MULTITROPHIC INTERACTIONS
IN TERRESTRIAL SYSTEMS

Multitrophic Interactions in Terrestrial Systems

THE 36TH SYMPOSIUM

OF THE BRITISH ECOLOGICAL SOCIETY

ROYAL HOLLOWAY COLLEGE

UNIVERSITY OF LONDON

1995

EDITED BY A. C. GANGE

School of Biological Sciences,
Royal Holloway

V. K. BROWN

International Institute of Entomology,
(An Institute of CAB International),
London

CAMBRIDGE
UNIVERSITY PRESS

CAMBRIDGE UNIVERSITY PRESS
Cambridge, New York, Melbourne, Madrid, Cape Town, Singapore, São Paulo, Delhi

Cambridge University Press
The Edinburgh Building, Cambridge CB2 8RU, UK

Published in the United States of America by Cambridge University Press, New York

www.cambridge.org
Information on this title: www.cambridge.org/9780521839952

First published on behalf of the British Ecological Society by Blackwell Science Ltd 1997
First published on behalf of the British Ecological Society by Cambridge University Press 2008
This digitally printed version 2008

A catalogue record for this publication is available from the British Library

ISBN 978-0-521-83995-2 hardback
ISBN 978-0-521-10055-7 paperback

CONTENTS

PART 2:
PLANT–MICROBE–ANIMAL INTERACTIONS

PART 3: PLANT–ANIMAL INTERACTIONS

PREFACE

Unravelling the complexity of the natural world relies on an understanding of the interactions between organisms. Interactions within one or two trophic levels have featured prominently in previous symposia of the British Ecological Society. However, there is a growing interest in the nature and role of more complex interactions, involving three or even four trophic levels. Such multitrophic interactions were the subject of the 1995 BES Symposium. In the past, the study of these interactions has been thwarted by tradition and training. Students, although trained in ecology, typically focused their interests on a particular taxonomic group – thus ecological principles were applied to bacteriology, mycology, entomology or mammology. This situation was commonly exacerbated by entire departments or even institutes focusing on particular subjects. Indeed, scientific meetings often followed the same trend!

Fortunately, in recent years, the situation has changed and there is far more emphasis on interdisciplinary, collaborative research. We felt there was a growing need for a summary of the state of this research and a consideration of the critical avenues to be pursued in the future. In assembling this symposium, we deliberately attempted to draw from as many disciplines as possible, to show how topics may be integrated and to suggest areas which require more study. Indeed, many of our speakers had not attended a BES meeting before. We believe that the meeting was a great success, since apart from the excellence of individual presentations, we observed a number of contacts being made between virologists and plant ecologists, entomologists and mycologists and molecular biologists and population ecologists, to name a few.

This volume is intended as a reference for those seeking a summary of the information on multitrophic interactions and as a source of ideas for the future. We trust it will lead to more interdisciplinary research, and thus major advances in our understanding of the forces which structure natural communities and ecosystems.

A. C. Gange
V. K. Brown
March 1996

PART 1
INTERACTIONS BETWEEN PLANTS AND LOWER ORGANISMS

INTRODUCTORY REMARKS

N. J. FOKKEMA

DLO Research Institute for Plant Protection (IPO-DLO), PO Box 9060, 6700 GW Wageningen, The Netherlands

Microbial interactions in soil, rhizosphere and phyllosphere and their consequences for plants, crops or natural vegetation form the central theme of this session. Traditionally, plant pathologists have focused on reducing plant pathogen populations, infection and crop losses, whereas soil ecologists are concerned with soil-borne bacteria and mycorrhizal fungi improving the nutritional status of the plant. Research on mycorrhizas and biological control of root pathogens has developed largely independently. The discovery of plant growth promoting rhizobacteria (PGPRs) and disease-suppressing effects of mycorrhizas, however, illustrates that the functional borders are rather artificial. Moreover, rhizobacteria and mycorrhizas occupy the same niche: the rhizosphere. Therefore, it is time for a better exchange of expertise on both groups of organisms, which is likely to benefit further exploitation of their beneficial properties. This session offers, for the first time, an opportunity for mutual acquaintance.

Interactions in the phyllosphere show considerable similarity with interactions in the rhizosphere. In this session, interactions between bacteria and fungal pathogens are discussed. With respect to biological control, we should realize, however, that interactions between saprophytic and pathogenic bacteria or fungi are also important, not only on the living plant, but also on dead plant material and crop debris. Some of these aspects are mentioned briefly in the summary of the session.

Why are we interested in all these phenomena? Apart from a fully justifiable scientific curiosity, which is difficult to finance nowadays, but which certainly formed the basis of our current knowledge, it is the public demand for a low-input sustainable agriculture that urges us to exploit biological alternatives for chemical crop protection and plant nutrition. This session gives us examples of our achievements so far. The available knowledge is considerable. Insight into the complexity of interactions will hopefully contribute to a reliable exploitation in the near future.

1. BACTERIAL ANTAGONIST–FUNGAL PATHOGEN INTERACTIONS ON THE PLANT AERIAL SURFACE

B. SEDDON*, S. G. EDWARDS*[1],
E. MARKELLOU*† AND N. E. MALATHRAKIS†

*Department of Agriculture, MacRobert Building, University of Aberdeen, Aberdeen AB24 5UA, Scotland, UK and †School of Agriculture, Technological Education Institute, 71500 Heraklio, Crete, Greece

INTRODUCTION

Plant aerial surfaces support a vast array of microbial communities, the interactions of which are poorly understood (Blakeman 1981; Fokkema & van den Heuval 1986; Andrews & Hirano 1991; Blakeman 1991; Shaw & Peters 1994). These communities are very unstable due to fluctuating environmental parameters. In contrast, the rhizosphere is a more stable environment, less affected by chemicals (e.g. fungicides) and therefore more is known about it (Lynch 1990; Beemster et al. 1991; Schroth & Hendson 1995). Knowledge of microbial interactions in the phyllosphere is biased towards important crops and agricultural environments. Moreover, most aerial fungal pathogens have been controlled by fungicides and their effectiveness, to a large extent, is independent of biotic factors. This situation held back studies and knowledge of these microbial communities. It is only recently, with development of fungicide resistance, the desire to reduce residues, concern over toxicity and lack of specificity of these chemicals, that ecophysiological studies and alternative methods of disease control have been made. An understanding of microbial interactions in the crop environment is the basis of biological control and more emphasis should be placed on research determining the ecological para-meters and reasons why biocontrol has not been consistently successful in the past (Blakeman & Fokkema 1982; Windels & Lindow 1985; Gowdu & Balasubramanian 1988; Andrews 1992; Edwards et al. 1994; Fokkema et al. 1994). This chapter aims to promote biocontrol by an understanding of bacterial antagonist–fungal pathogen interactions on the plant aerial surface. Successful biocontrol relies on an understanding of the pathogen and antagonist including their behaviour and interaction in relation to the host, environment and microclimate. These interactions must be elucidated if biocontrol is to be exploited in disease-control programmes.

---

[1] Present address: Department of Biology, University of York, PO Box 373, York YO1 5YW, UK.

BACTERIAL ANTAGONISTS TO FUNGAL PATHOGENS

Fungal pathogens on aerial plant surfaces generally fall into one of two groups: biotrophs or necrotrophs (Fokkema *et al.* 1994). Rusts and mildews are examples of biotrophic pathogens whilst *Botrytis cinerea* and *Alternaria alternata* are well-studied necrotrophs. The leaf surface and stems, especially tender young tissue of newly developed seedlings and those associated with wound damage, are particularly susceptible to pathogens of an opportunistic and ubiquitous nature. Flowers and developing fruits provide environments with variable levels of available nutrients, and post-harvest environments of fruits and vegetables, where ripening and senescence make plant tissues more prone to pathogen attack, are all subjected to invasion, disease initiation and spread from fungi (Droby & Chalutz 1994; Fokkema 1995). Additionally, the plant aerial environment is subjected to a greater degree of environmental change (humidity, temperature, irradiation) than is the rhizosphere and therefore it is much more difficult to predict microbial interactions and to practise biological control.

Antagonism

Antagonism is generally the basis from which biological control systems are developed and can be demonstrated *in vitro* (Murray *et al.* 1986; Utkhede & Sholberg 1986; Fravel 1988; Janisiewicz 1988). It is more difficult, however, to demonstrate antagonism *in vivo* where the host plant, microflora, microclimate and fluctuating external environmental parameters have modifying effects on both the fungal pathogen and its antagonist (Edwards *et al.* 1994). Initial screening studies have therefore been made *in vitro*. These studies, whilst giving information on the direct effect of the antagonist on the pathogen (antibiosis, competition for nutrients), have been criticized on the grounds that observations *in vitro* do not readily translate to efficacy *in vivo* (Fravel 1988) and some modes of antagonism, effective *in planta*, cannot even be demonstrated in these *in vitro* tests (Renwick *et al.* 1991). Nevertheless, it is worth considering bacterial antagonist–fungal pathogen interactions *in vitro* as a prelude to interactions *in vivo*.

In vitro *interactions between bacterial antagonists and fungal pathogens*

Bacterial antagonists isolated from plant aerial surfaces and introduced antagonists for biocontrol purposes mainly belong to the genera *Bacillus, Pseudomonas, Serratia, Enterobacter, Lactobacillus, Streptomyces* and *Erwinia* (Droby & Chalutz 1994; Edwards *et al.* 1994; Elad *et al.* 1994; Sobiczewski & Bryk 1995). Antibiosis and/or competition are the commonest mechanisms demonstrated *in vitro*. In *Pseudomonas* spp., metal-chelating agents and antimicrobial metabolites (siderophores, pyrrolnitrin, etc.) are among a variety of metabolites produced showing *in vitro* inhibition to plant pathogens (Jayaswal *et al.* 1993; Pfender *et al.* 1993; Dowling & O'Gara 1994) (see also Keel & Défago, this volume). The iron-chelating agents work by sequestering and hence depriving the pathogen of iron, which is an essential trace element for growth (Leong 1986). Antibiotics act directly to inhibit and suppress the pathogen. Pyrrolnitrin and the phenazine antibiotics are among many of the pseudomonad secondary metabolites that work this way although the precise mode of action is not always clearly understood (Janisiewicz *et al.* 1991; McManus *et al.* 1993; Dowling & O'Gara 1994). With *Bacillus* spp., antagonism *in vitro* is, in most cases, linked to the production of antifungal compounds. *Bacillus subtilis, B. brevis, B. cereus* and *B. polymyxa*, among others, have all been shown to produce antibiotics which have antifungal activity (Edwards *et al.* 1994; Markellou *et al.* 1995). Antibiotic-negative mutants confirm that *in vitro* antagonism is mediated by production of these antifungal compounds (Andrews 1990; Edwards & Seddon 1992).

In many cases the chemical identity of the antibiotic has not been confirmed, its biosynthesis and regulation of production not determined and the effects of the antibiotic on the fungus at the biochemical and physiological level have not been ascertained. It is imperative that these studies are made at an early stage in the investigation if interactions between bacterial antagonists and fungal pathogens are to form the basis of future biocontrol systems (Edwards *et al.* 1994). The precise characterization of the bacterial antagonist to subspecies level is not always made and this can be crucial for the proper development of biocontrol systems. Some *Bacillus* antagonists are pathogenic to beneficial insects and man (Sneath 1986) and the possibility of effects on other systems should not be overlooked. Working with ill-defined antagonists and inappropriate antibiotics does not help to promote the use of biocontrol systems and care should be taken in all such studies. Antibiotics easily countered by the development of resistance in fungi and *Bacillus* antagonists that show risks to non-target organisms (especially man) should be avoided (Li & Leifert 1994; Markellou *et al.* 1995). Only with an in-depth characterization of these systems can this be assured. Problems have arisen with fungicides in this way and we must make sure that this is not repeated with biocontrol systems. In our own work with the isolation and identification of *Bacillus* antagonists from the biotic environment for use against *B. cinerea* on tomato crops grown under protection, a strong candidate for antagonism and hence biocontrol was isolated from manure compost and identified as a *Bacillus cereus* strain. Further detailed characterization indicated that the isolate produced a human enterotoxin under defined conditions and therefore was potentially harmful to humans on consumption (Markellou *et al.* 1995). If these detailed characterization studies had not been made this isolate would have been amongst the strongest candidates, judged on its biocontrol potential, for future development. Other composts with manures of animal origin could equally well harbour such pathogens and detailed characterization of the bacteria present is essential.

In a similar way, detailed *in vitro* studies with the well-characterized antibiotic gramicidin S produced by *Bacillus brevis* Nagano strain, which is antagonistic to *B. cinerea*, have shown that conidial germination is more sensitive to inhibition from gramicidin S than mycelial growth and that the cell membrane of the germinating conidiospore is the target site of action of gramicidin S (Edwards & Seddon 1992; Edwards 1993; Seddon *et al.* 1996). An immediate release of adenosine triphosphate (ATP) with collapse of the cell membrane occurs on addition of gramicidin S. Such *in vitro* studies allow target stages in the life-cycle of the pathogen to be identified and biocontrol strategies *in vivo* to be developed.

Interactions of bacterial antagonists and fungal pathogens in vivo

Much early *in vitro* work was done with the hope that antagonism would be equally effective in the plant environment. Such expectations were naive and unrealistic and many factors other than the direct interaction between antagonist and pathogen play a role *in vivo* (Edwards *et al.* 1994). The host plant, the microclimate of the infection court, other microflora and inhabitants of the phyllosphere (Blakeman 1988), environmental parameters and insults (solar radiation, fungicides, etc.) – all contribute and modify this interaction. It is little wonder that many of these earlier biocontrol attempts failed or were invariably inconsistent (Windels & Lindow 1985; Pusey 1990; Andrews 1992). Widely fluctuating environmental conditions unsuitable to the antagonist led to many of these failures. Field crop situations are the most extreme of these conditions whereas crops grown under protection provide a certain degree of environmental control, the post-harvest environment is even better controlled and experiments in plant growth chambers operate under the strictest environmental control.

In vivo studies have been made in all these environments and bacterial antagonists targeted to a range of foliar plant pathogens (Andrews 1992; Edwards *et al.* 1994; Droby & Chalutz 1994; Fokkema *et al.* 1994). These studies have been carried out with natural isolates from the biotic environment where disease occurs, introduced bacterial antagonists and, in one or two situations, genetically modified bacteria (Pfender *et al.* 1993; Dowling & O'Gara 1994). Almost always there is a lack of fundamental knowledge concerning the microbial interactions that take place *in vivo*; it is much more difficult to carry out *in vivo* than *in vitro* studies and the necessary methodology to perform analysis *in planta* is not yet fully in place. Nevertheless, detailed analyses must be made if biocontrol is ever to achieve its full potential in crop protection strategies. Mechanisms of interaction other than antibiosis and nutrient competition can play a role *in vivo* and the situation may be more complex than direct antagonism normally encountered *in vitro*. Population levels, the ability to colonize and competition for space become important parameters, together with the involvement of the host plant (e.g. nutrient supply, induced resistance) (Knudsen & Spurr 1988; Andrews 1992).

In vivo *antibiosis*

Antibiotics are secondary metabolites produced by micro-organisms and it has been speculated that they do confer some competitive advantage to the producer (Katz & Demain 1977) such as improved colonization of the plant surface and therefore niche exclusion of the pathogen (Mazzola *et al.* 1992) or have more subtle effects as well as direct antagonism (Dowling & O'Gara 1994). Even when antibiosis is demonstrated as a mode of antagonism *in vitro* this can be difficult to demonstrate *in vivo* and strong antagonism *in vitro* does not always translate to good biocontrol *in vivo* (Spurr & Knudsen 1985; Lynch & Ebben 1986; Edwards & Seddon 1992; Pfender *et al.* 1993). On the plant aerial surface, antibiotics may or may not be produced, or may be produced at levels too low for effective antagonism. Lack of detection and/or activity of the antibiotic may be the result of insensitive detection techniques, binding and inactivation of the antibiotic on the plant surfaces, lack of release of the antibiotic or interaction with the pathogen (Williams & Vickers 1986; Edwards & Seddon 1992). Demonstration that antibiotic-negative mutants do not support effective antagonism against a pathogen, whereas the wild-type antibiotic producing strain does, is clear evidence that antibiosis plays a role (Pfender *et al.* 1993). A *Ps. fluorescens* strain producing phenazine-1-carboxylic acid inhibited teliospore germination of *Tilletia laevis* (common bunt of wheat) and protected wheat seedlings against disease. A mutant deficient in phenazine-1-carboxylic acid production did not inhibit teliospore germination and did not protect against common bunt (McManus *et al.* 1993). Antibiotic-negative mutants have also been used with *Bacillus* antagonists to demonstrate that antagonism *in vivo* to pathogenic fungi involves antibiosis. Using vital stains and epifluorescence microscopy, control of *Botrytis cinerea* conidial germination on Chinese cabbage leaf discs could be demonstrated with the *B. brevis* wild-type gramicidin S-producing strain whereas an antibiotic-negative mutant showed no inhibition (Edwards & Seddon 1992). Inhibition *in vivo* required much higher levels of gramicidin S and *B. brevis* populations for a fixed population level of *B. cinerea* than *in vitro*. Edwards and Seddon (1992) stated that gramicidin S was strongly adsorbed to the leaf surface and in this form was inactive resulting in much higher levels necessary for inhibition. Gramicidin S is also associated with the surface of the bacterial spore (Lazaridis 1981) and the *B. brevis* spores and *B. cinerea* conidia may not be sufficiently close to allow direct interaction unless high levels are used. In contrast, *in vitro* studies were carried out in a liquid culture medium or in water droplets on glass slides and in these aqueous suspensions would be readily distributed with frequent direct interaction between the two leading to lower levels of antibiotic necessary for inhibition (Edwards *et al.* 1994). Other reports that antibiotic-negative mutants of antagonistic bacteria do give some degree of disease control suggest that other modes of antagonism exist and could be exploited for biocontrol purposes (Colyer & Mount 1984; Kempf & Wolfe 1989).

Perhaps the greatest degree of success with biocontrol systems has been with post-harvest control of fungal plant pathogens. In many instances this has proved to be with fungal antagonists (mainly *Trichoderma* spp. and yeast) and these appear to interact via hyperparasitism and competition for nutrients and space, although a role for antibiosis is not always excluded (Droby & Chalutz 1994). Bacterial antagonists have also been identified and it is here that antibiosis plays a more prominent role. Several *Bacillus* antagonists interact with post-harvest rot pathogens. With citrus fruits, *B. sphaericus* was effective in reducing disease of *Alternaria citri* and *Penicillium italicum* (Sharma 1993). The relative population levels of pathogen and antagonist were important to control of disease as measured by lesion diameter, as was the time of submersion of the fruit in the antagonist suspension. Brown rot of stone fruit caused by *Monilinia fructicola* could be controlled by *B. subtilis* (Pusey & Wilson 1984) and its polypeptide antibiotic iturin. Other *B. subtilis* isolates reduced severity of alternaria rot and brown rot on cherry fruits (Utkhede & Sholberg 1986). Post-harvest biological control of *B. cinerea* on gerbera flowers was achieved with *B. brevis* and it was presumed that this was due to the action of gramicidin S produced by this strain (Kerssies 1993). Control was also demonstrated with a strain of *Ps. aureofaciens*. *Pseudomonas* spp. have also shown antagonism to *B. cinerea* with reduction of losses of Dutch white cabbage in storage (Leifert *et al.* 1993). A *Serratia liquefaciens* strain was also effective in these studies. Other pseudomonads shown to be effective with grey mould and blue mould (*B. cinerea* and *Ps. expansum*) are *Ps. cepacia* (Janisiewicz & Roitman 1988), *Ps. syringae* (Janisiewicz 1987) and *Ps. gladioli* (Mao & Cappellini 1989). The secondary metabolite pyrrolnitrin was shown to be the antagonistic agent of *Ps. cepacia*. An interesting line of study is that of nutritional and nutrient analogue enhancement of activity of the antagonist with specific amino acids and sugars (Janisiewicz *et al.* 1992; Janisiewicz 1994). Exploitation of bacterial antagonist–fungal pathogen interactions for biological control of post-harvest diseases of fruits and vegetables is an area where success is achievable due to the degree of control over environmental conditions (Wilson *et al.* 1991; Droby & Chalutz 1994; Sobiczewski & Bryk 1995).

STRATEGIES FOR BIOCONTROL

To study microbial ecophysiology of bacterial antagonist–fungal pathogen interactions and develop biocontrol strategies, it is best to choose a fungal pathogen that is recognized as an important disease-causing organism with environmental requirements, epidemiology and the life-cycle, well-characterized, Edwards *et al.* 1994. Powdery mildews are obligate biotrophic pathogens that require different humidity levels in the environment from other pathogens. They are associated with dry periods and can germinate and grow in the absence of surface moisture (Butt 1978; Yarwood 1978). On the other hand most fungal plant pathogens show a requirement for periods of surface wetness for disease to occur. *B. cinerea* is a ubiquitous aerial fungal plant pathogen with such a moisture requirement (Coley-Smith *et al.* 1980). Newly initiated stems of young seedlings are affected by *B. cinerea*, which is a causal agent of damping-off disease (Walker *et al.* 1994), and damaged stems of mature crops (e.g. tomato, cucumber) can be infected (Verhoeff *et al.* 1992). Leaves are also infected and this can be important in spread of the disease through the crop canopy (Fokkema *et al.* 1994) and to the yield of vegetable crops such as lettuce and cabbage (Edwards *et al.* 1994). Flowers are susceptible to *B. cinerea* (Kerssies 1993) and this is important to floriculture and to those crops eventually producing high-value fruits from flowers. *B. cinerea* also infects post-harvest fruits, vegetables and flowers in storage and transit (Droby & Chalutz 1994). It is one of the most important disease-causing organisms and, considering its ability to rapidly develop fungicide resistance and the difficulty in producing disease-resistant plants, the use of bacterial antagonists to this pathogen for disease control purposes is an attractive proposition (Edwards *et al.* 1994; Fokkema 1995).

The life-cycle of a pathogen provides several target sites where interaction between antagonist and pathogen could effectively reduce the level of disease throughout a crop. In the case of *B. cinerea*, survival and germination of conidia are two sequentially linked but different stages, each of which could be blocked in some way or another. Infection and disease development in the plant are associated with specific enzyme activities, growth, multiplication and hyphal development and could be targeted for control. Finally, once growth is limited, sporulation takes place and the formation of conidia could also be a target point for microbial interactions whereby antagonists suppress sporulation (Fokkema *et al.* 1994).

Interactions affecting germination of conidia

For initiation of disease by water-dependent pathogens, conidia alight on plant surfaces and germinate in the presence of water. Germ tubes elongate until eventually penetration of the plant tissue leads to infection (both causal and latent) (Verhoeff 1980; Williamson 1994). It has been suggested that interaction of bacterial antagonists here with conidia, as they germinate prior to infection, may be too narrow a time window for effective biocontrol purposes (Fokkema *et al.* 1994). This need not necessarily follow. Conidial germination, however brief, is an essential and vulnerable step in the development of a pathogen. In the case of *B. cinerea*, the pathogen must progress from dormant conidia on the surface of the plant to actively growing hyphae in the penetrated plant tissue if disease is to ensue. If this step is not successful then the fungus will die before disease can be established. Therefore this stage is especially vulnerable to those interactions aimed at inhibiting this process. When water is available and other environmental factors are also favourable (e.g. suitable temperature) then *B. cinerea* conidia will commence germination. If the antagonist is active against the pathogen under the same conditions or can modify the conditions such that they become unsuitable or even inhibitory for germination of the pathogen then there is every likelihood that biocontrol will be effective. Additionally, the bacterium itself need not be metabolically active under the conditions as long as the antagonistic products (e.g. antibiotics) are present and active. What is required is some form of active antagonism at this stage.

Under conditions of water availability when conidia of *B. cinerea* germinate on the leaf surface of Chinese cabbage, the presence of spores of *B. brevis* (carrying gramicidin S on their outer surface) leads to an inhibition of conidial germination. This inhibition, however, is not 100% (maximum spore levels tested of *B. brevis* and gramicidin S gave only 50–60% inhibition). In polythene tunnels, spray treatments with *B. brevis* cultures gave effective disease control of *B. cinerea* grey mould on Chinese cabbage (about 70% disease control and comparable to the standard fungicide treatment). This was therefore unexpected and some other parameter was suspected of improving the biocontrol. It was discovered that the *B. brevis* bacterial cultures also produced a biosurfactant that was effective in reducing periods of surface wetness on leaves of Chinese cabbage (Edwards & Seddon 1992; Edwards 1993; Edwards *et al.* 1994). After overhead irrigation, leaves treated with *B. brevis* dried four times faster than untreated leaves. The period of surface wetness on the leaf is greatly reduced by treatment with this bacterial antagonist. *B. brevis* therefore operates two modes of interaction with *B. cinerea*, both of which are antagonistic to conidial germination. First, interaction of the antibiotic gramicidin S at the level of the cell membrane leads to disruption of membrane integrity, leakage of essential metabolites such as ATP from the cell and cell death (Seddon *et al.* in press). This is a direct effect that appears to be instantaneous and would be expected to operate equally effectively under all conditions of surface wetness and relative humidity. Second, the production of a biosurfactant reduces periods of surface wetness and has an indirect effect on germination by limiting water availability to the pathogen. Conidial germination takes place on the plant surface and once germination has commenced the pathogen is exposed and vulnerable to desiccation until it penetrates the plant tissue whereupon it is protected. Surface wetness from rain, irrigation/misting systems, condensation and transpiration all create environmental situations suitable for germination of *B. cinerea* conidia. The use of an antibiotic-negative mutant strain of *B. brevis* (E-1) that produces the biosurfactant but no antibiotic, together with the wild-type strain that produces both gramicidin S and biosurfactant, would allow the antagonism from these two modes of action to be assessed in different environmental and crop situations.

In one study, conidia of *B. cinerea* were germinated on petals of red raspberry under controlled environmental conditions (defined temperature, relative humidity and wind speed) in a precision-built laboratory humidity chamber (Harrison *et al.* 1994). The petals were not subjected to drying and the contribution of the biosurfactant to antagonism could be ignored. The role of the antibiotic therefore was investigated and it was found that introduction of *B. brevis* wild-type cultures prior to inoculation of petals with *B. cinerea* and subsequent germination led to collapse of conidia. Observations were made using low temperature scanning electron microscopy (LTSEM). The extent of conidial collapse in preparations treated with *B. brevis* wild-type strain (presence of gramicidin S) was clear in comparison to controls (McDonald 1993). These LTSEM studies provided only qualitative data of the infection court and characteristics of germination of *B. cinerea* not possible by other methods. The adaxial surface of the petal had dome-shaped cells with radiating ridges. Germ tubes of *B. cinerea* appeared to penetrate these cells directly without formation of swollen appressoria, and similar LTSEM studies of penetration of rose petals by *B. cinerea* after dry inoculation support these data (Williamson *et al.* 1995). *B. brevis* cells were deposited in close juxtaposition to *B. cinerea* conidia over the surface of the plant cell and collapsed conidia of *B. cinerea* were observed in these *B. brevis* treated tests. These observations suggest that there is direct interaction between *B. brevis* and *B. cinerea* under these conditions and that the antagonism may be due to the antibiotic gramicidin S.

In another experiment, tomato plants were grown in polythene greenhouses and the level of grey mould infection from *B. cinerea* was monitored (Markellou, Malathrakis & Seddon, unpublished data). Both the wild-type *B. brevis* Nagano strain (produces biosurfactant and antibiotic) and the gramicidin S-negative mutant E-1 (produces only biosurfactant) were used. Biocontrol of leaf infection was as good with E-1 as with *B. brevis* wild-type and the fungicide used (rovral plus vinclozolin). This indicated that it was the biosurfactant that is effective in this system. Monitoring of surface wetness of the leaves with resistance sensors and dew point analysis indicated that E-1 treatment reduced periods of surface wetness. These results are similar to those with *B. brevis* treatment of Chinese cabbage leaves (Edwards & Seddon 1992) and suggest that in both these situations the biosurfactant plays a major role in antagonism of *B. cinerea*. Presumably, treatment with *B. brevis* E-1 reduced the time period of water availability to conidia and inhibited germination and initiation of infection by spreading water as a thin film across the leaf surface. When the relative humidity of the ambient air is low, water vapour resulting from transpiration will move from the leaf to the surrounding air during periods of high solar radiation. Condensation will occur when the leaf surface is at a lower temperature than the surrounding ambient air and the dew point has been reached (Harrison *et al.* 1994). Surface wetness during periods of high solar radiation is 'invisible' to the naked eye but can be detected by continuous recording of the leaf surface temperature and resistance. This monitoring indicated that E-1 treatment did reduce surface wetness under these conditions. This biocontrol system could be integrated with environmental control (control of temperature, relative humidity, solar radiation, irrigation systems, etc.) for more effective disease control. Most fungal plant pathogens are water dependent in that they require free water or high relative humidities (Everts & Lacy 1990; Butler & Jadhar 1991; Harrison *et al.* 1994) and this novel mode of antagonism could operate against a wide range of them. At high ambient relative humidity (prolonged periods of cold, damp weather) periods of leaf drying will not occur and the biosurfactant should not be effective. Disease forecasting would then predict the integrated use of fungicides and/or other biocontrol antagonists that work by other modes of interaction (antibiosis, competition for nutrients, hyperparasitism, etc.).

Interactions affecting infection

Two strategies for disease control using bacterial antagonists to reduce infection from foliar fungal pathogens seem promising. The first is to exploit direct modes of antagonism and the target for application is wounds of stems and fruits. Infection is associated with the growing phase (the hyphal mycelium) of pathogens such as *B. cinerea* and some bacterial antagonists specifically inhibit this growth stage either by nutrient competition or the production of antifungal compounds (Sobiczewski & Bryk 1995). A *Pseudomonas* isolate was effective against *B. cinerea* on cut stems (wounds) of cucumber and tomato plants (Koning & Köhl 1995). *Ps. cepacia* produces pyrrolnitrin and this metabolite used as a dip treatment for apples and pears protected against grey mould infection (Janisiewicz *et al.* 1991). *Bacillus* antagonists that produce medically unimportant antibiotics are well characterized with no toxicity and do not readily raise resistance in the pathogen; their use is acceptable. Otherwise, it would be best to select for antagonism based on nutrient competition (Sobiczewski & Bryk 1995).

The second strategy does not directly involve the pathogen but is mediated by the host-plant induced resistance. Many biotic and abiotic agents can induce these defence responses. Fungi, bacteria, viruses (Défago *et al.* 1995) and even plant extracts (Daayf *et al.* 1995) can all be used to induce resistance. The accepted phenomenon is that there is no direct effect against the pathogen but the pathogen is antagonized indirectly by changes in the host plant. The plant becomes 'immunized' to future challenge by a range of pathogens (Kúc 1995). For protection against aerial fungal plant pathogens the inducer organism or extract can be applied to, or be active in, the rhizosphere inducing systemic resistance (Défago *et al.* 1995) (see also Keel & Défago, and West, this volume) although localized resistance is also induced. Bacteria that promote this response can be pathogens or non-pathogens. The use of non-pathogenic, non-toxic bacteria are the most suitable and *Bacillus* and *Pseudomonas* spp. have been used (Défago *et al.* 1995; Schönbeck & Kraska 1995). Barley powdery mildew (*Erysiphe graminis* f. sp. *hordei*) showed 50% reduction in formation of primary haustoria with impaired ability for nutrient uptake and a reduced level of growth and reproduction when *B. subtilis* extracellular metabolites were used to induce resistance. Yield enhancement exceeded that of the fungicide control and increased tolerance was also observed (Schönbeck & Kraska 1995). Induced resistance is only now beginning to be understood and this phenomenon will undoubtedly play an integrated role in future crop protection systems.

Interactions and suppression of sporulation

The production of secondary inocula of a pathogen within the crop due to sporulation on a leaf (or stems and fruits) leads to spread of the disease throughout the crop and to epidemic situations if this is not controlled. Suppression of sporulation is another target stage for biological control (Fokkema *et al.* 1994; Fokkema 1995). With both biotrophic and necrotrophic pathogens the biological control agents must match the pathogen in their ability to occupy and colonize the niche, or be hyperparasites, and most antagonists identified are fungi (Fokkema *et al.* 1994). Few studies have been made on bacterial antagonists that suppress sporulation. *Phytophthora infestans* causes blight on potato leaves and a strain of *Pseudomonas fluorescens* is parasitic to this fungal pathogen. The bacteria are thought to develop in sporangia where they grow and multiply (Lewosz 1995). *B. cinerea* sporulates on dead plant tissue and antagonists would be required to be active in this environment. Saprophytic bacteria have been shown to control this pathogen on bean and tomato plants (Elad *et al.* 1994). Suppression of sporulation was greatest when uninterrupted periods of high humidity were maintained. Selection of drought-resistant mutants of bacteria to withstand adverse dry conditions in a manner similar to those employed to select fungal antagonists (Fokkema *et al.* 1994) might be a way forward.

INTERACTIONS IN THE AGRICULTURAL ENVIRONMENT

Conditions governing interactions *in vitro* and *in vivo* under controlled environments can be far removed from the agricultural situation where the addition of fertilizers, fungicides, and rapidly changing environmental parameters can all play a role in influencing and modifying these interactions. Currently, contradictions in biocontrol results occur which are undoubtedly due to the temporal and spatial fluctuations of the macro- and micro-environmental conditions and their effects on the bacterial antagonist–fungal pathogen interactions. However, analysis of a specific antagonist–pathogen–host system, under varying environmental conditions, may lead to a model for predicting the performance of other antagonist–pathogen interactions. Such models have already been developed and successfully used in practice; for example biological control using *Pseudomonas cepacia* and *Bacillus thuringiensis* (Knudsen & Spurr 1988).

Interactions with fungicides

The presence of fungicides can have such dramatic effects on microbial communities (including beneficial micro-organisms) that studies in the presence and absence of fungicides can give completely different observations. Fungicides can even promote disease rather than protect (Griffith 1981). Agrochemicals can directly or indirectly affect population levels of the pathogenic and non-pathogenic microflora (Andrews 1981). Seasonal sprays may have a long-term impact on non-target microflora. Smolka (1993) showed that the specificities of each system need to be assessed with respect to each particular fungicide and agricultural environment if sensible disease-control measures are to be developed. Multisite fungicides such as dichlofluanid, and the thiocarbamates (mancozeb, metiram and propineb) were detrimental to phylloplane yeasts and bacteria with antagonistic activity against fungal pathogens. More specific fungicides, such as triadimefon and vinclozolin, were less inhibitory and propamocarb did not cause any significant inhibition. The bacterial antagonists inhibited were a *Bacillus* sp. (B9) and *Erwinia herbicola* (B129), clearly showing the lack of specificity of some of these fungicides.

Our own work with *Bacillus* antagonists agrees well with that of Smolka (1993). A selection of *Bacillus* antagonists to fungal pathogens (*B. subtilis, B. brevis, B. polymyxa, B. cereus*) was tested with known fungicides. Of these, dichlofluanid proved the least suitable for integration studies showing inhibition to all *Bacillus* spp. tested, whereas vinclozolin showed no inhibition and therefore would be appropriate for use with these biocontrol agents (Seddon *et al.* 1995). These observations indicate that much more needs to be known about the effects of chemicals on target and non-target organisms and their interactions if integrated chemical and biological control is to be successful.

Compost extracts and interactions

Organic fertilizers as a source of nutrients are well established in agriculture and organic farming is perhaps the best situation for exploitation of microbial interactions for biocontrol purposes since fungicide use is excluded. The use of compost extracts can have beneficial effects on disease control as well as crop growth and yield (Weltzien 1991). Both bacterial and fungal antagonists to aerial fungal plant pathogens have been identified in compost extracts and both may play a role in biocontrol (Ketterer *et al.* 1992; McQuilken *et al.* 1994; Malathrakis *et al.* 1995). Indeed, some synergistic effect from each trophic level may be essential for optimum disease control. Bearing this in mind, only bacterial antagonists are highlighted here.

Well-composted manures of organic materials can reach temperatures in excess of 70°C. Most bacteria cannot survive such elevated temperatures but many *Bacillus* spp. produce heat-resistant spores that do and some can even grow under these conditions (Sneath 1986). *Bacillus* spp. therefore are found as antagonists to fungal pathogens from compost extracts. Pseudomonads, Enterobacteriaceae and Actinomycetes have also been implicated as active participants in the compost extract mix. The effectiveness of compost extracts is closely related to the number of bacterial cfu's ml^{-1} of the compost before application and the bacterial cfu's cm^{-2} of plant surface after application (Ketterer *et al.* 1992; Malathrakis *et al.* 1995). Studies on complete compost extracts are difficult to interpret with respect to the mode of interaction, although direct inhibition of the pathogen and induced resistance have been suggested (Weltzien 1991). Bacterial antagonists isolated from compost extracts allow a greater degree of characterization of the interactions. Two such isolates, a *Bacillus* sp. (B39) and a *Pseudomonas fluorescens* (C148), were able to control *P. infestans* on potatoes grown under controlled environmental conditions (Fokkema *et al.* 1994). In field trials, both antagonists failed to prevent complete destruction of the crop by a severe, natural outbreak of *P. infestans* although the *Bacillus* isolate (B39) delayed this epidemic of *P. infestans* by 1 week. It was thought that low ecological competence led to the lack of sustained biocontrol. These workers also commented that the fungicide control used, Shirlan (fluazinan), appeared to be harmful to the phyllosphere bacteria (Jongebloed *et al.* 1995).

Bacillus antagonists isolated from compost extracts have been identified to species level in our laboratories, their mode of interaction with fungal pathogens determined, and their potential for biocontrol assessed (Markellou *et al.* 1995). In some cases, isolates within the same species (e.g. *B. cereus*) differed in their mode of interaction with the pathogen *B. cinerea*. Some isolates *in vitro* inhibited conidial germination and mycelial growth, others inhibited only one or other of these two processes and one isolate showed no inhibition at all, although suppression of infection was demonstrated in an *in vivo* bean leaf bioassay. For integrated control, introducing two *Bacillus* antagonists into the crop situation, each with a different mode of action (e.g. one at conidial germination and the other on hyphal growth), could lead to higher levels of disease control. This indicates the importance of characterizing the antagonists and their modes of action both *in vitro* and *in vivo* prior to test trials in the crop situation.

FUTURE PROSPECTS, LIMITATIONS AND DEVELOPMENTS

There are many reports concerning potential bacterial antagonists to fungal pathogens for biocontrol of aerial plant surfaces but detailed studies aimed at elucidating these interactions and an understanding of the ecological and environmental factors that predispose the interactions one way or another are lacking. In some cases, it appears to be the relative population levels of antagonist and how these are influenced by environmental factors that govern whether biocontrol is successful or not (Andrews & Hirano 1991; Yuen *et al.* 1994). In others, a greater knowledge of the mechanism of action of these bacterial antagonist–fungal pathogen interactions *in vivo* would help to perfect biocontrol (Edwards *et al.* 1994). Certainly, future observations should be made under environmentally realistic conditions and sound ecological principles must be applied throughout all stages of the research. This is not to say that *in vitro* studies will not be relevant (indeed at the moment they can sometimes be the only way to make detailed observations prior to further *in vivo* studies), but the final studies must be made in environments where these interactions are to take place.

One of the issues in studying bacterial antagonist–plant pathogen interactions on the aerial plant surface is the ability to monitor population levels. Difficulty can be experienced in enumerating a selected micro-organism (antagonist or pathogen) in the presence of other natural microflora. Selective media used must be appropriate to the studies carried out. With the antagonist, *B. brevis*, a selective medium based on tyrosine allowed differential counting between *B. brevis* and other *Bacillus* spp. recovered from leaf surfaces (Edwards 1993). For differential enumeration of *B. brevis* wild-type and gramicidin S-negative mutant strains this medium was not suitable and further selection based on gramicidin S detection was necessary. Similarly, levels of the pathogen *B. cinerea* were monitored in the greenhouse by use of a formulated medium selective for *B. cinerea* (Edwards 1993). Pathogen and antagonist levels could be monitored and associated with disease and successful biocontrol respectively. Other workers have used antibiotic-resistance markers to allow enumeration of bacterial populations on plant surfaces (Epton *et al.* 1994). In this way the levels of antagonist necessary for suppression of the pathogen for biocontrol purposes can be determined. Recombinant DNA technology and reporter gene activity are now being used for environmental monitoring and no doubt will allow other indicator systems to be used in the future (Beauchamp *et al.* 1992).

Population levels of antagonists on aerial plant surfaces are expected to change dramatically over widely fluctuating environmental conditions and the ability to survive in this environment under harsh conditions has been considered more important than the ability to grow under favourable conditions (Andrews 1992). Certainly, this will be true for sustained biocontrol. Antagonists therefore should be selected for survival characteristics and *Bacillus* spp. that produce resistant spores are strong candidates in the future for biocontrol. These characteristics are also suitable for formulation purposes (Edwards *et al.* 1994). Other candidates, such as *Pseudomonas* spp., will need to be formulated in a manner that preserves their viability and ways will need to be found of sustaining their viability on the plant surface under harsh conditions. Under favourable conditions, however, many pseudomonads can colonize plant surfaces whereas *Bacillus* antagonists tend not to be as environmentally competent and colonization can be difficult, although this is not always the case. Nutrient additives have been shown to promote colonization and this work is worth pursuing (Janisiewicz *et al.* 1992). The manner in which bacteria have been previously grown (in liquid broth, on solid agar or on plants) influences their survival and colonization characteristics (Wilson & Lindow 1993). This work was with *Ps. syringae* and it will be interesting to see if bacterial antagonists behave similarly. Activity of the antagonist *in situ* against the pathogen is normally required for successful biocontrol and production of antifungal compounds, demonstrated *in vitro*, is sometimes difficult to reproduce *in vivo*. Production of antifungal compounds has been studied in detail under laboratory fermentation conditions but little or nothing is known of their production on plant surfaces. The addition of nutrient supplements and their withdrawal could be useful here in stimulating a round of bacterial growth followed by antibiotic (antifungal) production. Spores of *Bacillus* spp. can be triggered to germinate by known metabolites (L-alanine, adenosine, etc.) (Gould & Hurst 1969). Once germinated, they can then be induced to undergo growth and then sporulation, conditions which lead to synthesis of secondary metabolites (Edwards *et al.* 1994). In *Bacillus brevis*, the spore carries gramicidin S on its outer surface and in the presence of nutrients the antibiotic is released to its surroundings on germination (Lazaridis *et al.* 1980). If these activities could be reproduced on plant surfaces then improved biocontrol is possible. Practically no studies have been carried out with these objectives whereas such work could lead to efficient practical management of biocontrol systems.

Observations *in situ* are not easy to make. Fluorescence microscopy and LTSEM can be carried out *in vivo* on plant surfaces and give an indication of the likely interactions in the crop environment. Experiments are carried out under controlled environmental conditions and observations related to these fixed parameters. With *B. cinerea* and the antagonist *B. brevis*, details of conidial germination and the effects of the antagonist on this process have been established. This confirmed *in vitro* studies showing a direct effect of the antagonist and gramicidin S (Edwards *et al.* 1994; Seddon *et al.* 1996). Both fluorescence microscopy and LTSEM (although particularly difficult to quantify) give a much-needed insight into the cytology of these interactions.

Finally, instrumentation for environmental monitoring (leaf wetness, humidity, solar radiation, temperature, etc.) is becoming more sophisticated, automated and computerized. Measurements and monitoring regimes are such that simultaneous monitoring of all these parameters can be made and computer programs used to amass and store this information. In this way, detailed information can be provided that allows the necessary important environmental parameters associated with disease and its control to be tested under realistic conditions. Tomorrow's biocontrol systems will no doubt result from these detailed, integrated ecophysiological studies.

ACKNOWLEDGEMENTS

Parts of our experimental work reported here were supported by the AFRC, Aberdeen University and the ISIS programme of BBSRC. Thanks are given to Mrs K. Emslie and Mrs T. McKay for experimental assistance in laboratory studies, to Dr L. Manousakis and Dr K. Vidalis for technical and experimental design concerning greenhouse monitoring and to Dr J. G. Harrison, Dr B. Williamson and Mr G. H. Duncan of the Scottish Crop Research Institute, Dundee for supervising the LTSEM and humidity studies carried out by S. McDonald.

REFERENCES

Andrews, J. H. (1981). Effects of pesticides on non-target micro-organisms on leaves. *Microbial Ecology of the Phylloplane* (Ed. by J. P. Blakeman), pp. 283–304. Academic Press, London.

Andrews, J. H. (1990). Biological control in the phyllosphere: realistic goal or false hope? *Canadian Journal of Plant Pathology,* **12,** 300–307.

Andrews, J. H. (1992). Biological control in the phyllosphere. *Annual Review of Phytopathology* **30,** 603–635.

Andrews, J. H. & Hirano, S. S. (1991). *Microbial Ecology of Leaves.* Springer, New York.

Beauchamp, C. J., Kloepper, J. W. & Lemke, P. A. (1992). Luminometric analysis of plant root colonisation by bioluminescent pseudomonads. *Canadian Journal of Microbiology,* **39,** 434–441.

Beemster, A. B. R., Bollen, G. J., Gerlagh, M., Ruissen, M. A., Schippers, B. & Tempel, A. (1991). *Biotic Interactions and Soil-borne Diseases.* Elsevier, New York.

Blakeman, J. P. (1981). *Microbial Ecology of the Phylloplane.* Academic Press, London.

Blakeman, J. P. (1988). Competitive antagonism of air-borne fungal pathogens. *Fungi in Biological Control Systems* (Ed. by M. N. Burge), pp. 141–160. Manchester University Press. Manchester.

Blakeman, J. P. (1991). Foliar plant pathogens: epiphytic growth and interactions on leaves. *Journal of Applied Bacteriology,* Symposium Supplement, **70,** 49S–59S.

Blakeman, J. P. & Fokkema, N. J. (1982). Potential for the biological control of plant diseases on the phylloplane. *Annual Review of Phytopathology,* **20,** 167–192.

Butt, D. J. (1978). Epidemiology of powdery mildews. *The Powdery Mildews* (Ed. by D. M. Spencer), pp. 51–81. Academic Press, London.

Butler, D. R. & Jadhar, D. R. (1991). Requirements of leaf wetness and temperature for infection of groundnut by rust. *Plant Pathology*, 40, 395–400.

Coley-Smith, J. R., Verhoeff, K. & Jarvis, W. R. (1980). *The Biology of* Botrytis. Academic Press, London.

Colyer, P. D. & Mount, M. S. (1984). Bacterization of potatoes with *Pseudomonas putida* and its influence on postharvest soft rot diseases. *Plant Disease*, 68, 703–706.

Daayf, F., Schmitt, A. & Bélanger, R. (1995). The effects of plant extracts of *Reynoutria sachalinensis* on powdery mildew development and leaf physiology of long English cucumber. *Plant Disease*, 79, 577–580.

Défago, G., Hase, C. & Maurhofer, M. (1995). Mechanisms of induced resistance through plant growth-promoting rhizobacteria. *Environmental Biotic Factors in Integrated Plant Disease Control* (Ed. by M. Manka), pp. 193–197. Polish Phytopathological Society, Poznan.

Dowling, D. N. & O'Gara, F. (1994). Metabolites of *Pseudomonas* involved in the biocontrol of plant disease. *Trends in Biotechnology*, 12, 133–141.

Droby, S. & Chalutz, E. (1994). Successful biocontrol of postharvest pathogens of fruits and vegetables. *Proceedings of the Brighton Crop Protection Conference – Pests and Diseases 1994*, Vol. 3, pp. 18 1265–1272. BCPC, Farnham.

Edwards, S. G. (1993). *Biological control of* Botrytis cinerea *by* Bacillus brevis *on protected Chinese cabbage.* PhD Thesis, University of Aberdeen, Aberdeen.

Edwards, S. G. & Seddon, B. (1992). *Bacillus brevis* as a biocontrol agent against *Botrytis cinerea* on protected Chinese cabbage. *Recent Advances in* Botrytis *Research* (Ed. by K. Verheoff, N. E. Malathrakis & B. Williamson), pp. 267–271. Pudoc Scientific, Wageningen.

Edwards, S. G., McKay, T. & Seddon, B. (1994). Interactions of *Bacillus* species with phytopathogenic fungi – methods of analysis and manipulation for biocontrol purposes. *Ecology of Plant Pathogens* (Ed. by J. P. Blakeman & B. Williamson), pp. 101–118. CAB International, Wallingford.

Elad, Y., Köhl, J. & Fokkema, N. J. (1994). Control of infection and sporulation of *Botrytis cinerea* on bean and tomato by saprophytic bacteria and fungi. *European Journal of Plant Pathology*, 100, 315–336.

Epton, H. A. S., Wilson, M., Nicholson, S. L. & Sigee, D. C. (1994). Biological control of *Erwinia amylovora* with *Erwinia herbicola*. *Ecology of Plant Pathogens* (Ed. by J. P. Blakeman & B. Williamson), pp. 335–352. CAB International, Wallingford.

Everts, K. L. & Lacy, M. L. (1990). The influence of dew duration, relative humidity and leaf senescence on conidial formation and infection of onion by *Alternaria porri*. *Phytopathology*, 80, 1203–1207.

Fokkema, N. J. (1995). Strategies for biocontrol of foliar fungal diseases. *Environmental Biotic Factors in Integrated Plant Disease Control* (Ed. by M. Manka), pp. 69–79. Polish Phytopathological Society, Poznan.

Fokkema, N. J. & van den Heuval, J. (1986). *Microbiology of the Phyllosphere*. Cambridge University Press, Cambridge.

Fokkema, N. J., Gerlagh, M., Köhl, J., Jongebloed, P. H. J. & Kessell, G. J. T. (1994). Prospects for biological control of foliar pathogens. *Proceedings of the Brighton Crop Protection Conference – Pests and Diseases 1994*, Vol. 3, pp. 1249–1258. BCPC, Farnham.

Fravel, D. R. (1988). Role of antibiosis in the biocontrol of plant diseases. *Annual Review of Phytopathology*, 26, 75–91.

Gould, G. A. & Hurst, A. (1969). *The Bacterial Spore*. Academic Press, London.

Gowdu, B. J. & Balasubramanian, R. (1988). Role of phylloplane microorganisms in the biological control of foliar plant diseases. *Zeitschrift für Pflanzenkrankheiten und Pflanzenschutz*, 95, 310–331.

Griffith, E. (1981). Iatrogenic plant diseases. *Annual Review of Phytopathology*, **19**, 69–82.

Harrison, J. G., Lowe, R. & Williams, N. A. (1994). Humidity and fungal diseases of plants – problems. *Ecology of Plant Pathogens* (Ed. by J. P. Blakeman & B. Williamson), pp. 79–97. CAB International, Wallingford.

Janisiewicz, W. J. (1987). Postharvest biological control of blue mould on apple. *Phytopathology*, **77**, 481–485.

Janisiewicz, W. J. (1988). Biocontrol of postharvest diseases of apples with antagonistic mixtures. *Phytopathology*, **78**, 194–98.

Janisiewicz, W. J. (1994). Enhancement of biocontrol of blue mould with the nutrient analog 2-deoxy-D-glucose on apples and pears. *Applied and Environmental Microbiology*, **60**, 2671–2676.

Janisiewicz, W. J. & Roitman, J. (1988). Biological control of blue mould and grey mould on apple and pear with *Pseudomonas cepacia*. *Phytopathology*, **78**, 1697–1700.

Janisiewicz, W. J., Yourman, L., Roitman, J. & Mahoney, N. (1991). Postharvest control of blue mould and grey mould of apples and pears by dip treatment with pyrrolnitrin, a metabolite of *Pseudomonas cepacia*. *Plant Disease*, **75**, 490–494.

Janisiewicz, W. J., Usall, J. & Bors, B. (1992). Nutritional enhancement of biocontrol of blue mould on apples. *Phytopathology*, **82**, 1364–1370.

Jayaswal, R. K., Fernandez, M., Upadhyay, R. S., Visintin, L., Kurz, M., Webb, J. & Rinehart, K. (1993). Antagonism of *Pseudomonas cepacia* against phytopathogenic fungi. *Current Microbiology*, **26**, 17–22.

Jongebloed, P. H. J., Kessel, G. J. T., van der Plas, C. H., Molhoek, W. M. L. & Fokkema, N. J. (1995). Possibilities for biocontrol of *Phytophthora infestans* in potato with two bacterial antagonists. *Environmental Biotic Factors in Integrated Plant Disease Control* (Ed. by M. Manka), pp. 277–282. Polish Phytopathological Society, Poznan.

Katz, E. & Demain, A. L. (1977). The peptide antibiotics of *Bacillus*: chemistry, biogenesis, and possible functions. *Bacteriological Reviews*, **41**, 449–474.

Kempf, H. J. & Wolfe, G. (1989). *Erwinia herbicola* as a biocontrol agent of *Fusarium culmorum* and *Puccinia recondita* f.s.p. *tritici* on wheat. *Phytopathology*, **79**, 990–994.

Kerssies, A. (1992). Epidemiology of *Botrytis cinerea* in gerbera and rose grown in glasshouses. *Recent Advances in* Botrytis *Research* (Ed. by K. Verhoeff, N. E. Malathrakis & B. Williamson), pp. 159–166. Pudoc Scientific, Wageningen.

Kerssies, A. (1993). Postharvest biological control of *Botrytis cinerea* on gerbera. *Biological Control of Foliar and Postharvest Diseases IOBC/WPRS Bulletin* (Ed. by N. J. Fokkema, J. Köhl & Y. Elad), pp. 131–135. Wageningen.

Ketterer, N., Fisher, B. & Weltzien, H. C. (1992). Biological control of *Botrytis cinerea* on grapevine by compost extracts and their microorganisms in pure culture. *Recent Advances in* Botrytis *Research* (Ed. by K. Verhoeff, N. E. Malathrakis & B. Williamson), pp. 179–186. Pudoc Scientific, Wageningen.

Knudsen, G. R. & Spurr, H. W. (1988). Management of bacterial populations for foliar disease control. *Biocontrol of Plant Diseases*, Vol. 1 (Ed. by K. G. Mukerji & K. L. Garg), pp. 83–92. CRC Press, Boca Raton, FL.

Koning, G. P. & Köhl, J. (1995). Wound protection by antagonists against *Botrytis* stem rot in cucumber and tomato. *Environmental Biotic Factors in Integrated Plant Disease Control* (Ed. by M. Manka), pp. 313–316. Polish Phytopathological Society, Poznan.

Kúc, J. (1995). Systemic induced resistance as part of integrated plant disease control. *Environmental Biotic Factors in Integrated Plant Disease Control* (Ed. by M. Manka), pp. 129–136. Polish Phytopathological Society, Poznan.

Lazaridis, I. (1981). *Spore characteristics of wild-type and gramicidin S-negative mutants of* Bacillus brevis *Nagano*. Ph D Thesis. University of Aberdeen, Aberdeen.

Lazaridis, I., Frangou-Lazaridis, M., MacCuish, F. C., Nandi, S. & Seddon, B. (1980). Gramicidin

S content and germination and outgrowth of *Bacillus brevis* Nagano spores. *FEMS Microbiology Letters*, **7**, 229–232.

Leifert, C., Sigee, D. C., Stanley, R., Knight, C. & Epton, H. A. S. (1993). Biocontrol of *Botrytis cinerea* and *Alternaria brassicicola* on Dutch white cabbage by bacterial antagonists at cold-storage temperatures. *Plant Pathology*, **42**, 270–279.

Leong, J. (1986). Siderophores: their biochemistry and possible role in the biocontrol of plant pathogens. *Annual Review of Phytopathology*, **24**, 187–209.

Lewosz, J. (1995). Bacteria parasitic towards the mycelium and sporangia of *Phytophthora infestans*. *Environmental Biotic Factors in Integrated Plant Disease Control* (Ed. by M. Manka), pp. 351–356. Polish Phytopathological Society, Poznan.

Li, H. & Leifert, C. (1994). Development of resistance in *Botryotinia fuckeliana* (de Bary) Whetzel against the biological control agent *Bacillus subtilis* CL27. *Zeitschrift für Pflanzenkrankheiten und Pflanzenschutz*, **101**, 414–418.

Lynch, J. M. (1990). *The Rhizosphere*. J.H. Wiley & Sons, Chichester.

Lynch, J. M. & Ebben, H. (1986). The use of microorganisms to control plant disease. *Journal of Applied Bacteriology Symposium*, Supplement, **61**, 115S–126S.

Malathrakis, N. E., Goumas, D., Markellou, E., Marazakis, M. & Panagiotakis, G. (1995). Biological control of tomato grey mould (*Botrytis cinerea*) by watery compost extracts. *Environmental Biotic Factors in Integrated Plant Disease Control* (Ed. by M. Manka), pp. 379–382. Polish Phytopathological Society, Poznan.

Mao, G. H, & Cappellini, R. A. (1989). Postharvest biocontrol of grey mould of pear by *Pseudomonas gladioli*. *Phytopathology*, **79**, 1153.

Markellou, E., Malathrakis, N .E., Walker R., Edwards, S. G., Powell, A. A. & Seddon, B. (1995). Characterisation of bacterial antagonists to *Botrytis cinerea* from the biotic environment and its importance with respect to *Bacillus* species and biocontrol considerations. *Environmental Biotic Factors in Integrated Plant Disease Control* (Ed. by M. Manka), pp. 385–389. Polish Phytopathological Society, Poznan.

Mazzola, M., Cook, R. J., Thomashow, L. S., Weller, D. M. & Pierson, L. S. (1992). Contribution of phenazine antibiotic biosynthesis to the ecological competence of fluorescent pseudomonads in soil habitats. *Applied and Environmental Microbiology*, **58**, 2616–2624.

McDonald, S. (1993). *Behaviour of* Botrytis cinerea *and three bacterial antagonists on plant surfaces as a prelude to successful biocontrol of grey mould*. MSc Thesis, University of Aberdeen, Aberdeen.

McManus, P. S., Ravenscroft, A. V. & Fulbright, D. W. (1993). Inhibition of *Tilletia laevis* teliospore germination and suppression of common bunt of wheat by *Pseudomonas fluorescens* 2–79. *Plant Disease*, **77**, 1012–1015.

McQuilken, M. P., Whipps, J. M. & Lynch, J. M. (1994). Effect of water extracts of a composted manure on the plant pathogen *Botrytis cinerea*. *World Journal of Microbiology and Biotechnology*, **10**, 20–26.

Murray, T., Leighton, F. C. & Seddon, B. (1986). Inhibition of fungal spore germination by gramicidin S and its potential use as a biocontrol agent against fungal plant pathogens. *Letters in Applied Microbiology*, **3**, 5–7.

Pfender, W. F., Kraus, J. & Loper, J. E. (1993). A genomic region from *Pseudomonas fluorescens* Pf-5 required for pyrrolnitrin production and inhibition of *Pyrenophora tritici-repentis* in wheat straw. *Phytopathology*, **83**, 1223–1228.

Pusey, P. L. (1990). Control of pathogens on aerial plant surfaces by antagonistic microorganisms. *Biological and Cultural Tests for Control of Plant Diseases*, Vol. 5 (Ed. by W. F. Wilcox), pp. v–vii. American Phytopathological Society Press, St Paul, MN.

Pusey, P. L. & Wilson, C. L. (1984). Postharvest biological control of stone fruit brown rot by *Bacillus subtilis*. *Plant Disease*, **68**, 753–756.

Renwick, A., Campbell, R. & Coe, S. (1991). Assessment of an *in vivo* screening system for potential biocontrol agents of *Gaeumannomyces graminis*. *Plant Pathology*, **40**, 524–532.

Schönbeck, F. & Kraska, T. (1995). Induced resistance: mechanisms and application. *Environmental Biotic Factors in Integrated Plant Disease Control* (Ed. by M. Manka), pp. 495–499. Polish Phytopathological Society, Poznan.

Schroth, M. N. & Hendson, M. (1995). Metabolic, genetic and ecological considerations of rhizobacteria as biological control agents. *Environmental Biotic Factors in Integrated Plant Disease Control* (Ed. by M. Manka), pp. 37–44. Polish Phytopathological Society, Poznan.

Seddon, B., Edwards, S. G. & Rutland, L. (1996). Development of *Bacillus* species as antifungal agents in crop protection. *Modern Fungicides and Antifungal Compounds* (Ed. by H. Lyr, P. E. Russell & H. D. Sisler), pp. 555–560. BCPC, Intercept, Andover.

Sharma, N. (1993). Post harvest biological control of citrus fruit rot. *Journal of Biological Control*, 7, 84–86.

Shaw, M. W. & Peters, J. C. (1994). The biological environment and pathogen population dynamics: uncertainty, coexistence and competition. *Ecology of Plant Pathogens* (Ed. by J. P. Blakeman & B. Williamson), pp. 17–37. CAB International, Wallingford.

Smolka, S. E. (1993). Assessment of pesticide side effects on beneficial microorganisms from the phylloplane. *Biological Control of Fungal & Bacterial Pathogens. Bulletin Vol. 16(11)* (Ed. by N. J. Fokkema, J. Köhl & Y. Elad), pp. 155–160, IOBC/WPRS, Montfavet.

Sneath, P. H. A. (1986). Endospore-forming Gram-positive rods and cocci, *Bergey's Manual of Systematic Bacteriology*, Vol. 2 (Ed. by P. H. A. Sneath, N. S. Mair, M. E. Sharpe & J. G. Holt), pp. 1104–1207. Williams & Wilkins, Baltimore.

Sobiczewski, P. & Bryk, H. (1995). Bacteria for biocontrol of postharvest diseases of fruits. *Environmental Biotic Factors in Integrated Plant Disease Control* (Ed. by M. Manka), pp. 105–111. Polish Phytopathological Society, Poznan.

Spurr, H. W. & Knudsen, G. R. (1985). Biological control of leaf diseases with bacteria. *Biological Control on the Phylloplane* (Ed. by C. E. Windels & S. E. Lindow), pp. 45–62. American Phytopathological Society, St Paul, MN.

Utkhede, R. S. & Sholberg, P. L. (1986). *In vitro* inhibition of plant pathogens by *Bacillus subtilis* and *Enterobacter aerogenes* and *in vivo* control of two postharvest cherry diseases. *Canadian Journal of Microbiology*, 32, 963–967.

Verhoeff, K. (1980). The infection process and host-pathogen interactions. *The Biology of* Botrytis (Ed. by J. R. Coley-Smith, K. Verhoeff & W. R. Jarvis), pp. 153–180. Academic Press, London.

Verhoeff, K., Malathrakis, N. E. & Williamson, B. (1992). *Recent Advances in* Botrytis *Research*. Pudoc Scientific, Wageningen.

Walker, R., Powell, A. A. & Seddon, B. (1994). Tests for biological control of seed and seedling damping-off diseases of peas and beans using *Bacillus* species. *Seed Treatment: Progress and Prospects* (Ed. by T. Martin), pp. 333–338. BCPC, Farnham.

Weltzien, H. C. (1991). Biocontrol of foliar fungal diseases with compost extracts. *Microbial Ecology of Leaves* (Ed. by J. H. Andrews & S. S. Hirano), pp. 430–450. Springer, New York.

Williams, S. T. & Vickers, J. C. (1986). The ecology of antibiotic production. *Microbial Ecology*, 12, 43–52.

Williamson, B. (1994). Latency and quiescence in survival and success of fungal plant pathogens. *Ecology of Plant Pathogens* (Ed. by J. P. Blakeman & B. Williamson), pp. 187–207. CAB International, Wallingford.

Williamson, B., Duncan, G. H., Harrison, J. G., Harding, L. A., Elad, Y. & Zimand, G. (1995). Effect of humidity on infection of rose petals by dry-inoculated conidia of *Botrytis cinerea*. *Mycological Research*, 99, 1303–1310.

Wilson, M. & Lindow, S. (1993). Effect of phenotypic plasticity on epiphytic survival and colonisation by *Pseudomonas syringe*. *Applied and Environmental Microbiology*, 59, 410–416.

Wilson, C. L., Wisniewski, M. E., Biles, C. L., McLaughlin, R., Chalutz, E. & Droby, S. (1991).

Biological control of post-harvest diseases of fruits and vegetables: Alternatives to synthetic fungicides. *Crop Protection*, **10**, 172–177.

Windels, C. E. & Lindow, S. E. (1985). *Biological Control on the Phylloplane.* American Phytopathological Society Press, St Paul, MN.

Yarwood, C. E. (1978). Water and the infection process. *Water Deficits and Plant Growth* (Ed. by T. T. Kozlowski), pp. 143–173. Academic Press, New York.

Yuen, G. Y., Craig, M. L., Kerr, E. D. & Steadman, J. R. (1994). Influences of antagonistic population levels, blossom development stage and canopy temperature on the inhibition of *Sclerotinia sclerotiorum* on dry edible bean by *Erwinia herbicola*. *Phytopathology*, **84**, 495–501.

2. INTERACTIONS BETWEEN BENEFICIAL SOIL BACTERIA AND ROOT PATHOGENS: MECHANISMS AND ECOLOGICAL IMPACT

C. KEEL AND G. DÉFAGO

Phytopathology, Institute of Plant Sciences, Swiss Federal Institute of Technology, CH-8092 Zurich, Switzerland

INTRODUCTION

Roots can be attacked by major pathogens which cause well-known diseases with characteristic symptoms such as root rots and vascular wilts. Prominent examples include the pathogenic fungi *Gaeumannomyces graminis* var. *tritici*, the causative agent of take-all of wheat; *Rhizoctonia solani* and *Pythium ultimum* causing root rot and damping-off diseases of a variety of plants; *Thielaviopsis basicola* which induces black root rot of tobacco; and *Fusarium oxysporum* f. spp. causing vascular wilts. In many cases damage to the plant is due to minor pathogens – that is, saprophytes or parasites which are harmful to plant growth but do not produce obvious disease symptoms (Weller 1988).

One of the most effective techniques to reduce population densities of certain soil-borne pathogens is crop rotation with poor or non-host crops. However, many farmers tend to reduce crop rotation because it has become impracticable for economic reasons. Such reductions often lead to increased problems with soil-borne pathogens. Today, soil-borne diseases are controlled by fungicides and soil fumigants such as methyl bromide. The use of these chemical pesticides will be restricted in the future since many of them have created severe problems for the environment and public health, for example contamination of soil, groundwater and food; the broad-spectrum soil fumigant methyl bromide, which has been used extensively for soil-borne disease and pest control for decades, affects not only the ecological balance in soils but also contributes to stratospheric ozone depletion and therefore is likely to be banned in the near future (Noling & Becker 1994). These problems have led to a clear need for alternative, environmentally more sound methods to control soil-borne plant diseases.

Biological control of root diseases with micro-organisms is a forward-looking concept in the overall trend towards a more sustainable agriculture. Examples of 'natural' biological control of root-infecting fungi include disease-suppressive soils, effects of organic amendments such as composts, and crop rotation, all based on the exploitation or management of indigenous communities of micro-organisms (Cook & Baker 1983; Hoitink & Fahy 1986). In the case of naturally suppressive soils, the biocontrol effect has been attributed mostly to resident populations of beneficial root-colonizing bacteria and to their interaction with certain soil edaphic factors (Cook & Baker 1983; Défago & Haas 1990; Défago & Keel 1995). The disease-suppressive effects of these soils may be mimicked to a certain extent in other soils by introduction of single strains or consortia of selected biocontrol agents. Among a range of bacteria with the capacity to suppress soil-borne diseases, including *Agrobacterium, Arthrobacter, Alcaligenes, Bacillus, Enterobacter, Erwinia* and *Serratia*, root-colonizing fluorescent *Pseudomonas* spp. in particular are targets of extensive research. However, even though the beneficial effects of these bacteria are amply documented and the underlying mechanisms partially understood, data on their actual behaviour and impact in the soil ecosystem are still scarce. In this review, some principal aspects of the complex interactions among beneficial *Pseudomonas* strains, pathogen(s), the plant and the biotic/abiotic soil environment, and some safety concerns associated with the release of these bacteria, are highlighted.

USE OF ROOT-COLONIZING PSEUDOMONADS TO CONTROL SOIL-BORNE PLANT DISEASES

During the past 20 years, numerous reports have illustrated the beneficial, plant growth-promoting effects of fluorescent pseudomonads (Weller 1988; Défago & Haas 1990; Cook 1993). Under controlled conditions – that is, in growth chamber and greenhouse experiments – *Pseudomonas* strains, added as a seed treatment, soil drench or granules, effectively colonized underground plant organs and consistently promoted plant growth and reduced disease incidence caused by a range of soil-borne pathogenic fungi. The beneficial effect of pseudomonads has also been observed in many field experiments, involving soils naturally infested with major (e.g. *G. g. tritici, Pythium* spp.) or minor pathogens (e.g. deleterious rhizobacteria) and in most cases is described as a general enhancement of plant growth. Yields are increased commonly by 15–30%; but there is considerable variation (e.g. Weller 1988; Défago & Haas 1990; Défago *et al.* 1990). The reasons for the variation of the antagonistic effects under field conditions are manifold; they are discussed below.

In many cases the antagonistic effects of the *Pseudomonas* strains are not restricted to a specific plant host–pathogen system. In our studies, we have focused on *P. fluorescens* CHA0, an isolate from a soil in the Morens region of Switzerland that is naturally suppressive to black root rot of tobacco, caused by *T. basicola* (Stutz *et al.* 1986). In growth chamber and greenhouse experiments, strain CHA0 was not only effective against diseases caused by *T. basicola* but also against various other diseases caused by *Aphanomyces euteiches, F. oxysporum* f. spp., *G. g. tritici, P. ultimum* and *R. solani* (Défago *et al.* 1990). In most, but not all, field trials conducted over several years, the strain also improved the health and yield of wheat grown in plots naturally or artificially infested with *G. g. tritici* (Défago *et al.* 1990).

It is generally accepted that fluorescent pseudomonads promote plant growth mainly by suppressing major or minor pathogens. It has also been suggested that pseudomonads may exhibit their plant growth-promoting effect indirectly by improving the beneficial action of other root-associated micro-organisms such as mycorrhiza. Growth promotion in the absence of other micro-organisms has been attributed to the increased availability of mineral nutrients, such as phosphate or nitrogen, to the bacterial production of root growth-promoting phytohormones or to the degradation of the precursor of root ethylene (Glick 1995).

MECHANISMS OF DISEASE SUPPRESSION

There is a diversity of possible mechanisms by which beneficial pseudomonads suppress disease (Table 2.1). Major mechanisms proposed include the inhibition of the pathogen by antimicrobial compounds, competition for iron, competition for colonization sites and nutrients supplied by roots, and induction of plant defence mechanisms (Weller 1988; Défago & Haas 1990; Paulitz 1990; Loper & Buyer 1991). Inhibition of the pathogen by production of antimicrobial and iron-chelating metabolites is considered to be the primary mechanism of biocontrol. Other mechanisms that have been named are the degradation of pathogenicity factors of the pathogen, for example toxins (Toyoda *et al.* 1988), the inactivation of pathogen germination factors present in seed or root exudates (Nelson 1992), and the production of extracellular enzymes, for example chitinase, laminarase and glucanase, that can lyse fungal cell walls (Lim *et al.* 1991; Fridlender *et al.* 1993).

Root colonization is a prerequisite for the functioning of all other biocontrol mechanisms exhibited by pseudomonads. Effective root colonization is linked directly to successful competition with other micro-organisms for colonization sites and nutrients available from root exudates (Paulitz 1990; O'Sullivan & O'Gara 1992; Kluepfel 1993). To exhibit its disease-suppressive effect, a biocontrol agent needs to become distributed along the root, multiply and survive for several weeks in the presence of competition from the indigenous microbiota. Besides the indigenous microbiota, many other factors may influence root colonization (Weller 1988; Parke 1990; Weller & Thomashow 1994): (i) characteristics of the introduced antagonist, such as cell surface properties, production of siderophores or antibiotics, pili, flagellae and chemotaxis towards root or seed exudates; (ii) the species, cultivar and growth stage of the host plant; and (iii) physical and chemical characteristics of the soil, for example, moisture, temperature, pH, soil texture, and mineral nutrients.

Particularly exciting is the recent finding that some pseudomonads that are closely associated with the roots can induce systemic plant defence mechanisms

TABLE 2.1. Proposed mechanisms of suppression of soilborne plant diseases by beneficial pseudomonads.

Mechanism	Mode of action	Bacterial metabolites involved
Antibiosis	Inhibition of the pathogen by diffusible or volatile antimicrobial compounds	Phenazines, 2,4-diacetylphloroglucinol, pyoluteorin, pyrrolnitrin, oomycin A, hydrogen cyanide
Iron competition	Inhibition of the pathogen by limitation of the iron supply through iron chelators which sequester iron in a form that is unavailable to the pathogen	Siderophores (pyoverdines, pseudobactins, pyochelin)
Induced resistance	Induction of systemic plant defence mechanisms against the pathogen	Metal chelators (?)
Root colonization	Exclusion of the pathogen by competition for suitable niches on the roots	Antibiotics, siderophores
Nutrient competition	Exclusion of the pathogen by competition for available nutrients, e.g. root and seed exudates	
Inactivation of pathogenicity and germination factors	Degradation of pathogen toxins and inactivation of pathogen-stimulatory molecules from seed or root exudates	Extracellular enzymes

against fungal or bacterial pathogens (Alström 1991; Van Peer *et al.* 1991; Wei, Kloepper & Tuzun 1991). In our studies, colonization of tobacco roots by *P. fluorescens* CHA0 reduced leaf necrosis caused by tobacco necrosis virus and induced physiological changes (pathogenesis related (PR) proteins) in the plant, commonly associated with induced resistance (Maurhofer *et al.* 1994). Since strain CHA0 can also be found in the root cortex, it seems possible that bacterial metabolites (e.g. metal chelators or antibiotics) might be delivered into the plant and induce the stress necessary to activate defence mechanisms (Voisard *et al.* 1989; Défago *et al.* 1990; Maurhofer *et al.* 1994). However, the bacterial components that trigger systemic induced resistance still need to be elucidated.

It is evident that the mechanisms listed (Table 2.1) are complementary. The diversity of compounds secreted by pseudomonads (Leisinger & Margraff 1979; Défago & Haas 1990; Dowling & O'Gara 1994) is the major reason why disease suppression by these bacteria is considered to be the product of several mechanisms which vary from strain to strain. Even for a single strain of *Pseudomonas*, different mechanisms or combinations of mechanisms may be important in the suppression of different plant diseases.

BACTERIAL METABOLITES INVOLVED IN DISEASE SUPPRESSION

It is generally assumed that *Pseudomonas* strains with biocontrol ability use root exudates to synthesize metabolites, some of which may play a key role in disease suppression. During the past decade, antibiotic and metal-chelating compounds in particular, have been recognized as being essential for biocontrol activity of many *Pseudomonas* strains (Table 2.2).

Antimicrobial metabolites

During the stationary growth phase, fluorescent pseudomonads produce an array of low molecular weight metabolites, which belong to many groups of organic compounds (Leisinger & Margraff 1979; Défago & Haas 1990; Dowling & O'Gara 1994). Many of them have antibiotic activities against plant pathogens. Important antimicrobial metabolites which have been shown to be involved in the control of plant diseases include phenazine 1-carboxylic acid (PCA), 2,4-diacetylphloroglucinol (Phl), pyrrolnitrin, pyoluteorin (Plt), oomycin A and hydrogen cyanide (HCN) (Table 2.2). Some *Pseudomonas* strains produce a considerable spectrum of antibiotics, for example *P. fluorescens* CHA0, which excretes Phl, Plt and pyrrolnitrin in addition to HCN (Voisard *et al.* 1994). In most cases, the importance of a given antibiotic metabolite in pathogen control has been investigated by the use of genetically or chemically induced mutants deficient in the production of the metabolite (for

TABLE 2.2. Antimicrobial and metal-chelating metabolites of *Pseudomonas* involved in disease suppression.

Metabolite(s)	Properties	Contribution to the suppression of	Reference
Phenazines	Antibiotic	*G. g. tritici* on wheat	Thomashow *et al.* 1990
2,4-diacetylphloroglucinol	Antibiotic	*G. g. tritici* on wheat *P. ultimum* on sugar beet *T. basicola* on tobacco	Keel *et al.* 1992 Fenton *et al.* 1992 Keel *et al.* 1992
Pyoluteorin	Antibiotic	*P. ultimum* on cotton and cress	Kraus & Loper 1995
Pyrrolnitrin	Antibiotic	*R. solani* on cotton and radish *Pyrenophora tritici-repentis* on wheat straw	Homma & Suzui 1989; Hill *et al.* 1994 Pfender *et al.* 1993
Oomycin A	Antibiotic	*P. ultimum* on cotton	Gutterson 1990; Howie & Suslow 1991
Hydrogen cyanide	Biocide, metal chelator	*P. ultimum* on cress and cucumber *T. basicola* on tobacco	Keel, unpublished data Voisard *et al.* 1989
Pyoverdine Pseudobactin	Fluorescent siderophore Metal chelator	*P. ultimum* on cotton and wheat *F. oxysporum* f. spp. on carnation and tomato Deleterious rhizobacteria	Becker & Cook 1988; Loper 1988 Lemanceau *et al.* 1992; Scher 1986 Bakker *et al.* 1986
Pyochelin	Siderophore, metal chelator	*P. ultimum* on tomato	Buysens *et al.* 1994

references, see Table 2.2). If a mutant showed a reduced disease-suppressive capacity, the attempt was then made to restore the desired trait by complementation of the mutant with the corresponding wild-type genes. If a complementing fragment was found, a further step was to identify and sequence the gene(s) involved in the production of the compound, and to study their regulation. Additional evidence may then be obtained by heterologous expression of complementing genes in strains that naturally do not produce the antibiotic, thereby improving their biocontrol activity (e.g. Voisard *et al.* 1989; Fenton *et al.* 1992; Hara *et al.* 1994; Hill *et al.* 1994). Further support may also come from the *in situ* detection of the metabolite in the rhizosphere of plants treated with the producing strain, as demonstrated for PCA (Thomashow *et al.* 1990), Phl (Keel *et al.* 1992; Maurhofer *et al.* 1995), Plt (Maurhofer *et al.* 1995), and pyrrolnitrin (Kempf *et al.* 1994).

To be effective in disease suppression, production of antimicrobial metabolites must be induced at the correct time and in the appropriate location, so a complex regulatory network is necessary. Genetic analysis of *Pseudomonas* mutants with pleiotropic defects in the synthesis of antimicrobial metabolites caused by single mutations has provided evidence that antibiotic production is under global control of a two-component regulatory system. This system is based on two protein components, an environmental sensor (presumably a transmembrane protein) and a cytoplasmic response regulator that mediates changes in response to sensor signals. A response regulator, GacA (for global activator), was identified first in *P. fluorescens* strain CHA0 (Laville *et al.* 1992). Strains with a mutation in the *gacA* gene have lost their ability to produce Phl, Plt and HCN (Laville *et al.* 1992) and two exoenzymes, protease and phospholipase C (Sacherer *et al.* 1994). Genes corresponding to *gacA* of strain CHA0 have also been identified in *P. fluorescens* BL915 (Gaffney *et al.* 1994) and in *P. syringae* (Rich *et al.* 1994). The cognate sensor component is encoded by the *lemA* gene in *P. syringae* (Hrabak & Willis 1992) and the *lemA*-like *apdA* gene in *P. fluorescens* Pf-5 (Corbell & Loper 1995). A mutation in the *lemA* or *apdA* genes of the respective strains causes multiple defects, similar to those induced by a mutation in *gacA*; *lemA* mutants do not produce syringomycin and protease and fail to form lesions on host plants, and *apdA* mutants are unable to produce HCN, Plt and pyrrolnitrin (Hrabak & Willis 1992; Corbell & Loper 1995). Genetic evidence for *LemA/GacA* forming a two-component system in *P. syringae* has been provided recently by Rich *et al.* (1994). However, the environmental signals that trigger the *LemA/GacA* two-component system are not known. Particularly intriguing is the recent finding of a possible involvement of cell-density-dependent autoregulatory mechanisms in antibiotic synthesis. In *P. aureofaciens* 30–84 phenazine antibiotic production is regulated by the transcriptional activator PhzR in response to cell density, and the possible involvement of an auto-inducing compound has been suggested (Pierson *et al.* 1994). Auto-inducing compounds, identified as N-acylhomoserine lactones, are diffusable, low molecular weight compounds, which are accumulated in batch cultures at high cell densities; auto-inducers have been reported to be important signals of secondary metabolism and virulence factors in many bacterial species (Salmond *et al.* 1995).

Siderophores

During growth under iron-low conditions fluorescent pseudomonads produce characteristic fluorescent, yellow-green siderophores, termed pyoverdines or pseudobactins, which have a very high affinity for ferric ion (Abdallah 1991). These potent bacterial iron chelators are thought to sequester the limited supply of iron available in the rhizosphere to a form that is unavailable to pathogenic fungi and other harmful micro-organisms, thereby restricting their growth (Loper & Buyer, 1991; O'Sullivan & O'Gara 1992). Evidence for an involvement of siderophores in disease suppression has come from studies on siderophore-deficient mutants and on the effects of purified siderophores or synthetic iron chelators; siderophores in particular have been reported to contribute to the suppression of fusarium wilt diseases, pythium damping-off and minor pathogens (Table 2.2). Nevertheless, in a number of other *Pseudomonas* strains, siderophores have a minimal role in disease suppression and other bacterial metabolites, especially antibiotics, have been suggested to be the determinant factors involved (Keel *et al.* 1989; Hamdan *et al.* 1991; Paulitz & Loper 1991). Little is known about the role of two other siderophores, pyochelin and its precursor salicylate, produced by certain *Pseudomonas* strains (Meyer *et al.* 1992; Voisard *et al.* 1994). Pyochelin has been reported to contribute to the protection of tomato plants from *Pythium* (Buysens *et al.* 1994). Salicylate may act as a signal in plants inducing systemic resistance against pathogen attack (Métraux *et al.* 1990).

ENVIRONMENTAL FACTORS AFFECTING BACTERIAL METABOLITE PRODUCTION

The production and activity of bacterial metabolites involved in pathogen suppression may be significantly affected by environmental conditions that prevail in the rhizosphere, for example available sources of carbon, nitrogen and micronutrients, temperature and availability of oxygen. Laboratory studies have identified several factors that influence the production of biocontrol metabolites; for example, iron-deplete conditions induce the production of siderophores (Loper & Buyer 1991). Iron-replete conditions are necessary for HCN production and for efficient suppression of black root rot of tobacco by strain CHA0 (Keel *et al.* 1989; Voisard *et al.* 1989). Zinc was reported to enhance PCA production by *P. fluorescens* 2–79 *in vitro* and to improve the suppression of take-all of wheat (Weller & Thomashow

1994). Production of oomycin A is stimulated by glucose and good aeration (Gutterson 1990). An elegant approach to identify biological and physico-chemical factors that influence the synthesis of biocontrol metabolites is the use of reporter genes fused to promoters of the corresponding genes. Reporter genes based on β-galactosidase or ice-nucleation activity have allowed determination of the *in situ* expression of genes required for the synthesis of phenazines, oomycin A and Plt and pyoverdine on roots or seeds (Howie & Suslow 1991; Georgakopoulos *et al.* 1994; Loper & Lindow 1994; Kraus & Loper 1995). Reporter gene fusions have also been used to assess the availability of certain nutrients such as iron and phosphate in soil habitats (Loper & Lindow 1994; de Weger *et al.* 1994).

APPROACHES TO IMPROVE THE PERFORMANCE OF BIOCONTROL AGENTS

Although there are numerous examples of micro-organisms that are capable of reducing soil-borne plant diseases, only few biocontrol agents have been commercialized (Cook 1993). A major impediment to the commercial use of beneficial bacteria is their inconsistent performance under field conditions often varying from site to site and from year to year. Given that disease suppression depends on multiple interactions between the biocontrol agent, the host plant, the pathogen and the soil environment, there may be many reasons for this inconsistency. Variability in root colonization and in the production of metabolites which contribute to disease control, site specificity of the beneficial effect (i.e. adaptation to a specific crop and local soil environmental factors), limited persistence and activity after release, occurrence of non-target pathogens, and genetic instability during maintenance in the laboratory and during scale-up are some possible reasons that have been suggested (Weller 1988; Voisard *et al.* 1994; Weller & Thomashow 1994; Défago & Keel 1995).

Several strategies are being followed to overcome the problem of inconsistent performance. One approach is the application of mixtures of beneficial micro-organisms highly adapted to control specific diseases in specific environments. One rationale behind using consortia of micro-organisms is to profit from the synergistic effect obtained by combining strains which exert their beneficial activity based on diverse mechanisms. In greenhouse and field experiments, combinations of certain *Pseudomonas* strains provided better control of take-all of wheat than did the individual strains alone (Pierson & Weller 1994). *Pseudomonas* strains have also been used to increase the beneficial effect of mycorrhiza or of non-pathogenic fungi used as biocontrol agents (Meyer & Lindermann 1986; Park *et al.* 1988; Lemanceau *et al.* 1992); for example, non-pathogenic isolates of *Fusarium oxysporum* have been successfully combined with fluorescent *Pseudomonas* strains to control *Fusarium* wilt of tomato and cucumber (Park *et al.* 1988; Lemanceau *et al.* 1992; Défago & Keel 1995).

Another means to improve performance and reliability of biocontrol agents may be to enhance their biocontrol and root-colonizing ability by genetic engineering. This strategy is promising since a series of genes relevant for biocontrol capacity in the bacteria has been identified (see above) and is now amenable to manipulation either at the level of regulation or by transferral to new hosts. One approach to construct strains with improved biocontrol activity is to overexpress gene(s) involved in antibiotic biosynthesis; for example, production of the antibiotic oomycin A in *P. fluorescens* HV37a was enhanced by placing the corresponding structural genes under the control of the constitutive *tac* promoter from *E. coli*. The antibiotic overproducer had an improved biocontrol efficacy against Pythium damping-off of cotton (Gutterson 1990). In *P. fluorescens* CHA0, the disease suppressive capacity could be enhanced by the introduction of the recombinant plasmid pME3090, which carries a 22-kb insert of CHA0 DNA (Maurhofer *et al.* 1992). Production of the antibiotics Phl and Plt was increased several-fold *in vitro* and in the rhizosphere of wheat (Maurhofer *et al.* 1992, 1995). The recombinant strain showed an improved protection of cucumber against *P. ultimum*, *Phomopsis sclerotioides* and *F. oxysporum* f. sp. *cucumerinum* and of tobacco against *T. basicola*. Interestingly, the same amplification was deleterious to the growth of some other plant species, presumably because of a phytotoxic effect of the antibiotics (Maurhofer *et al.* 1992, 1995). By subcloning experiments, transposon mutagenesis and sequence analysis, the *rpoD* gene (encoding the house-keeping sigma factor σ^{70}) was identified as the gene on pME3090 being responsible for enhanced antibiotic production and biocontrol activity (Schnider *et al.* 1995). In *P. fluorescens* Pf-5, another biocontrol strain which produces multiple antibiotics, inactivation of *rpoS* (encoding the stationary phase sigma factor σ^{38}) stops pyrrolnitrin production, but enhances production of Phl and Plt and improves disease control (Sarniguet *et al.* 1994).

Improvements of the protective effect may also be achieved through transfer of relevant biosynthetic genes, in particular those involved in the synthesis of antimicrobial metabolites, into heterologous hosts; for example, a naturally HCN-negative *P. fluorescens* strain was capable of producing HCN after chromosomal insertion of the cloned *hcn*-region from strain CHA0 and gave improved protection of tobacco against black root rot (Voisard *et al.* 1989). Other examples include the heterologous expression, respectively, of phenazine and Phl biosynthetic genes from *P. fluorescens* 2–79 and *P. aureofaciens* Q2-87. Introduction of either of these regions into a series of other *Pseudomonas* strains conferred on them the ability to produce PCA or Phl and enhanced their capacity to suppress take-all of wheat (Hara *et al.* 1994). Similarly, a Phl biosynthetic DNA region of *Pseudomonas* sp. F113 was cloned upstream of a constitutive tetracycline resistance promoter and introduced into *P. aureofaciens* strain M114 that is unable to produce Phl; the engineered strain synthesized Phl and improved the emergence of sugar beet seeds, subject to damping-off caused by *P. ultimum* (Fenton *et al.* 1992). Transfer of a gene region involved in pyrrolnitrin synthesis from *P. fluorescens* strain BL915 into two other *P. fluorescens* strains conferred upon them the ability to produce this antibiotic and to suppress damping-off of cotton, caused by *R. solani* (Hill *et al.* 1994).

Further improvements may be achieved by strict quality control during scale-up for mass production of inoculum for field application and by optimizing the delivery systems for the bacterial inoculants (Weller 1988; Van Elsas *et al.* 1992; Voisard *et al.* 1994; Défago & Keel 1995). For example, a well-known problem is the genetic instability of some *Pseudomonas* strains during *in vitro* cultivation which may result in the loss of their suppressive effect. In the case of *P. fluorescens* CHA0, prolonged exposure to rich media under laboratory conditions leads to the formation of spontaneous mutants having pleiotropic defects. The phenotypes of these mutants interestingly match with those of mutants defective in the regulatory genes *gacA* or *lemA*, and are characterized by the loss of production of antimicrobial and other metabolites and by the lack of biocontrol activity (Défago & Keel 1995 and unpublished data). The possible loss of phenotypes which are critical for biocontrol due to laboratory maintenance accentuates the need for quality control of bacterial inocula to ensure their biocontrol effectiveness in the field (Voisard *et al.* 1994).

ECOLOGICAL BEHAVIOUR AND IMPACT OF BENEFICIAL BACTERIA

Commercial application of beneficial bacteria to control plant diseases requires the release of large numbers of wild-type or genetically engineered strains into the environment. Although the plant-beneficial effects are amply documented and the mechanisms involved partially understood, the behaviour of released bacteria in the environment is largely unknown. Concern about the ecological safety of such release has been raised, especially in the context of using genetically engineered strains with enhanced biological activity. The assessment of potential ecological risks requires information on: (i) the persistence of introduced micro-organisms and their possible dissemination to non-target sites; (ii) possible negative effects on indigenous organisms and biological processes; and (iii) the likelihood of the horizontal transfer of introduced genes with possible adverse effects.

Persistence

Following introduction into field soil the biocontrol agent is faced with numerous biotic and abiotic factors that may affect its survival. It is not surprising, therefore, that any introduced population of bacteria (natural or genetically modified) tends to decrease gradually to a low level (Défago *et al.* 1990; Drahos *et al.* 1992; Keel *et al.* 1994; Ryder *et al.* 1994; Weller & Thomashow 1994); for example, after 9 months, the number of culturable cells of *P. fluorescens* CHA0 in the surface soil of large field lysimeters planted with wheat declined from an initial concentration of 10^8 cells to about 10^3 cells per gram (Keel *et al.* 1994; Keel & Défago unpublished data). Similarly, the population of a *P. corrugata* biocontrol strain, applied at 10^7 bacteria per seed, had dropped by three orders of magnitude within 4 months (Ryder *et al.* 1994). Two years after introduction, the strain still remained detectable at very low levels (seven culturable cells per 100 g of soil), indicating that introduced *Pseudomonas* strains may persist over a considerable period of time. Populations of introduced bacteria decline more rapidly in non-sterile soil than in sterile soils (Mazzola *et al.* 1992; Natsch *et al.* 1994), showing that competition with established organisms and predation (mainly by protozoa) are important factors.

In view of a potential use of bacteria genetically engineered for enhanced biological activity, it is important to know whether production of metabolites that are relevant for the enhanced activity may also contribute to the survival and ecological competence of a released strain. For *P. fluorescens* CHA0, the regulatory gene *gacA*, which controls the synthesis of antimicrobial and other secondary metabolites, was found to be important not only for disease suppression but also for the strain's survival in soil (Natsch *et al.* 1994). Similarly, phenazine antibiotics contributed positively to the long-term persistence of two *Pseudomonas* strains in soil and in the rhizosphere (Mazzola *et al.* 1992).

The finding that, after release into the environment, a significant fraction of the bacteria may enter a viable-but-non-cultivable or dormant state, and as such become undetectable with traditional plating methods (Roszak & Colwell 1987; Van Overbeek *et al.* 1990), is particularly intriguing and adds a new dimension to the monitoring of population dynamics of introduced biocontrol strains. In our studies, we found that 9 months after application to the field-scale lysimeters, < 1% of the total number of *P. fluorescens* CHA0 cells (detected by fluorescent antibody staining) in soil remained in a cultivable state, while < 10% were viable-but-non-culturable and over 90% were in a dormant state (Keel *et al.* 1994; Keel & Défago unpublished data). Recent experiments set up in a wheat field indicated that the proportion of the different physiological states of strain CHA0 seems to depend on the habitat examined. Two months after release, most of the cells in sandy and loamy parts of deeper soil layers (40- to 200-cm depth) were in a dormant state; on the plough-pan about 50% of the cells were dormant and the rest viable-but-non-culturable; in the rhizosphere about half the cells were culturable and half dormant; on the root surface and in the root interior, only viable cells could be detected, but 70–90% were non-culturable. Subsequent experiments therefore should focus on conditions that favour resuscitation of non-culturable states, for example soil conditions and nutrient-availability from roots. However, the occurrence of non-culturable cells is a future challenge for safety assessment, since virtually nothing is known about their ecological significance.

Dissemination

One concern with the release of wild-type or genetically modified bacteria is that, once they are established in the rhizosphere, they might eventually be disseminated to non-target sites, for example leading to the contamination of groundwater or other water sources. Bacterial dissemination occurs mainly through passive transport by percolating or run-off water, air, animals and human activities. However, literature about the transport of micro-organisms released for biocontrol under field conditions is scarce. Although the lateral and vertical dissemination of released rhizobacteria has been reported to be limited largely to the first few centimetres from the point of application (e.g. Drahos *et al.* 1992; Weller & Thomashow 1994), these findings must be considered with care.

Percolating water is the major factor acting upon the vertical distribution of introduced bacteria in soil. Percolating water was shown to be crucial for the distribution of a *Pseudomonas* biocontrol strain along the root system (Parke *et al.* 1986). Bacterial transport by percolating water is influenced by soil type and structure, by the occurrence of roots, and by bacterial adhesion properties (Gammack *et al.* 1992; Kluepfel 1993). The presence of earthworms seems to increase bacterial root colonization and movement in soil (Stephens *et al.* 1993). Nevertheless, it has been suggested that the major fraction of bacteria is effectively retained in soil by the mechanisms of straining in the soil matrix or adsorption on soil particles. However, vertical translocation of bacteria is substantially increased in intact soil columns compared with repacked soil columns (Natsch *et al.* 1994). Based on this observation it is likely that a significant fraction of introduced bacteria may be transported by percolating water to deeper soil layers, by following preferential flowpaths (e.g. macropores, cracks, earthworm channels, root channels), rather than being translocated uniformly through the matrix soil.

For a more thorough investigation of this question, we have examined patterns of vertical transport of *P. fluorescens* CHA0 under field conditions in large outdoor lysimeters exposed to natural rainfall and in field plots exposed to a simulated rainstorm, both planted with wheat. Strain CHA0 was applied at 10^8 cfu g^{-1} of surface soil. In the lysimeters, CHA0 was detected for the first time after 200 days in the effluent water at 2.5-m depth, following a heavy natural rainfall. CHA0 continued to leach from soil at a concentration of 10^1–10^4 cfu ml^{-1} effluent until day 230. In the wheat field plots, immediately after bacterial application, heavy precipitation was simulated by the sprinkling, over a period of 8 h, of 40 mm of water containing a blue dye to identify channels of preferential flow. After 1 day, CHA0 counts along the blue-stained macropores in the subsurface soil between

10-cm and 150-cm depth were constant at about 2.5×10^6 cfu g^{-1}. In the matrix soil the number of CHA0 cells was substantially fewer. Two months later, the number of CHA0 cells had decreased sharply in the surface soil. However, CHA0 was recovered at relatively high numbers (10^3–10^5 cfu g^{-1} soil) in deeper soil layers (20- to 200-cm depth) (Keel & Défago unpublished data). It seems, therefore, that after heavy rainfall, released bacteria may be transported in considerable numbers through the channels of preferential flow to deeper soil layers and thereby possibly to the groundwater level.

Approaches to evaluate perturbations
on the indigenous microbiota

The establishment of introduced biocontrol agents in the rhizosphere is likely to affect not only the target pathogen but also many other micro-organisms. Of relevance to environmental safety is whether they negatively affect beneficial micro-organisms, such as mycorrhiza or nitrogen-fixing bacteria, and relevant biological processes in soil. Although positive interactions with other beneficial micro-organisms, for example mycorrhiza, have been reported, improved information about the ecological impact of introduced biocontrol agents on the indigenous microbial community is needed. A number of approaches have been taken to assess quantitative and qualitative changes in the composition of microbial communities, including selective plating, assessment of metabolic patterns (e.g. carbon source utilization), DNA fingerprinting and analysis of fatty acids (Gilbert *et al.* 1993; de Leij *et al.* 1994; Ellis *et al.* 1995; Lemanceau *et al.* 1995). In some cases an influence on native microbial populations could be detected, for example, in a very detailed study, based on the analysis of over 2600 individual isolates and using up to 50 physiological attributes, Gilbert *et al.* (1993) found that introduction of the antagonist *Bacillus cereus* into the rhizosphere caused significant quantitative and qualitative changes in the bacterial communities on roots of soybean. However, further research is needed to assess the ecological significance of possible changes in microbial communities in response to the introduction of a non-resident biocontrol agent. This requires a combination of population diversity and functional analysis. This means that possible changes should be evaluated in relation to actual effects on plant growth or certain key biological processes in soil (e.g. nitrogen fixation, phosphorus solubilization) and to traits of the introduced antagonist (e.g. enhanced biocontrol activity).

Gene transfer

Natural gene exchange among micro-organisms in the environment is an established fact (Levy & Miller 1989; Van Elsas & Smit 1994). With respect to known gene transfer mechanisms (transformation, transduction, conjugation), most data are available on the conjugal transfer of plasmids. Conjugal transfer of broad-host-range plasmids between soil or rhizosphere inhabiting bacteria is amply documented (Van Elsas & Smit 1994). In view of a potential application of biocontrol strains carrying genetically manipulated genes inserted in their chromosomes (rather than on plasmids), it is of interest to see whether chromosomal genes of soil bacteria can also be transmitted. In our laboratory, a conjugation system has been established which consists of donor strains carrying a conjugative plasmid with chromosome mobilizing ability (Cma) (Reimmann & Haas 1993) and auxotrophic recipient strains. Transfer of chromosomal genes transfer of *P. fluorescens* CHA0 and *P. aeruginosa* PAO has been detected *in vitro* and in the rhizosphere of wheat. Conjugative plasmids with Cma occur naturally in environmental isolates of Gram-negative bacteria (Reimmann & Haas 1993). The finding that chromosomal recombinants can be detected readily, illustrates the potential for gene flux in the rhizosphere. The probability of such an event being recovered depends largely on the selective advantage of the recombinants (prototrophic recombinants are better root colonizers than auxotrophic mutants).

ACKNOWLEDGEMENTS

We thank M. S. Wolfe and D. Haas for carefully reading the manuscript and P. Azelvandre, C. Bull, B. Duffy, A. Natsch, U. Schnider, J. Troxler and M. Zala for contributing unpublished data. Support from the Swiss National Foundation for Scientific Research projects 5002-035142 (Priority Programme Biotechnology) and 31-32473.91, the Swiss Federal Office for Environment FE/OFEFP/310.92.46 project and the Swiss Federal Office for Education and Science (EU IMPACT – project PL 920053) is gratefully acknowledged.

REFERENCES

Abdallah, M. A. (1991). Pyoverdins and pseudobactins. *Handbook of Microbial Iron Chelates* (Ed. by G. Winkelmann), pp. 139–153. CRC Press, Boca Raton, FL.

Alström, S. (1991). Induction of disease resistance in common bean susceptible to halo blight bacterial pathogen after seed bacterization with rhizosphere pseudomonads. *Journal of General and Applied Microbiology*, **37**, 495–501.

Bakker, P. A. H. M., Lamers, J. G., Bakker, A. W., Marugg, J. D., Weisbeck, P. J. & Schippers, B. (1986). The role of siderophores in potato tuber yield increase by *Pseudomonas putida* in a short rotation of potato. *Netherlands Journal of Plant Pathology*, **92**, 249–256.

Becker, J. O. & Cook, R. J. (1988). Role of siderophores in suppression of *Pythium* species and production of increased growth response of wheat by fluorescent pseudomonads. *Phytopathology*, **78**, 778–782.

Buysens, S., Poppe, J. & Höfte, M. (1994). Role of siderophores in plant growth stimulation and antagonism by *Pseudomonas aeruginosa* 7NSK2. *Improving Plant Productivity with Rhizosphere*

Bacteria (Ed. by M. H. Ryder, P. M. Stephens & G. D. Bowen), pp. 139–141. CSIRO Division of Soils, Adelaide.

Cook, R. J. (1993). Making greater use of introduced microorganisms for biological control of plant pathogens. *Annual Review of Phytopathology*, **31**, 53–80.

Cook, R. J. & Baker, K. R. (1983). *The Nature and Practice of Biological Control of Plant Pathogens.* American Phytopathological Society, St Paul, MN.

Corbell, N. & Loper, J. E. (1995). A global regulator of secondary metabolite production in *Pseudomonas fluorescens* Pf-5. *Journal of Bacteriology*, **177**, 6230–6236.

de Leij, F. A. A. M., Sutton, E. J., Whipps, J. M. & Lynch, J. M. (1994). Effect of genetically modified *Pseudomonas aureofaciens* on indigenous microbial populations of wheat. *FEMS Microbiology Ecology*, **13**, 249–258.

de Weger, L. A., Dekkers, L. C., Van der Bij, A. J. & Lugtenberg, B. J. J. (1994). Use of phosphate-reporter bacteria to study phosphate limitation in the rhizosphere and in bulk soil. *Molecular Plant–Microbe Interactions*, **7**, 32–38.

Défago, G. & Haas, D. (1990). Pseudomonads as antagonists of soilborne plant pathogens: modes of action and genetic analysis. *Soil Biochemistry*, **6**, 249–291.

Défago, G. & Keel, C. (1995). Pseudomonads as biocontrol agents of diseases caused by soil-borne pathogens. *Benefits and Risks of Introducing Biocontrol Agents* (Ed. by H. M. T. Hokkanen & J. M. Lynch), pp. 137–148. Cambridge University Press, Cambridge.

Défago, G., Berling, C-H., Burger, U., Haas, D., Kahr, G., Keel, C., Voisard, C., Wirthner, P. & Wüthrich, B. (1990). Suppression of black root rot of tobacco and other root diseases by strains of *Pseudomonas fluorescens*: potential application and mechanisms. *Biological Control of Soil-Borne Plant Pathogens* (Ed. by D. Hornby, R. J. Cook, Y. Henis, W. H. Ko, A. D. Rovira, B. Schippers & P. R. Scott), pp. 93–108. CAB International, Wallingford.

Dowling, D. N. & O'Gara, F. (1994). Metabolites of *Pseudomonas* involved in the biocontrol of plant disease. *Trends in Biotechnology*, **12**, 133–141.

Drahos, D. J., Barry, G. F., Hemming, B. C., Brandt, E. J., Kline, E. L., Skipper, H. D., Kluepfel, D. A., Gooden, D. T. & Hughes, T. A. (1992). Spread and survival of genetically marked bacteria in soil. *Release of Genetically Engineered and Other Micro-organisms* (Ed. by J. C. Fry & M. J. Day), pp. 147–159. Cambridge University Press, Cambridge.

Ellis, R. J., Thompson, I. P. & Bailey, M. J. (1995). Metabolic profiling as a means of characterizing plant-associated microbial communities. *FEMS Microbiology Ecology*, **16**, 9–18.

Fenton, A. M., Stephens, P. M., Crowley, J., O'Callaghan, M. & O'Gara, F. (1992). Exploitation of gene(s) involved in 2,4-diacetylphloroglucinol biosynthesis to confer a new biocontrol capability to a *Pseudomonas* strain. *Applied and Environmental Microbiology*, **58**, 3873–3878.

Fridlender, M., Inbar, J. & Chet, I. (1993). Biological control of soilborne pathogens by a (β-1,3 glucanase-producing *Pseudomonas cepacia*. *Soil Biology and Biochemistry*, **25**, 1211–1221.

Gaffney, T. D., Lam, S. T., Ligon, J., Gates, K., Frazelle, A., di Maio, J., Hill, S., Goodwin, S., Torkewitz, N., Allshouse, A. M., Kempf, H.J. & Becker, J.O. (1994). Global regulation of expression of antifungal factors by a *Pseudomonas fluorescens* biological control strain. *Molecular Plant–Microbe Interactions*, **7**, 455–463.

Gammack, S. M., Paterson, E., Kemp, J. S., Cresser, M. S. & Killham, K. (1992). Factors affecting the movement of microorganisms in soil. *Soil Biochemistry*, **7**, 263–305.

Georgakopoulos, D. G., Hendson, M., Panopoulos, N. J. & Schroth, M. N. (1994). Analysis of expression of a phenazine biosynthesis locus of *Pseudomonas aureofaciens* PGS12 on seeds with a mutant carrying a phenazine biosynthesis locus-ice nucleation reporter gene fusion. *Applied and Environmental Microbiology*, **60**, 4573–4579.

Gilbert, G. S., Parke, J. L., Clayton, M. K. & Handelsman, J. (1993). Effects of an introduced bacterium on bacterial communities on roots. *Ecology*, **74**, 840–854.

Glick, B. R. (1995). The enhancement of plant growth by free-living bacteria. *Canadian Journal of Microbiology*, **41**, 109–117.

Gutterson, N. (1990). Microbial fungicides: recent approaches to elucidating mechanisms. *Critical Review in Biotechnology*, **10**, 69–91.

Hamdan, H., Weller, D. M. & Thomashow, L. S. (1991). Relative importance of fluorescent siderophores and other factors in biological control of *Gaeumannomyces graminis* var. *tritici* by *Pseudomonas fluorescens* 2–79 and M4-80R. *Applied and Environmental Microbiology*, **57**, 3270–3277.

Hara, H., Bangera, M., Kim, D.-S., Weller, D. M. & Thomashow, L. S. (1994). Effect of transfer and expression of antibiotic biosynthesis genes on biological control activity of fluorescent pseudomonads. *Improving Plant Productivity with Rhizosphere Bacteria* (Ed. by M. H. Ryder, P. M. Stephens & G. D. Bowen), pp. 247–249. CSIRO Division of Soils, Adelaide.

Hill, D. S., Stein, J. I., Torkewitz, N. R., Morse, A. M., Howell, C. R., Pachlatko, J. P., Becker, J. O. & Ligon, J. M. (1994). Cloning of genes involved in the synthesis of pyrrolnitrin from *Pseudomonas fluorescens* and role of pyrrolnitrin synthesis in biological control of plant disease. *Applied and Environmental Microbiology*, **60**, 78–85.

Hoitink, H. A. J. & Fahy, P. C. (1986). Basis for the control of soilborne plant pathogens with composts. *Annual Review of Phytopathology*, **24**, 93–114.

Homma, Y. & Suzui, T. (1989). Role of antibiotic production in suppression of radish damping-off by seed bacterization with *Pseudomonas cepacia*. *Annals of the Phytopathological Society of Japan*, **55**, 643–652.

Howie, W. J. & Suslow, T. V. (1991). Role of antibiotic biosynthesis in the inhibition of *Pythium ultimum* in the cotton spermosphere and rhizosphere by *Pseudomonas fluorescens*. *Molecular Plant–Microbe Interactions*, **4**, 393–399.

Hrabak, E. M. & Willis, D. K. (1992). The *lemA* gene required for pathogenicity of *Pseudomonas syringae* pv. *syringae* on beam is a member of a family of two-component regulators. *Journal of Bacteriology*, **174**, 3011–3020.

Keel, C., Voisard, C., Berling, C. H., Kahr, G. & Défago, G. (1989). Iron sufficiency, a prerequisite for the suppression of tobacco black root rot by *Pseudomonas fluorescens* strain CHA0 under gnotobiotic conditions. *Phytopathology*, **79**, 584–589.

Keel, C., Schnider, U., Maurhofer, M., Voisard, C., Laville, J., Burger, U., Wirthner, P., Haas, D. & Défago, G. (1992). Suppression of root diseases by *Pseudomonas fluorescens* CHA0: importance of the bacterial secondary metabolite 2,4-diacetylphloroglucinol. *Molecular Plant–Microbe Interactions*, **5**, 4–13.

Keel, C., Zala, M., Troxler, J., Natsch, A., Pfirter, H. A. & Défago, G. (1994). Application of biocontrol strain *Pseudomonas fluorescens* CHA0 to standard soil-columns and field-scale lysimeters – I. Survival and vertical translocation. *Improving Plant Productivity with Rhizosphere Bacteria* (Ed. by M. H. Ryder, P. M. Stephens & G. D. Bowen), pp. 255–257. CSIRO Division of Soils, Adelaide.

Kempf, H.-J., Sinterhauf, S., Müller, M. & Pachlatko, P. (1994). Production of two antibiotics by a biocontrol bacterium in the spermosphere of barley and in the rhizosphere of cotton. *Improving Plant Productivity with Rhizosphere Bacteria* (Ed. by M. H. Ryder, P. M. Stephens & G. D. Bowen), pp. 114–116. CSIRO Division of Soils, Adelaide.

Kluepfel, D. (1993). The behaviour and tracking of bacteria in the rhizosphere. *Annual Review of Phytopathology*, **31**, 441–472.

Kraus, J. & Loper, J. E. (1995). Characterization of a genomic region required for production of the antibiotic pyoluteorin by the biological control agent *Pseudomonas fluorescens* Pf-5. *Applied and Environmental Microbiology*, **61**, 849–854.

Laville, J., Voisard, C., Keel, C., Maurhofer, M., Défago, G. & Haas, D. (1992). Global control in *Pseudomonas fluorescens* mediating antibiotic synthesis and suppression of black root rot of tobacco. *Proceedings of the National Academy of Science, USA*, **89**, 1562–1566.

Leisinger, T. & Margraff, R. (1979). Secondary metabolites of the fluorescent pseudomonads. *Microbiological Reviews*, **43**, 422–442.

Lemanceau, P., Bakker, P. A. H. M., de Kogel, W. J., Alabouvette, C. & Schippers, B. (1992). Effect of pseudobactin 358 production by *Pseudomonas putida* WCS358 on suppression of fusarium wilt of carnations by nonpathogenic *Fusarium oxysporum* F047. *Applied and Environmental Microbiology*, **58**, 2978–2982.

Lemanceau, P., Corberand, T., Gardan, L., Latour, X., Laguerre, G., Boeufgras, J.-M. & Alabouvette, C. (1995). Effect of two plant species, flax (*Linum usitatissinum* L.) and tomato (*Lycopersicon esculentum* Mill.), on the diversity of soilborne populations of fluorescent pseudomonads. *Applied and Environmental Microbiology*, **61**, 1004–1012.

Levy, S. B. & Miller, R. (1989). *Gene Transfer in the Environment.* McGraw-Hill, New York.

Lim, H-S., Kim, Y-S. & Kim, S-D. (1991). *Pseudomonas stutzeri* YPL-1 genetic transformation and antifungal mechanisms against *Fusarium solani*, an agent of plant root rot. *Applied and Environmental Microbiology*, **57**, 510–516.

Loper, J. E. (1988). Role of fluorescent siderophore production in biological control of *Pythium ultimum* by a *Pseudomonas fluorescens* strain. *Phytopathology*, **78**, 166–172.

Loper, J. E. & Buyer, J. S. (1991). Siderophores in microbial interactions on plant surfaces. *Molecular Plant–Microbe Interactions*, **4**, 5–13.

Loper, J. E. & Lindow, S. E. (1994). A biological sensor for iron available to bacteria in their habitats on plant surfaces. *Applied and Environmental Microbiology*, **60**, 1934–1941.

Maurhofer, M., Keel, C., Schnider, U., Voisard, C., Haas, D. & Défago, G. (1992). Influence of enhanced antibiotic production in *Pseudomonas fluorescens* strain CHA0 on its disease suppressive capacity. *Phytopathology*, **82**, 190–195.

Maurhofer, M., Hase, C., Meuwly, Ph., Métraux, J-P. & Défago, G. (1994). Induction of systemic resistance of tobacco to tobacco necrosis virus by the root colonizing *Pseudomonas fluorescens* strain CHA0: influence of the *gacA* gene and of pyoverdin production. *Phytopathology*, **89**, 139–146.

Maurhofer, M., Keel, C., Haas, D. & Défago, G. (1995). Influence of plant species on disease suppression by *Pseudomonas fluorescens* strain CHA0 with enhanced antibiotic production. *Plant Pathology*, **44**, 40–50.

Mazzola, M., Cook, R. J., Thomashow, L. S., Weller, D. M. & Pierson, L. S. III (1992). Contribution of phenazine antibiotic biosynthesis to ecological competence of fluorescent pseudomonads in soil habitats. *Applied and Environmental Microbiology*, **58**, 2616–2624.

Métraux, J. P., Signer, H., Ryals, J., Ward, E., Wyss-Benz, M., Gaudin, J., Raschdorf, K., Schmid, E., Blum, W. & Inverardi, B. (1990). Increase in salicylic acid at the onset of systemic acquired resistance in cucumber. *Science*, **250**, 1004–1006.

Meyer, J. M., Azelvandre, P. & Georges, C. (1992). Iron metabolism in *Pseudomonas*: salicylic acid, a siderophore of *Pseudomonas fluorescens* CHA0. *BioFactors*, **4**, 23–27.

Meyer, J. R. & Linderman, R. G. (1986). Response of subterranean clover to dual inoculation with vesicular–arbuscular mycorrhizal fungi and a growth-promoting bacterium, *Pseudomonas putida*. *Soil Biology and Biochemistry*, **18**, 185–190.

Natsch, A., Keel, C., Pfirter, H. A., Haas, D. & Défago, G. (1994). Contribution of the global regulator gene gacA to persistence and dissemination of *Pseudomonas fluorescens* biocontrol strain CHA0 introduced into soil microcosms. *Applied and Environmental Microbiology*, **60**, 2553–2560.

Nelson, E. B. (1992). Bacterial metabolism of propagule stimulants as an important trait in the biocontrol of *Pythium* seed infections. *Biological Control of Plant Diseases, Progress and Challenges for the Future* (Ed. by E. C. Tjamos, G. C. Papavizas & R. J. Cook), pp. 353–357. Plenum Press, New York.

Noling, J. W. & Becker, J. O. (1994). The challenge of research and extension to define and implement alternatives to methyl bromide. *Journal of Nematology*, **26(4S)**, 573–586.

O'Sullivan, D. J. & O'Gara, F. (1992). Traits of fluorescent *Pseudomonas* spp. involved in suppression of plant root pathogens. *Microbiological Reviews*, **56**, 662–676.

Park, C. S., Paulitz, T. C. & Baker, R. (1988). Biocontrol of fusarium wilt of cucumber resulting from interactions between *Pseudomonas putida* and non-pathogenic isolates of *Fusarium oxysporum. Phytopathology*, **78**, 194–199.

Parke, J. L. (1990). Root colonization by indigenous and introduced microorganisms. *The Rhizosphere and Plant Growth* (Ed. by D. L. Keister & P. B. Gregan), pp. 33–42. Kluwer Academic Publishers, Dordrecht.

Parke, J. L., Moen, R., Rovira, A. D. & Bowen, G. D. (1986). Soil water flow affects the rhizosphere distribution of a seed-borne biological control agent, *Pseudomonas fluorescens. Soil Biology and Biochemistry*, **18**, 583–588.

Paulitz, T. C. (1990). Biochemical and ecological aspects of competition in biological control. *New Directions in Biological Control: Alternatives for Suppressing Agricultural Pests and Diseases* (Ed. by R. R. Baker & P. E. Dunn), pp. 713–724. Alan R. Liss, New York.

Paulitz, T. C. & Loper, J. E. (1991). Lack of a role for fluorescent siderophores production in the biological control of *Pythium* damping-off of cucumber by a strain of *Pseudomonas putida. Phytopathology*, **81**, 930–935.

Pfender, W. F., Kraus, J. & Loper, J. E. (1993). A genomic region from *Pseudomonas fluorescens* Pf-5 required for pyrrolnitrin production and inhibition of *Pyrenophora tritici-repentis* in wheat straw. *Phytopathology*, **84**, 940–947.

Pierson, E. A. & Weller, D. M. (1994). Use of mixtures of fluorescent pseudomonads to suppress take-all and improve the growth of wheat. *Phytopathology*, **84**, 940–947.

Pierson, L. S. III, Keppenne, V. D. & Wood, D. W. (1994). Phenazine antibiotic biosynthesis in *Pseudomonas aureofaciens* 30-84 is regulated by PhzR in response to cell density. *Journal of Bacteriology*, **176**, 3966–3974.

Reimmann, C. & Haas, D. (1993). Mobilization of chromosomes and nonconjugative plasmids by cointegrative mechanisms. *Bacterial Conjugation* (Ed. by D. B. Clewell), pp. 137–187. Plenum Press, New York.

Rich, J. J., Kinscherf, T. G., Kitten, T. & Willis, D. K. (1994). Genetic evidence that the *gacA* gene encodes the cognate response regulator for the *lemA* sensor in *Pseudomonas syringae. Journal of Bacteriology*, **176**, 7468–7475.

Roszak, D. B. & Colwell, R. R. (1987). Survival strategies of bacteria in the natural environment. *Microbiological Reviews*, **51**, 365–379.

Ryder, M. H., Pankhurst, C. E., Rovira, A. D., Correll, R. L. & Ophel Keller, K. M. (1994). Detection of introduced bacteria in the rhizosphere using marker genes and DNA probes. *Molecular Ecology of Rhizosphere Microorganisms* (Ed. by F. O'Gara, D. Dowling & B. Boesten), pp. 29–47. VCH, Weinheim.

Sacherer, P., Défago, G. & Haas, D. (1994). Extracellular protease and phospholipase C are controlled by the global regulatory gene *gacA* in the biocontrol strain *Pseudomonas fluorescens* CHA0. *FEMS Microbiology Letters*, **16**, 155–160.

Salmond, G. P. C., Bycroft, B. W., Stewart, G. S. A. & Williams, P. (1995). The bacterial 'enigma': cracking the code of cell-cell communication. *Molecular Microbiology*, **16**, 615–624.

Sarniguet, A., Kraus, J. & Loper, J. E. (1994). An *rpoS* homolog affects antibiotic production, ecological fitness and suppression of plant diseases by *Pseudomonas fluorescens* Pf-5. *Molecular Ecology*, **3**, 607.

Scher, F. M. (1986). Biological control of fusarium wilts by *Pseudomonas putida* and its enhancement by EDDHA. *Iron, Siderophores and Plant Diseases* (Ed. by T. R. Swinburne), pp. 109–117. Plenum Press, New York.

Schnider, U., Keel, C., Blumer, C., Troxler, J., Défago, G. & Haas, D. (1995). Amplification of the house-keeping sigma factor in *Pseudomonas fluorescens* CHA0 enhances antibiotic production and improves biocontrol abilities. *Journal of Bacteriology*, **177**, 5387–5392.

Stephens, P. M., Davoren, C. W., Ryder, M. H. & Doube, B. M. (1993). Influence of the lumbricid earthworm *Aporrectodea trapezoides* on the colonization of wheat roots by *Pseudomonas corrugata* strain 2140R in soil. *Soil Biology and Biochemistry*, **25**, 1719–1724.

Stutz, E. W., Défago, G. & Kern, H. (1986). Naturally occurring fluorescent pseudomonads involved in suppression of black root rot of tobacco. *Phytopathology*, 76, 181–185.

Thomashow, L. S., Weller, D. M., Bonsall, R. F. & Pierson, L. S. III (1990). Production of the antibiotic phenazine-1-carboxylic acid by fluorescent *Pseudomonas* species in the rhizosphere of wheat. *Applied and Environmental Microbiology*, 56, 908–912.

Toyoda, H., Hashimoto, H., Utsumi, R., Kobayashi, H. & Ouchi, S. (1988). Detoxification of fusaric acid by a fusaric acid-resistant mutant of *Pseudomonas solanacearum* and its application to biological control of fusarium wilt of tomato. *Phytopathology*, 78, 1307–1311.

Van Elsas, J. D. & Smit, E. (1994). Some considerations on gene transfer between bacteria in soil and rhizosphere. *Molecular Ecology of Rhizosphere Microorganisms* (Ed. by F. O'Gara, D. Dowling & B. Boesten), pp. 151–164. VCH, Weinheim.

Van Elsas, J. D., Trevors, J. T., Jain, D., Wolters, A. C., Heijnen, C. E. & Van Overbeek, L. S. (1992). Survival of, and root colonization by, alginate encapsulated *Pseudomonas fluorescens* cells following introduction into soil. *Biology and Fertility of Soils*, 14, 14–22.

Van Overbeek, L. S., Van Elsas, J. D., Trevors, J. T. & Starodub, M. E. (1990). Long-term survival of and plasmid stability in *Pseudomonas* spp. and *Klebsiella* spp. and appearance of nonculturable cells in agricultural drainage water. *Microbial Ecology*, 19, 239–250.

Van Peer, R., Niemann, G. J. & Schippers, B. (1991). Induced resistance and phytoalexin accumulation in biological control of fusarium wilt of carnation by *Pseudomonas* sp. WCS417r. *Phytopathology*, 81, 728–734.

Voisard, C., Keel, C., Haas, D. & G. Défago, G. (1989). Cyanide production by *Pseudomonas fluorescens* helps suppress black root rot of tobacco under gnotobiotic conditions. *European Molecular Biology Organisation Journal*, 8, 351–358.

Voisard, C., Bull, C., Keel, C., Laville, J., Maurhofer, M., Schnider, U., Défago, G. & Haas, D. (1994). Biocontrol of root diseases by *Pseudomonas fluorescens* CHA0: current concepts and experimental approaches. *Molecular Ecology of Rhizosphere Microorganisms* (Ed. by F. O'Gara, D. Dowling & B. Boesten), pp. 67–89. VCH, Weinheim.

Wei, G., Kloepper, J. W. & Tuzun, S. (1991). Induction of systemic resistance of cucumber to *Colletotrichum orbiculare* by selected strains of plant growth-promoting rhizobacteria. *Phytopathology*, 81, 1508–1512.

Weller, D. M. (1988). Biological control of soilborne plant pathogens in the rhizosphere with bacteria. *Annual Review of Phytopathology*, 26, 379–407.

Weller, D. M. & Thomashow, L. S. (1994). Current challenges in introducing beneficial microorganisms into the rhizosphere. *Molecular Ecology of Rhizosphere Microorganisms* (Ed. by F. O'Gara, D. Dowling & B. Boesten), pp. 1–18. VCH, Weinheim.

3. INTERACTIONS BETWEEN FUNGI AND PLANT PATHOGENS IN SOIL AND THE RHIZOSPHERE

J. M. WHIPPS

Department of Plant Pathology and Microbiology, Horticulture Research International, Wellesbourne, Warwick CV35 9EF, UK

INTRODUCTION

Historically, soil fungi and soil-borne plant pathogens have been studied by disparate groups of researchers with different aims. Soil ecologists have been concerned with the study of the behaviour of fungi in soil and the rhizosphere, whereas plant pathologists have been more interested in the effects of fungal pathogens on their hosts and methods for their control. However, recently the need to combine these different approaches has been stimulated with the increase in interest in biological disease control, brought about through environmental concerns over the use of pesticides in general. Because of problems involving poor reproducibility of control, the significance of applying sound ecological principles to all stages of the research and development procedure for any biological control agent is gradually being recognized. Antagonists are being selected from environments where they will be required to work as well as being screened for activity under environmentally realistic conditions. Indeed, several new commercial biologial control products for use against soil-borne plant pathogens have been selected in this way (Whipps 1996 in press). Consequently, in this review, ecological aspects of fungal–fungal interactions in soil, on roots and seeds are considered against a background of developing successful or reproducible biological disease control. Only ecologically obligate mycoparasites or saprophytic fungi with potential for antagonistic behaviour towards plant pathogens are considered here.

FUNGAL–FUNGAL INTERACTIONS

There is a huge literature concerning the study of interactions between fungal pathogens and other fungi but it is largely based on experiments carried out *in vitro*. Frequently, the conditions used bear little or no relation to the soil or rhizosphere are so the results may be viewed as laboratory oddities. However, when *in vitro* studies are designed to mimic the normal environment, they can provide evidence for potential modes of action in the natural situation. In addition, such experiments can provide physiological information on the fungi which can be useful if they

47

reach the developmental stage of biological disease control where inoculum production is required.

Several modes of action have been identified during fungal–fungal interactions involving fungi attacking or controlling plant pathogenic fungi, none of which is mutually exclusive. These include direct antagonism involving mycoparasitism, where contact and nutrient transfer from host to parasite occurs through combat between individual hyphae or cells; antibiosis, where antagonistic fungi secrete metabolites harmful to other micro-organisms; and competition, where demand exceeds immediate supply of materials or space. In some interactions, hyphal interference may occur following hyphal contact when cell contents of the host hyphae rupture or become disorganized in a localized region of the affected mycelium. This localized effect contrasts with generalized antibiotic production or non-specific enzyme release which do not constitute part of true mycoparasitism in most cases, although induction of specific degradative enzymes during hyphal penetration may do so. In the field, however, antibiotic effects may occur first and continue during a mycoparasitic phase.

A further indirect mode of action termed cross-protection or induced resistance is known. Here, exposure of a plant to a species or strain of fungus that is non-pathogenic, or only mildly so, results in a subsequent decrease in disease after exposure to a pathogen but the inducer organism may not be directly involved in combat with the pathogen. Here, effectively, the plant can become the antagonist. Alternatively, the plant can respond to the presence of some fungi by increasing growth, thus diminishing the effects of any subsequent pathogen attack.

All aspects of interactions have been widely reviewed in depth (e.g. Burge 1988; Whipps & Lumsden 1989, 1991; Lumsden 1992; Jeffries & Young 1994) and so emphasis here is placed, wherever possible, on studies that relate directly to observations in the field or in microcosm studies relevant to the field situation.

DIRECT ANTAGONISM

Mycoparasitism

There are relatively few detailed studies examining the significance of mycoparasitism in soil or the rhizosphere. Most relate to the recovery of natural or baited pathogen propagules which are then found to contain mycelium or spores of putative mycoparasites. The viability of the pathogen propagules at the time of infection is unknown and whether the fungi isolated act as mycoparasites or merely secondary invaders of already damaged tissue is often unclear, limiting the value of such observations in some cases. Nevertheless, in ecological terms, the greatest opportunity for mycoparasitism of propagules is likely to occur within the rhizosphere, and in and around decomposing plant residues, where these reproductive and resting structures are produced in greatest numbers. The presence of these nutrient-rich, persistent structures is likely to have encouraged the development of mycoparasitism, reflected by the occurrence of some highly specialized fungi which have become ecologically dependent on them for survival.

Oospores of *Aphanomyces*, *Phytophthora*, *Pythium* and *Sclerospora* species have been found to be infected by a wide range of fungi from different taxonomic groups including *Dactylella* spp., *Fusarium mesmoides*, *Hyphochytrium catenoides* and *Trinacrium subtile* (Drechsler 1938; Rao & Pavgi 1976; Sneh, Humble & Lockwood 1977; Hoch & Abawi 1979; Wynn & Epton 1979; Humble & Lockwood 1981; Daft & Tsao 1984). The level of parasitism can be extremely high in some cases; for instance, depending on the soil, over 70% of the oospores of *Phytophthora cinnamomi* and *P. nicotiana* incorporated as bait could become infected by *Catenaria anguillae* (Daft & Tsao 1984), suggesting that such mycoparasitism may be involved in the suppressiveness of some soils to these pathogens.

Similar approaches have also been used in relation to sclerotial mycoparasites and extensive lists of species isolated from sclerotia have been compiled (e.g. Whipps 1991; Jeffries & Young 1994). Species of *Trichoderma* and *Gliocladium* are frequently reported to occur in sclerotia of many different pathogens including *Phymatotrichum omnivorum*, *Rhizoctonia* spp., *Sclerotium* spp. and *Verticillium dahliae* (Gladders & Coley-Smith 1980; Howell 1982; Artigues & Davet 1984; Zazzerini & Tosi 1985; Kenerley & Stack 1987; Keinath *et al.* 1991; Van den Boogert & Saat 1991) and *Talaromyces flavus* has been repeatedly isolated from sclerotia of *Rhizoctonia* spp., *Sclerotinia* spp. and *Verticillium dahliae* (Marois *et al.* 1984; McLaren *et al.* 1986). Depending on the species and strain, cell-wall degrading enzymes and other lytic enzymes, such as proteases and oxidases, and antibiotics have all been implicated as having a role in the activity of these mycoparasites against plant pathogenic fungi (Elad *et al.* 1982; Brückner & Przybylski 1984; Howell 1987; Ridout *et al.* 1988; Barak & Chet 1990; Ghisalberti & Sivasithamparam 1991; Fravel & Roberts 1991; Ordentlich *et al.* 1991; Lumsden *et al.* 1992; Benhamou & Chet 1993). *In vitro*, synergism between different hydrolytic enzymes and between hydrolytic enzymes and antibiotics for inhibition of spore germination of several fungi has been demonstrated (Di Pietro *et al.* 1993; Lorito *et al.* 1993a, b, 1994; Schirmböck *et al.* 1994) suggesting a concert of enzyme and antibiotic activities may be involved during mycoparasitism in nature. Lectins may also be involved in the recognition process between mycoparasite and host (Barak & Chet 1990).

These facultative mycoparasites have a wide host range and are not ecologically dependent on their hosts for long-term survival. In contrast, some mycoparasites are generally only found in association with their hosts; for example, *Coniothyrium minitans* and *Sporidesmium sclerotivorum* are mycoparasites of sclerotia of *Sclerotinia* species and some ascomycetous *Sclerotium* species (Adams & Ayers

1979; Whipps & Gerlagh 1992). Similarly, *Verticillium biguttatum* is associated with sclerotia and mycelia of *Rhizoctonia solani* (Van den Boogert & Saat 1991; Morris *et al.* 1992). In the absence of a suitable host, these ecologically obligate mycoparasites remain inactive as spores in the soil; or as spores or mycelium within parasitized organs or associated with plant material colonized by the host (Adams & Ayers 1979; Van den Boogert & Velvis 1992; Whipps & Gerlagh 1992).

There is growing evidence that these ecologically obligate mycoparasites can impose a major constraint on disease caused by sclerotium-forming pathogens. In a 7-year study in Canada, two sunflower fields naturally infested with *Sclerotinia sclerotiorum* showed a rapid decline in incidence of sclerotinia wilt associated with sunflower (Huang & Kozub 1991). In one field, the wilt declined from more than 90% to less than 10% during the 6-year period after the disease reached its peak, and was related to a high frequency of indigenous mycoparasites, notably *C. minitans*. *C. minitans* is now becoming recorded in sclerotinia-suppressive soils from other parts of North America and Europe (McLaren 1989; Pfeffer & Lüth 1990; Gerlagh & Vos 1991; Whipps *et al.* 1993). Similarly, *Sporidesmium sclerotivorum* has also been associated with natural biological control of *Sclerotinia minor* on lettuce in the USA (Adams & Ayers 1981). The specificity of this mycoparasite is significant in that in soil, spores will only germinate in close proximity to host sclerotia but following colonization, it can grow from sclerotium to sclerotium through the soil (Adams *et al.* 1984). The production of haustoria within the sclerotium cells (Bullock *et al.* 1989) and the stimulation of host glucanases resulting in degradation of glucans to glucose, which *S. sclerotivorum* utilizes as a carbon source for growth (Adams & Ayers 1983), highlights the closeness of this relationship.

Most recently, population studies have provided evidence that *Verticillium biguttatum* can control the development of *Rhizoctonia solani* on potato in the long term (Van den Boogert & Velvis 1992). In fields infested with *R. solani*, the pathogen colonized the roots and developing potato tubers rapidly. Subsequently, depending on the initial soil population density, temperature and soil type, the populations of *V. biguttatum* on the roots and tubers increased. In the absence of potato, no increase in *V. biguttatum* occurred as contact between the plant pathogen and mycoparasite could not be established. Consequently, consecutive cropping would serve to increase total populations of *V. biguttatum* and could be one of the main reasons for the decline in *Rhizoctonia* disease observed under monoculture of potato in The Netherlands. Significantly, observations *in vitro* have shown that parasitism of *R. solani* by *V. biguttatum* involves the production of haustoria by *V. biguttatum* in the hyphae of *R. solani* (Van den Boogert & Deacon 1994), possibly accompanied by release of low molecular weight antibiotics (Morris *et al.* 1995) and this may have a long-term weakening effect on the pathogen by limiting sclerotium formation and inoculum carryover.

Antibiosis

Antibiotics have often been considered to be relatively low molecular weight materials (< 1 kDa), often secondary metabolites produced when nutrients become limiting, which can inhibit microbial growth (Lewis *et al.* 1991). However, much the same effect can be achieved *in vitro* by fungi that freely release larger moieties such as oxidases or cell-wall degrading enzymes (Lynch 1990). Consequently, the significance of both groups of inhibitory compounds in control of plant pathogenic fungi is considered here. By virtue of the difficulty of recovery of such materials from soil or the rhizosphere due to rapid breakdown or binding to clay particles, direct evidence for their involvement in interactions in the field is scanty. This is in marked contrast to the situation *in vitro* where most studies are done.

A role for antibiotic production *in vivo* has been suggested by the use of various antibiotic-minus mutants of *Gliocladium virens* lacking ability to produce the non-enzymic metabolites glioverin (steroid-like antibiotic) and gliotoxin (an epipolythiodioxopiperazine) (Howell & Stipanovic 1983; Howell 1987, 1991). A strain of *G. virens* that suppressed *Pythium*-induced damping-off in cotton seedlings was shown to produce the antibiotic glioverin in culture. Antibiotic-minus mutants failed to give biological control and were overgrown by *P. ultimum in vitro*. An overproducing mutant of *G. virens* was more inhibitory in culture than the wild-type and showed similar disease-suppressive activity, even though its growth was slower than the wild-type. Similarly, mutants of another isolate of *G. virens*, (GL-21), lacking ability to produce gliotoxin, exhibited only 54% of the disease-suppressive ability of the wild-type isolate *in vivo* and displayed a near total loss of antagonistic activity *in vitro* toward *P. ultimum* (Wilhite *et al.* 1994). Gliotoxin could be detected in soil following addition of inocula of *G. virens* GL-21 and was correlated with disease-suppression activity towards *P. ultimum* and *Rhizoctonia solani* in non-sterile growing media (Lumsden *et al.* 1992). Even though culture supernatants from *G. virens* GL-21 grown in 5% bran extract contained laminarinase, amylase, carboxymethylcellulase, chitinase and protease activity in addition to gliotoxin, the culture filtrates still remained inhibitory to sporangial germination and mycelial growth of *P. ultimum* when the enzyme activities were inactivated by heating or removed by ultra-filtration (Roberts & Lumsden 1990). This further emphasizes the significance of gliotoxin production for biocontrol activity with this isolate of *G. virens*.

Production of the metabolite, chaetomin (an epidithiadiketopiperazine), by *Chaetomium globosum* was also correlated with efficacy in suppressing *Pythium* damping-off of sugar beet in heat-pasteurized soil (Di Pietro *et al.* 1992). Moreover, chaetomin was extracted from pasteurized soil inoculated with strain Cg-13, an effective biocontrol strain, but not from pasteurized soil inoculated with a spontaneous variant unable to suppress *Pythium* damping-off.

Much the same approach has been used to demonstrate a role for glucose oxidase in the biocontrol of verticillium wilt by *Talaromyces flavus* (Fravel & Roberts 1991). *Talaromyces flavus* Tf-1 produces glucose oxidase which suppresses radial growth of *V. dahliae in vitro* and kills microsclerotia of *V. dahliae in vitro* and in soil. Glucose oxidase exhibits antibiotic activity in the presence of glucose due to the production of the reaction product, hydrogen peroxide (Kim, Fravel & Papavizas 1988). A single-spore variant, Tf-1-np, which produced 2% of the level of glucose oxidase activity of the wild-type, did not control verticillium wilt of eggplant in non-sterile field soil in a greenhouse experiment, whilst the wild-type significantly reduced the incidence of wilt (Fravel & Roberts 1991). Importantly, purified glucose oxidase of Tf-1 significantly reduced the growth rate of *V. dahliae* in the presence, but not in the absence, of eggplant roots, suggesting that a supply a glucose from the roots was of major importance.

The significance of the generalized production of lytic enzymes such as chitinases, β-1,3 glucanases, proteases and lipases, capable of degrading fungal tissues during biocontrol in the field, is still a moot point. Although the ability to produce these enzymes *in vitro* is easily demonstrated (Elad *et al.* 1985; Ridout *et al.* 1988; Lewis, Whipps & Cooke 1989; Tweddell *et al.* 1994), it would seem energetically wasteful to produce quantities of these compounds in excess of those required, for example, during hyphal–hyphal combat and resource capture. Consequently, such lytic enzymes may play a key role during the mycoparasitic process, especially during penetration of cell walls where production may be regulated, but may not be a normal ecological feature. Therefore, they are not considered further.

Competition

Competition occurs between micro-organisms when space or nutrients are limited. Evidence for the role of competition between fungi in biocontrol is always indirect and frequently comes from experiments involving additions of nutrients or manipulation of the environment; for example, additions of dried bean leaf material to Nunn sandy loam soil decreased propagule density of plant pathogenic *Pythium ultimum* and this was related to an increase in the antagonistic species, *P. nunn* (Paulitz & Baker 1987a, b). The increase in *P. nunn* was dependent on the food base and, in the absence of *P. nunn*, addition of the bean leaf material resulted in an increase in disease (Paulitz & Baker 1987b). This indicated that *P. nunn* and *P. ultimum* occupied overlapping environmental niches (Paulitz & Baker 1987a). The situation was somewhat different when *P. nunn* was used as a biological control agent of *Phytophthora* species (Fang & Tsao 1995). Although addition of *P. nunn* at 300 propagules per gram to a peat/sand mix containing 1% ground rolled oats resulted in a reduction of population densities of *P. cinnamomi*, *P. citrophthora* and one of two isolates of *P. parasitica*, root rot of azalea caused by

P. cinnamomi or *P. parasitica* and root rot of sweet orange caused by *P. parasitica*, was not suppressed. However, when *P. nunn* was applied at 1000 propagules per gram peat/sand mix under similar conditions, disease suppression in sweet orange caused by *P. nunn* was increased, indicating the importance of relative inoculum levels in determining the outcome of any interactions and, thus, biological control.

Other examples of competition for substrates are known. For instance, in the San Joachim Valley of California in soils with raised chloride levels, *Pythium oligandrum* was able to colonize leaf debris more successfully than *P. ultimum* (Martin & Hancock 1985, 1986). Here, *P. ultimum* was less tolerant to chloride than *P. oligandrum* which resulted in the suppression of saprophytic activity of *P. ultimum* and suppression of disease.

Competition for seed or root exudates may also be important for control of several pathogens; for example, chlamydospore germination of *Fusarium oxysporum* in melon and cotton rhizosphere soil was significantly inhibited after soil or seed application with *Trichoderma harzianum* (Sivan & Chet 1989). However, addition of an excess of seedling exudates, glucose or asparagine eliminated the inhibition due to *T. harzianum* and reduced the disease-control capacity of the antagonist. Interestingly, in the presence of a sterile red fungus (SRF) known to control take-all caused by *Gaeumannomyces graminis* var. *tritici*, foliar applications of thiamine on wheat resulted in an increase in severity of take-all (Shankar *et al.* 1994). It was suggested that competition for thiamine occurred naturally in the rhizosphere where it is present at low levels and that addition of thiamine alleviated the normal competitive effect. The SRF required levels of 0.6 µg thiamine l^{-1} for growth in culture whereas the take-all fungus required 20 µg l^{-1}. This competition, combined with antibiotic production and parasitism by SRF may result in the overall suppressive effect observed in nature.

General soil suppressiveness to *Fusarium* diseases may also act, albeit in part, through competition for nutrients (Alabouvette *et al.* 1985). In a suppressive soil from Chateaurenard in France, a greater microbial biomass was present compared with conducive soils and it was suggested that this led to greater nutrient competition and subsequent inhibition of *Fusarium* species. Recently, it was demonstrated that suppression of pathogenic *Fusarium* species by non-pathogenic *Fusarium* isolates involved competition for carbon (Couteaudier 1992) and that this was related to iron availability of the soil (Lemanceau *et al.* 1993). It was suggested that siderophores produced by *Pseudomonas* species lowered overall iron availability which affected pathogenic strains of *F. oxysporum* to a greater extent than non-pathogenic ones. This again demonstrates the complexity of the interactions that may occur in suppressive soils.

Within the rhizosphere, competition for space as well as nutrients can occur. Occupation by antagonists of the root surface, senescing cortical cells or zones

within the roots, thus denying access to pathogens, could be a key feature for biocontrol activity; for instance, rhizosphere competent isolates of *Trichoderma* have
been obtained by mutagenesis and their ability to utilize cellulosic substrates
associated with the root have been correlated with rhizosphere competence (Ahmad
& Baker 1987). Similarly, ability to occupy the rhizosphere of potato, cotton and
aubergine is viewed as an important attribute in the biological control of *Verticillium
dahliae* by *Talaromyces flavus* (Marois *et al.* 1984). Rhizosphere competence is
also thought to be important in the control of *Phytophthora cinnamomi* by *Myrothecium roridum* on avocado (Gees & Coffey 1989), *Gaeumannomyces graminis*
by *Phialophora graminicola* and *Idriella bolleyi* on wheat (Kirk & Deacon
1987; Lascaris & Deacon 1991; Hemens *et al.* 1992) and *Rhizoctonia solani*
by hypovirulent isolates of *R. solani* on radish and cotton (Sneh *et al.* 1989a, b).
Occupation of the vascular tissue by non-pathogenic isolates of *Fusarium* may
also be important for the control of *Fusarium oxysporum* f. sp. *dianthi* on carnation
(Postma & Rattink 1991).

INDIRECT INTERACTIONS

Cross-protection and induced resistance

Inoculation of plants with avirulent or non-pathogenic isolates of fungal pathogens
can prevent or reduce disease expression when the same plants are subsequently
challenged with a virulent isolate of the same or closely related pathogen. This
phenomenon of cross-protection is widely reported for *Fusarium* species; for instance, fusarium wilt of sweet potato has been decreased by dipping freshly cut
ends of cuttings into suspensions of bud cells of non-pathogenic isolates of *F.
oxysporum* before planting (Ogawa & Komada 1985), a non-pathogenic strain of
F. oxysporum introduced into soil reduced wilt of cucumber caused by *F. oxysporum*
f. sp. *cucumerinum* (Mandeel & Baker 1991), an avirulent strain of *F. oxysporum*
f. sp. *niveum* inoculated on to the roots of watermelon provided good control of
wilt in the field (Martyn *et al.* 1991) and a non-pathogenic isolate of *F. oxysporum*
drenched on to steamed soil before application of a pathogenic isolate of *F.
oxysporum* f. sp. *dianthi* virtually eliminated carnation wilt (Rattink 1989). Similar
effects have been reported with other pathogens; for example, control of verticillium
wilt of tomato has been obtained by dipping roots in a suspension of an avirulent
strain of *Verticillium albo-atrum* prior to planting (Matta & Garibaldi 1977) and
dipping shoot-tip cuttings of peppermint in conidial suspensions of *V. nigrescens*
protected the cuttings against wilt caused by *V. dahliae* (Melouk & Horner 1975).
Rhizoctonia damping-off of cotton, radish and wheat has been depressed with
avirulent strains of *Rhizoctonia solani* (Ichielevich-Auster *et al.* 1985), and avocado
could be protected from *Phytophthora citricola* and *P. cinnamomi* by prior inoculation

with *P. parasitica*, a non-pathogen on avocado (Dolan *et al.* 1986). Rather than using only avirulent strains of the same species or genus, cross-protection can also be carried out using fungi from different genera; for instance, take-all of wheat caused by *G. graminis* f. sp. *tritici* may be prevented from growing within the stele by prior inoculation with *Phialophora graminicola* (Speakman & Lewis 1978).

Importantly, in all the cases cited above, there is evidence that the cross-inoculation involves some degree of induced resistance in the host, and this mechanism is different to those systems where an inoculated micro-organism occupies the surface of the root or enters the root and directly competes or antagonizes the pathogen as mentioned earlier. Thus, *Phialophora graminicola* induces changes in the stele of wheat such that it becomes resistant to the take-all pathogen (Speakman & Lewis 1978), and phytoalexins are synthesized both in carnation, in response to inoculation with a non-pathogenic isolate of *Fusarium oxysporum*, and in cotton, in response to inoculation with mildly virulent *Verticillium albo-atrum* (Rattink 1989; Zaki *et al.* 1972).

Other indirect evidence for the role of induced systemic resistance comes from experiments where inoculation with the avirulent isolate is physically separate from the site of subsequent challenge with the pathogen; for example, sweet potato cuttings prior inoculated with a non-pathogenic isolate of *Fusarium oxysporum* were resistant to injection with a pathogenic isolate 10 cm from the inoculated cut ends (Ogawa & Komada 1985). A similar result was also obtained in cucumber when roots were treated with a non-pathogenic isolate of *F. oxysporum* to protect from stem infection by *F. oxysporum* f. sp. *cucumerinum* (Mandeel & Baker 1991) and in avocado when the stem was challenged with a drop of water containing zoospores of either *Phytophthora citricola* or *P. cinnamomi* 4 days after root treatment with *P. parasitica* (Dolan *et al.* 1986). An interval between prior inoculation and challenge is always required for this induced resistance to be successful, presumably to allow for the signal initiating the response to be transmitted. Perhaps the best evidence for systemic induced resistance involving soil-borne pathogens comes from experiments with split root systems. Here, half root systems are exposed to the cross-inoculating strain and subsequently, the other half exposed to the pathogen. Using this system in cucumber, a non-pathogenic strain of *Fusarium oxysporum* induced resistance to *F. oxysporum* f.sp. *cucumerinum* (Mandeel & Baker 1991).

This area of work is extremely topical at the moment, for not only can fungi induce resistance in plants, but viruses, bacteria and a range of inorganic and organic moieties can have the same effect (Fulton 1986; Van Peer & Schippers 1992; Wilson *et al.* 1994; Keel & Défago, this volume). Although developed to combat aerial pathogens and post-harvest diseases, control of pathogens in the rhizosphere and root environment in this way is only just beginning to be addressed and holds great promise for the future.

Plant growth promotion

In recent years there have been growing numbers of reports of growth promotion in plants in response to incorporation of both bacteria and fungi in soil or the rhizosphere. This has stemmed from work on biological control agents used in systems lacking pathogens. A massive literature exists on this topic for bacteria, and this, together with the closely related areas of endophytes and mycorrhizas, is covered elsewhere in this volume (Barea *et al.*; Clay; Keel & Défago, this volume). But, except for a few reports involving the effects of sterile fungi isolated from the rhizosphere of some members of the Gramineae (Dewan & Sivasithamparam 1989; Shivana *et al.* 1994), and of non-pathogenic isolates of *Rhizoctonia solani* (Sneh *et al.* 1986), work on plant growth promotion by fungi has almost exclusively focused on the effect of *Trichoderma* species.

Increased plant growth in response to *Trichoderma* has been reported in a large range of plants including vegetables such as aubergine, bean, cucumber, lettuce, pea, pepper, radish and tomato (Baker *et al.* 1984; Chang *et al.* 1986; Paulitz *et al.* 1986; Ahmad & Baker 1988; Baker 1988; Coley-Smith *et al.* 1991; Kleifeld & Chet 1992;, Ousley *et al.* 1993, 1994b; Inbar *et al.* 1994) and flowers such as alyssum, carnation, chrysanthemum, French marigold, moss rose, marigold, periwinkle, petunia, snapdragon, tobacco and verbena (Baker *et al.* 1984; Chang *et al.* 1986; Paulitz *et al.* 1986; Baker 1988; Ousley *et al.* 1994a). The increased growth responses which have been measured differ between experiments and crops, but for vegetables they have been expressed in terms of shorter germination times and increased percentage germination, plant height, fresh weight, dry weight and leaf area. For flowers, additional parameters are available. These have included, for petunia and periwinkle, increased number of blooms and earlier flowering, and for chrysanthemum and carnation, enhanced rooting of cuttings.

Although there are substantial numbers of observations of this growth-promotion phenomenon, the interactions between *Trichoderma* isolates, inoculum form and growth conditions are highly complex. Consequently, it is extremely difficult to make generalizations or predictions of the effect of a single *Trichoderma* isolate taken from one system and used in another; for example, there are inherent differences in ability of strains to increase plant growth, some working only in one system and not in another, others lacking stability of effect or exhibiting potential to become phytotoxic (Chang *et al.* 1986; Ousley *et al.* 1993, 1994b). Inoculum form, age and rate of application of some strains can influence growth promotion (Baker *et al.* 1984; Kleifeld & Chet 1992; Ousley *et al.* 1993) and the medium used for growing the plants can influence the growth response (Paulitz *et al.* 1986; Kleifeld & Chet 1992). Nevertheless, some strains such as *T. harzianum* WT and *T. harzianum* T-95 from Colorado (Chang *et al.* 1986; Ousley *et al.* 1994b), *T. harzianum* T-203 from Israel (Chang *et al.* 1986; Inbar *et al.* 1994) and several

Trichoderma isolates from the UK (Ousley *et al.* 1994b) do appear to retain a general growth promotion capability across a large range of systems. Nevertheless, the key growth-promoting properties of these strains are still unclear.

A variety of mechanisms have been proposed to account for the growth promotion. Most relate to work carried out with bacteria but could be equally applicable to fungal antagonists. These include control of minor pathogens in the rhizosphere, production of plant hormones or vitamins, conversion of materials to a form useful to the plant, nutrient release from soil or organic matter, and increased uptake and translocation of minerals (Windham *et al.* 1986; Baker 1989; Kleifeld & Chet 1992). Importantly, the growth-promotion effect is over and above that in response to the simple presence of nutrients in the fungal inoculum. For instance, in general, systems with nutrient levels optimal for normal plant growth still give increases in plant growth (Chang *et al.* 1986) and control treatments consisting of autoclaved inoculum or the original, uncolonized food-base itself, if used, fail to provide the same level of plant growth promotion compared with incorporation of live inoculum (Chang *et al.* 1986; Baker *et al.* 1984) although there are exceptions (Ousley *et al.* 1993). In the latter case, where autoclaved inoculum still promoted plant growth, the *Trichoderma* inoculum may have contained a heat-stable metabolite which promoted plant growth or it may have acted to remove a toxic material from the growing medium (Ousley *et al.* 1994b). Such possible mechanisms deserve further study.

ACKNOWLEDGEMENTS

This work was carried out through the support of the Biotechnology and Biological Sciences Research Council (BBSRC).

REFERENCES

Adams, P. B. & Ayers, W. A. (1979). Mycoparasitism of sclerotia of *Sclerotinia* and *Sclerotium* species by *Sporidesmium sclerotivorum*. *Canadian Journal of Microbiology*, **25**, 17–23.

Adams, P. B. & Ayers, W. A. (1981). *Sporidesmium sclerotivorum*: Distribution and function in natural biological control of sclerotial fungi. *Phytopathology*, **71**, 90–93.

Adams, P. B. & Ayers, W. A. (1983). Histological and physiological aspects of infection of sclerotia of two *Sclerotinia* spp. by two mycoparasites. *Phytopathology*, **73**, 1072–1076.

Adams, P. B., Marois, J. J. & Ayers, W. A. (1984). Population dynamics of the mycoparasite *Sporidesmium sclerotivorum* and its host, *Sclerotinia minor*, in soil. *Soil Biology and Biochemistry*, **16**, 627–633.

Ahmad, J. S. & Baker, R. (1987). Implications of rhizosphere competence of *Trichoderma harzianum*. *Canadian Journal of Microbiology*, **34**, 229–234.

Ahmad, J. S. & Baker, R. (1988). Rhizosphere competence of *Trichoderma harzianum*. *Phytopathology*, **77**, 182–189.

Alabouvette, C., Couteaudier, Y. & Louvet, J. (1985). Soils suppressive to *Fusarium* wilt. Mechanisms and management of suppressiveness. *Ecology and Management of Soilborne Plant Pathogens* (Eds. C. A. Parker, A. D. Rovira, K. J. Moore, P. T. W. Wong & J. F. Kollmorgen), pp. 101–106. American Phytopathological Society, St Paul, MN.

Artigues, M. & Davet, P. (1984). Comparaison des aptitudes parasitaires de clones de *Trichoderma* vis-à-vis de quelques champignons a sclérotes. *Soil Biology and Biochemistry*, **16**, 413–417.

Baker, R. (1988). *Trichoderma* spp. as plant-growth stimulants. *CRC Critical Reviews in Biotechnology*, **7**, 97–106.

Baker, R. (1989). Improved *Trichoderma* spp. for promoting crop productivity. *Trends in Biotechnology*, **7**, 34–38.

Baker, R., Elad, Y. & Chet, I. (1984). The controlled experiment in the scientific method with special emphasis on biological control. *Phytopathology*, **74**, 1019–1021.

Barak, R. & Chet, I. (1990). Lectin of *Sclerotium rolfsii*: its purification and possible function in fungal–fungal interaction. *Journal of Applied Bacteriology*, **69**, 101–112.

Benhamou, N. & Chet, I. (1993). Hyphal interactions between *Trichoderma harzianum* and *Rhizoctonia solani*: Ultrastructure and gold cytochemistry of the mycoparasitic process. *Phytopathology*, **83**, 1062–1071.

Brückner, H. & Przybylski, M. (1984). Isolation and structural characterization of polypeptide antibiotics of the peptaibol class by hplc with field desorption and fast atom bombardment mass spectrometry. *Journal of Chromatography*, **296**, 263–275.

Bullock, S., Willetts, H. J. & Adams, P. B. (1989). Morphology, histochemistry and germination of conidia of *Sporidesmium sclerotivorum*. *Canadian Journal of Botany*, **67**, 313–317.

Burge, M. N. (1988). *Fungi in Biological Control Systems*. Manchester University Press, Manchester.

Chang, Y-C., Chang, Y-C., Baker, R., Kleifeld, O. & Chet, I. (1986). Increased growth of plants in presence of the biological control agent *Trichoderma harzianum*. *Plant Disease*, **70**, 145–148.

Coley-Smith, J. R., Ridout, C.J., Mitchell, C.M. & Lynch, J.M. (1991). Control of bottom rot disease of lettuce (*Rhizoctonia solani*) using preparations of *Trichoderma viride*, *T. harzianum* or tolclofos-methyl. *Plant Pathology*, **40**, 359–366.

Couteaudier, Y. (1992). Competition for carbon in soil and rhizosphere, a mechanism involved in biological control of fusarium wilts (Ed. by E. C. Tjamos, G. C. Papavizas & R. M. Cook), pp. 99–104. Plenum Press, New York.

Daft, G.C. & Tsao, P.H. (1984). Parasitism of *Phytophthora cinnamomi* and *P. parasitica* spores by *Catenaria anguillulae* in a soil environment. *Transactions of the British Mycological Society*, **82**, 485–490.

Dewan, M. M. & Sivasithamparam, K. (1989). Growth promotion of rotation crop species by a sterile fungus from wheat and effect of soil temperature and water potential on its suppression of take-all. *Mycological Research*, **93**, 156–160.

Di Pietro, A., Gut-Rella, M., Pachlatko, J. P. & Schwinn, F. J. (1992). Role of antibiotics produced by *Chaetomium globosum* in biocontrol of *Pythium ultimum*, a causal agent of damping-off. *Phytopathology*, **82**, 131–135.

Di Pietro, A., Lorito, M., Hayes, C. K., Broadway, R. M. & Harman, G. E. (1993). Endochitinase from *Gliocladium virens*: Isolation characterization, and synergistic antifungal activity in combination with gliotoxin. *Phytopathology*, **83**, 308–313.

Dolan, T. E., Cohen, Y. & Coffey, M. D. (1986). Protection of *Persea* species against *Phytophthora cinnamomi* and *P. citricola* by prior inoculation with a citrus isolate of *P. parasitica*. *Phytopathology*, **76**, 194–198.

Drechsler, C. (1938). Two hyphomycetes parasitic on oospores of root-rotting Oomycetes. *Phytopathology*, **28**, 81–103.

Elad, Y., Chet, I. & Henis, Y. (1982). Degradation of plant pathogenic fungi by *Trichoderma harzianum*. *Canadian Journal of Microbiology*, **28**, 719–725.

Elad, Y., Lifshitz, R. & Baker, R. (1985). Enzymatic activity of the mycoparasite *Pythium nunn* during interaction with host and non-host fungi. *Physiological Plant Pathology*, **27**, 131–148.

Fang, J. G. & Tsao, P. H. (1995). Evaluation of *Pythium nunn* as a potential biocontrol agent against phytophthora root rots of Azalea and Sweet Orange. *Phytopathology,* **85,** 29–36.

Fravel, D. R. & Roberts, D. P. (1991). *In situ* evidence for the role of glucose oxidase in the biocontrol of *Verticillium* wilt by *Talaromyces flavus. Biocontrol Science and Technology,* **1,** 91–99.

Fulton, R.W. (1986). Practices and precautions in the use of cross protection for plant virus disease control. *Annual Review of Phytopathology,* **24,** 67–81.

Gees, R. & Coffey, M. D. (1989). Evaluation of a strain of *Myrothecium roridum* as a potential biocontrol agent against *Phytophthora cinnamomi. Phytopathology,* **79,** 1079–1084.

Gerlagh, M. & Vos, I. (1991). Enrichment of soil with sclerotia to isolate antagonists of *Sclerotinia sclerotiorum. Biotic Interactions and Soil-borne Diseases* (Ed. by A. B. R. Beemster, G. J. Bollen, M. Gerlagh, M. A. Ruissen, B. Schipper & A. Tempel), pp. 165–171. Elsevier, Amsterdam.

Ghisalberti, E. L. & Sivasithamparam, K. (1991). Antifungal antibiotics produced by *Trichoderma* spp. *Soil Biology and Biochemistry,* **23,** 1011–1020.

Gladders, P. & Coley-Smith, J. R. (1980). Interactions between *Rhizoctonia tuliparum* sclerotia and soil micro-organisms. *Transactions of the British Mycological Society,* **74,** 579–586.

Hemens, E., Steiner, U. & Schönbeck, F. (1992). Infektionsstrukturen von *Microdochium bolleyi* an Wurzeln und Koleoptilen von Gerste. *Journal of Phytopathology,* **136,** 57–66.

Hoch, H. C. & Abawi, G. S. (1979). Mycoparasitism of oospores of *Pythium ultimum* by *Fusarium merismoides. Mycologia,* **71,** 621–625.

Howell, C. R. (1982). Effect of *Gliocladium virens* on *Pythium ultimum, Rhizoctonia solani,* and damping-off of cotton seedlings. *Phytopathology,* **72,** 496–498.

Howell, C. R. (1987). Relevance of mycoparasitism in the biological control of *Rhizoctonia solani* by *Gliocladium virens. Phytopathology,* **77,** 992–994.

Howell, C.R. (1991). Biological control of *Pythium* damping-off of cotton with seed-coating preparations of *Gliocladium virens. Phytopathology,* **81,** 738–741.

Howell, C. R. & Stipanovic, R. D. (1983). Gliovirin, a new antibiotic from *Gliocladium virens* and its role in the biological control of *Pythium ultimum. Canadian Journal of Microbiology,* **29,** 321–324.

Huang, H. C. & Kozub, G. C. (1991). Monocropping to sunflower and decline of sclerotinia wilt. *Botanical Bulletin of Academia Sinica,* **32,** 163–169.

Humble, S. J. & Lockwood, J. L. (1981). Hyperparasitism of oospores of *Phytophthora megasperma* var. *sojae. Soil Biology and Biochemistry,* **13,** 355–360.

Ichielevich-Auster, M., Sneh, B., Koltin, Y. & Barash, I. (1985). Suppression of damping-off caused by *Rhizoctonia* species by a non-pathogenic isolate of *R. solani. Phytopathology,* **75,** 1080–1084.

Inbar, J., Abramsky, M., Cohen, D. & Chet, I. (1994). Plant growth enhancement and disease control by *Trichoderma harzianum* in vegetable seedlings grown under commercial conditions. *European Journal of Plant Pathology,* **100,** 337–346.

Jeffries, P. & Young, T. W. K. (1994). *Interfungal Parasitic Relationships.* CAB International, Wallingford.

Keinath, A. P., Fravel, D. R. & Papavizas, G. C. (1991). Potential of *Gliocladium roseum* for biocontrol of *Verticillium dahliae. Phytopathology,* **81,** 644–648.

Kenerley, C. M. & Stack, J. P. (1987). Influence of assessment methods on selection of fungal antagonists of the sclerotium-forming fungus *Phymatotrichum omnivorum. Canadian Journal of Microbiology,* **33,** 632–635.

Kim, K. K., Fravel, D. R. & Papavizas, G. C. (1988). Identification of a metabolite produced by *Talaromyces flavus* as glucose oxidase and its role in the biocontrol of *Verticillium dahliae. Phytopathology,* **78,** 488–492.

Kirk, J. J. & Deacon, J. W. (1987). Control of the take-all fungus by *Microdochium bolleyi,* and interactions involving *M. bolleyi, Phialophora graminicola* and *Periconia macrospinosa* on cereal roots. *Plant and Soil,* **98,** 231–237.

Kleifeld, O. & Chet, I. (1992). *Trichoderma harzianum* – interaction with plants and effect on growth response. *Plant and Soil,* **144,** 267–272.

Lascaris, D. & Deacon, J. W. (1991). Colonization of wheat roots from seed-applied spores of *Idriella* (*Microdochium*) *bolleyi*: a biocontrol agent of take-all. *Biocontrol Science and Technology*, 1, 229–240.

Lemanceau, P., Bakker, P. A. H. M., de Kogel, W. J., Alabouvette, C. & Schippers, B. (1993). Antagonistic effect of nonpathogenic *Fusarium oxysporum* Fo47 and pseudobactin 358 upon pathogenic *Fusarium oxysporum* f. sp. *dianthi*. *Applied & Environmental Microbiology*, 59, 74–82.

Lewis, J. A., Roberts, D. P. & Hollenbeck, M. D. (1991). Induction of cytoplasmic leakage from *Rhizoctonia solani* hyphae by *Gliocladium virens* and partial characterization of leakage factor. *Biocontrol Science and Technology*, 1, 21–29.

Lewis, K., Whipps, J. M. & Cooke, R. C. (1989). Mechanisms of biological disease control with special reference to the case study of *Pythium oligandrum* as an antagonist, in *Biotechnology of Fungi for Improving Plant Growth* (Ed. by J.M. Whipps & R.D. Lumsden), pp. 191–217. Cambridge University Press, Cambridge.

Lorito, M., Harman, G. E., Hayes, C. K., Broadway, R. M., Tronsmo, A., Woo, S. L. & Di Pietro, A. (1993a). Chitinolytic enzymes produced by *Trichoderma harzianum*: Antifungal activity of purified endochitinase and chitobiosidase. *Phytopathology*, 83, 302–307.

Lorito, M., Di Pietro, A., Hayes, C. K., Woo, S. L. & Harman, G. E. (1993b). Antifungal synergistic interaction between chitinolytic enzymes from *Trichoderma harzianum* and *Enterobacter cloacae*. *Phytopathology*, 83, 721–728.

Lorito, M., Hayes, C. K., Di Pietro, A., Woo, S. L. & Harman, G. E. (1994). Purification characterization and synergistic activity of a glucan 1,3-β-glucosidase and an *N*-acetyl-β-glucosaminidase from *Trichoderma harzianum*. *Phytopathology*, 84, 398–405.

Lumsden, R. D. (1992). Mycoparasitism of soilborne plant pathogens. *The Fungal Community: Its Organisation and Role in the Ecosystem* (Ed. by G. C. Carroll & D. T. Wicklow), pp. 275–293. Marcel Dekker, New York.

Lumsden, R. D., Locke, J. C., Adkins, S. T., Walter, J. F. & Ridout, C. J. (1992). Isolation and localization of the antibiotic gliotoxin produced by *Gliocladium virens* from alginate prill in soil and soilless media. *Phytopathology*, 82, 230–235.

Lynch, J. M. (1990). Microbial metabolites. *The Rhizosphere* (Ed. by J. M. Lynch), pp. 177–206. John Wiley & Sons, Chichester.

Mandeel, Q. & Baker, R. (1991). Mechanisms involved in biological control of fusarium wilt on cucumber with strains of nonpathogenic *Fusarium oxysporum*. *Phytopathology*, 81, 462–469.

Marois, J. J., Fravel, D. R. & Papavizas, G. C. (1984). Ability of *Talaromyces flavus* to occupy the rhizosphere and its interaction with *Verticillium dahliae*. *Soil Biology and Biochemistry*, 16, 387–390.

Martin, F. N. & Hancock, J. G. (1985). Chemical factors in soils suppressive to *Pythium ultimum*. *Ecology and Management of Soilborne Plant Pathogens* (Ed. by C. A. Parker, A. D. Rovira, K. J. Moore, P. T. W. Wong & J. F. Kollmorgen), pp. 113–116. American Phytopathological Society, St Paul, MN.

Martin, F. N. & Hancock, J. G. (1986). Association of chemical and biological factors in soils suppressive to *Pythium ultimum*. *Phytopathology*, 76, 1221–1231.

Martyn, R. D., Biles, C. L. & Dillard, E. A. (1991). Induced resistance to fusarium wilt of watermelon under simulated field conditions. *Plant Disease*, 75, 874–877.

Matta, A. & Garibaldi, A. (1977). Control of *Verticillium* wilt to tomato by preinoculation with a virulent fungi. *Netherlands Journal of Plant Pathology*, 83 (Supplement 1), 457–462.

McLaren, D. L. (1989). *Biocontrol of sclerotinia diseases* (Sclerotinia sclerotiorum) *of sunflower and bean by* Talaromyces flavus *and* Coniothyrium minitans. PhD Thesis, University of Manitoba.

McLaren, D. L., Huang, H. C. & Rimmer, S. R. (1986). Hyperparasitism of *Sclerotinia sclerotiorum* by *Talaromyces flavus*. *Canadian Journal of Plant Pathology*, 8, 43–48.

Melouk, H. A. & Horner, C. E. (1975). Cross protection in mints by *Verticillium nigrescens* against *V. dahliae*. *Phytopathology*, **65**, 767–769.

Morris, R. A. C., Coley-Smith, J. R. & Whipps, J. M. (1992). Isolation of mycoparasite *Verticillium biguttatum* from sclerotia of *Rhizoctonia solani* in the United Kingdom. *Plant Pathology*, **41**, 513–516.

Morris, R. A. C., Ewing, D., Whipps, J. M. & Coley-Smith, J. R. (1995). Antifungal hydroxymethylphenols from the mycoparasite *Verticillium biguttatum*. *Phytochemistry*, **39**, 1043–1048.

Ogawa, K. & Komada, H. (1985). Biological control of fusarium wilt of sweet potato with crossprotection by nonpathogenic *Fusarium oxysporum*. *Ecology & Management of Soilborne Plant Pathogens* (Ed. by C. A. Parker, A. D. Rovira, K. J. Moore, P. T. W. Wong & J. F. Kollmorgen), pp. 121–123. *APS Press, The American Phytopathological Society*, St Paul, MN.

Ordentlich, A., Migheli, Q. & Chet, I. (1991). Biological control activity of three *Trichoderma* isolates against fusarium wilts of cotton and muskmelon and lack of correlation with their lytic enzymes. *Phytopathology*, **133**, 177–186.

Ousley, M. A., Lynch, J. M. & Whipps, J. M. (1993). Effect of *Trichoderma* on plant growth: a balance between inhibition and growth promotion. *Microbial Ecology*, **26**, 277–285.

Ousley, M. A., Lynch, J. M. & Whipps, J. M. (1994a). The effects of addition of *Trichoderma* inocula on flowering and shoot growth of bedding plants. *Scientia Horticulturae*, **59**, 147–155.

Ousley, M. A., Lynch, J. M. & Whipps, J. M. (1994b). Potential of *Trichoderma* spp. as consistent plant growth stimulators. *Biology and Fertility of Soils*, **17**, 85–90.

Paulitz, T. C. & Baker, R. (1987a). Biological control of pythium damping-off of cucumbers with *Pythium nunn*: population dynamics and disease suppression. *Phytopathology*, **77**, 335–340.

Paulitz, T. C. & Baker, R. (1987b). Biological control of pythium damping-off of cucumbers with *Pythium nunn*: influence of soil environment and organic amendments. *Phytopathology*, **77**, 341–346.

Paulitz, T., Windham, M. & Baker, R. (1986). Effect of peat: vermiculite mixes containing *Trichoderma harzianum* on increased growth response of radish. *Journal of the American Society of Horticultural Science*, **111**, 810–814.

Pfeffer, H. & Lüth, P. (1990). Der Einfluss einer Rotkleemonokultur auf das Antiphytopathogene. Potential des Bodens in Bezug auf *Sclerotinia trifoliorum* Erikss. *Nachrichtenblatt Pflazenschutz DDR*, **44**, 214–216.

Postma, J. & Rattink, H. (1991). Biological control of fusarium wilt of carnation with nonpathogenic *Fusarium* isolates. *Mededelingen-van-de-Faculteit-Landbouwwetenschappen-Universiteit-Gent*, **56**, 179–183.

Rao, N. N. R. & Pavgi, M. S. (1976). A mycoparasite on *Sclerospora graminicola*. *Canadian Journal of Botany*, **54**, 220–223.

Rattink, H. (1989). Possibilities of biological control of fusarium wilt of carnation. *Mededelingen-van-de-Faculteit-Landbouwwetenschappen-Universiteit-Gent*, **54**, 517–524.

Ridout, C. J., Coley-Smith, J. R. & Lynch, J. M. (1988). Fractionation of extracellular enzymes from a mycoparasitic strain of *Trichoderma harzianum*. *Enzyme and Microbial Technology*, **10**, 180–187.

Roberts, D. P. & Lumsden, R. D. (1990). Effect of extracellular metabolites from *Gliocladium virens* on germination of sporangia and mycelial growth of *Pythium ultimum*. *Phytopathology*, **80**, 461–465.

Schirmböck, M., Lorito, M., Wang, Y.-L., Hayes, C. K., Arisan-Atac, I., Scala, F., Harman, G. E. & Kubicek, C. P. (1994). Parallel formation and synergism of hydrolytic enzymes and peptaibol antibiotics, molecular mechanisms involved in the antagonistic action of *Trichoderma harzianum* against phytopathogenic fungi. *Applied & Environmental Microbiology*, **60**, 4364–4370.

Shankar, D. I., Kurtböke, Gillespie-Sasse, L. M. J., Rowland, C. Y. & Sivasithamparam, K. (1994). Possible roles of competition for thiamine, production of inhibitory compounds, and hyphal interactions in suppression of the take-all fungus by a sterile red fungus. *Canadian Journal of Microbiology*, **40**, 478–483.

Shivana, M. B., Meera, M. S. & Hyakumachi, M. (1994). Sterile fungi from zoysiagrass rhizosphere as plant growth promoters in spring wheat. *Canadian Journal of Microbiology*, **40**, 637–644.

Sivan, A. & Chet, I. (1989). The possible role of competition between *Trichoderma harzianum* and *Fusarium oxysporum* on rhizosphere colonization. *Phytopathology*, **79**, 198–203.

Sneh, B., Humble, S. J. & Lockwood, J. L. (1977). Parasitism of oospores of *Phytophthora megasperma* var. *sojae*, *P. cactorum*, *Pythium* sp. and *Aphanomyces euteiches* by oomycetes, chytridiomycetes, hyphomycetes, actinomycetes, and bacteria. *Phytopathology*, **67**, 622–628.

Sneh, B., Zeidan, M., Ichielevich-Auster, M., Barash, I. & Koltin, Y. (1986). Increased growth responses induced by a nonpathogenic *Rhizoctonia solani*. *Canadian Journal of Botany*, **64**, 2372–2378.

Sneh, B., Ichielevich-Auster, M. & Plaut, Z. (1989a). Mechanisms of seedling protection induced by a hypovirulent isolate of *Rhizoctonia solani*. *Canadian Journal of Botany*, **67**, 2135–2141.

Sneh, B., Ichielevich-Auster, M., & Shomer, I. (1989b). Comparative anatomy of colonization of cotton hypocotyls and roots by virulent and hypovirulent isolates of *Rhizoctonia solani*. *Canadian Journal of Botany*, **67**, 2142–2149.

Speakman, J. B. & Lewis, B. G. (1978). Limitation of *Gaeumannomyces graminis* by wheat root responses to *Phialophora radicicola*. *New Phytologist*, **80**, 373–380.

Tweddell, R. J., Jabaji-Hare, S. H. & Charest, P. M. (1994). Production of chitinases and β-1,3-glucanases by *Stachybotrys elegans*, a mycoparasite of *Rhizoctonia solani*. *Applied and Environmental Microbiology*, **60**, 489–495.

Van den Boogert, P. H. J. F. & Deacon, J. W. (1994). Biotrophic mycoparasitism by *Verticillium biguttatum* on *Rhizoctonia solani*. *European Journal of Plant Pathology*, **100**, 137–156.

Van den Boogert, P. H. J. F. & Saat, T. A. W. M. (1991). Growth of the mycoparasitic fungus *Verticillium biguttatum* from different geographical origins at near minimum temperatures. *Netherlands Journal of Plant Pathology*, **97**, 115–124.

Van den Boogert, P. H. J. F. & Velvis, H. (1992). Population dynamics of the mycoparasite *Verticillium biguttatum* and its host, *Rhizoctonia solani*. *Soil Biology and Biochemistry*, **24**, 157–164.

Van Peer, R. & Schippers, B. (1992). Lipopolysaccharides of plant-growth promoting *Pseudomonas* sp. strain WCS417r induce resistance in carnation to fusarium wilt. *Netherlands Journal of Plant Pathology*, **98**, 129–139.

Whipps, J. M. (1991). Effects of mycoparasites on sclerotia-forming fungi. *Biotic Interactions and Soil-borne Diseases* (Ed. by A. B. R. Beemster, G. J. Bollen, M. Gerlagh, M. A. Ruissen, B. Schippers & A. Tempel), pp. 129–140. Elsevier, Amsterdam.

Whipps, J. M. (1994). Advances in biological control in protected crops. *Brighton Crop Protection Conference – Pests and Diseases*, pp. 1259–1263. British Crop Protection Council, Farnham.

Whipps, J. M. (1996). Ecological considerations involved in commercial development of biological control agents for soil-borne diseases. *Modern Soil Microbiology* (Ed. by J. D. van Elsas, E. M. H. Wellington & J. T. Trevors) (in press).

Whipps, J. M., Budge, S. P. & Mitchell, S. J. (1993). Observations on sclerotial mycoparasites of *Sclerotinia sclerotiorum*. *Mycological Research*, **97**, 697–700.

Whipps, J. M. & Gerlagh, M. (1992). Biology of *Coniothyrium minitans* and its potential for use in disease biocontrol. *Mycological Research*, **96**, 897–907.

Whipps, J. M. & Lumsden, R. D. (Eds) (1989). *Biotechnology of Fungi for Improving Plant Growth*. Cambridge University Press, Cambridge.

Whipps, J. M. & Lumsden, R. D. (1991). Biological control of *Pythium* species. *Biocontrol Science and Technology*, **1**, 75–90.

Wilhite, S. E., Lumsden, R. D. & Straney, D. C. (1994). Mutational analysis of gliotoxin production by the biocontrol fungus *Gliocladium virens* in relation to suppression of pythium damping-off. *Phytopathology*, **84**, 816–821.

Wilson, C. L., Ghaouth, A. E., Chalutz, E., Droby, S., Stevens, C., Lu, J. Y., Khan, V. & Arul, J.

(1994). Potential of induced resistance to control postharvest diseases of fruits and vegetables. *Plant Disease*, **78**, 837–844.

Windham, M. T., Elad, Y. & Baker, R. (1986). A mechanism for increased plant growth induced by *Trichoderma* spp. *Phytopathology*, **76**, 518–521.

Wynn, A. R. & Epton, H. A. S. (1979). Parasitism of oospores of *Phytophthora erythroseptica* in soil. *Transactions of the British Mycological Society*, **73**, 255–259.

Zaki, A. I., Keen, N. T. & Erwin, D. C. (1972). Implication of vergosin and hemigossypol in the resistance of cotton to *Verticillium albo-atrum*. *Phytopathology*, **62**, 1402–1406.

Zazzerini, A. & Tosi, I. (1985). Antagonistic activity of fungi isolated from sclerotia of *Sclerotinia sclerotiorum*. *Plant Pathology*, **34**, 415–421.

4. INTERACTIONS BETWEEN MYCORRHIZAL FUNGI AND RHIZOSPHERE MICRO-ORGANISMS WITHIN THE CONTEXT OF SUSTAINABLE SOIL–PLANT SYSTEMS

J. M. BAREA, C. AZCON-AGUILAR AND R. AZCON

Departamento de Microbiología del Suelo y Sistemas Simbióticos, Estación Experimental del Zaidín, CSIC, Professor Albareda 1, 18008 Granada, Spain

INTRODUCTION

Degradation of natural and agro-ecosystems is associated with disturbances in the vegetation cover, loss of soil structure, increase in soil erosion, loss of available nutrients and organic matter, loss of microbial propagules, and/or a decrease in microbial activity (Bethlenfalvay & Linderman 1992; Barea & Jeffries 1995). These determinants may lead to negative feedback causing a downward spiral in soil fertility and vegetation cover. The 'cause–consequence' fluxes involved in eco-system degradation are outlined in Fig. 4.1.

A sustainable ecosystem can be achieved with the rational use of natural re-sources to maintain environmental quality in general, and in particular, the structure and diversity of natural plant communities (Barea & Jeffries 1995). Microbial populations are key components of the soil–plant systems where they are involved in a framework of interactions (Lynch 1990). The aim of this chapter is to describe such interactions and to discuss how the activity of specific members of soil microbiota is crucial for the sustainability of natural and agro-ecosystems.

THE MICROBIAL COMPONENT IN SUSTAINABLE SOIL–PLANT SYSTEMS

Sustainability, either in natural ecosystems or in agro-ecosystems, is dependent on a biological balance in the soil (Barea & Jeffries 1995), which is mainly governed by the activity of microbial communities, some of which can be managed as a nat-ural resource (Bethlenfalvay & Linderman 1992). Many of the soil-borne microbes are bound to the surface of soil particles or found in soil aggregates, while others interact specifically with the plant root system (Glick 1995). A large number of micro-organisms live in the soil–plant interface where a microcosm system develops. This microcosm is known as the rhizosphere (Lynch 1990; Azcón-Aguilar & Barea 1992; Linderman 1992).

65

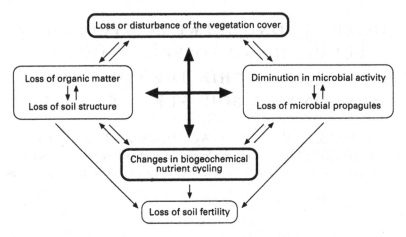

FIG. 4.1. Cause–consequence fluxes involved in the degradation of soil–plant systems.

Soil microbial dynamics largely govern ecosystem functioning (Kennedy & Smith 1995) through a number of activities, carried out mainly by rhizosphere microbiota constituents, which are known to enhance soil and plant quality (Bethlenfalvay & Schüepp 1994). The functions involved include improvement of plant establishment, increased availability of plant nutrients, enhancement of nutrient uptake, protection against cultural and environmental stresses and improvement of soil structure (Barea & Jeffries 1995). Certainly, plant health and productivity depend on soil quality which, in turn, is dependent on the diversity and effectiveness of its microbiota (Bethlenfalvay & Schüepp 1994). However, little research has been done to unite all components of the agro-ecosystems and to evaluate the beneficial relationships between microbial diversity, soil quality, plant productivity and health, and ecosystem sustainability (but see Bethlenfalvay & Schüepp 1994; Kennedy & Smith 1995).

Certain micro-organisms can be used as inoculants to benefit plant growth and health. These can be integrated into two main groups: (i) saprophytes; and (ii) mutualistic symbionts. Among other microbial types, the saprophytes include the so-called plant growth promoting rhizobacteria (PGPR). These bacteria participate in many key ecosystem processes such as those involved in the biological control of plant pathogens, nutrient cycling and seedling establishment (Kloepper 1992; Glick 1995). Mycorrhizal fungi and nitrogen (N_2)-fixing bacteria are among the most important members within the influential group of mutualistic symbionts. The mycorrhizal fungi develop an external mycelium which is a bridge connecting the root with the surrounding soil microhabitats. Symbiotic N_2-fixing associations, involving *Rhizobium*, *Frankia* and cyanobacteria, are also relevant to sustainable systems.

MYCORRHIZAL SYMBIOSIS

This fungal–plant (root) association plays a key role in nutrient cycling in eco-systems. Additionally, the external mycorrhizal mycelium, in co-operation with other soil organisms, forms water-stable aggregates necessary for good soil tilth. Mycorrhizas improve plant health through increased protection against biotic and abiotic stresses (Bethlenfalvay & Linderman 1992).

Mycorrhizal associations can be found in nearly all ecological situations, and most plant species are able to form this symbiosis naturally, the arbuscular mycor-rhiza (AM) being the most common type involved in the normal cropping systems and natural ecosystems (Harley & Smith 1983). The responsible symbiotic fungi belong to the order Glomales in the Zygomycetes (Rosendahl, *et al.* 1994).

The role of AM in determining plant community structure and successional processes (Read 1993; Francis & Read 1994), in crop production (Barea *et al.* 1993; Thompson 1994), and in revegetation (Jasper 1994) has been reviewed.

The role of AM in sustainable systems derives from the ability of this symbiotic association to contribute to: (i) nutrient cycling; (ii) establish bridges between plants; (iii) the stability and diversity of natural ecosystems; (iv) biotic and abiotic stress alleviation; and (v) in soil structure improvement (Barea & Jeffries 1995).

THE MYCORRHIZOSPHERE

Mycorrhiza establishment is known to change the mineral nutrient composition, hormonal balance, carbon (C) allocation patterns, and other aspects of plant phy-siology (Azcón-Aguilar & Bago 1994; Smith *et al.* 1994). The nutrient transport attributes of mycorrhizal fungi, when active, and the release of materials, when senescing, represent a supply of nutrients into the rhizosphere (Harley & Smith 1983). Thus, the AM symbiotic status affects the chemical composition of root exudates while the development of a mycorrhizal soil mycelium also introduces physical modifications in the environment surrounding the roots. These changes affect the rhizospheric microbial communities in the so-called 'mycorrhizosphere', and thus the 'mycorrhizosphere' effect (Linderman 1992), takes place.

These changes affect both quantitatively and qualitatively the microbial populations in either the rhizosphere or the rhizoplane (Azcón-Aguilar & Barea 1992; Linderman 1992). A typical function of the AM soil mycelium is to serve as a carbon source to microbial communities; even outside the limit of the proper rhizosphere, it results in an important contribution of the mycorrhizosphere to improve soil quality (Bethlenfalvay & Schüepp 1994).

Conversely, soil micro-organisms can affect mycorrhiza formation. In fact, 'specific' soil micro-organisms are capable of benefiting mutualistic symbiont establishment in plant roots, and in the corresponding rhizosphere (Azcón-Aguilar

& Barea 1992; Fitter & Garbaye 1994; Puppi *et al.* 1994). It appears that microbial-induced changes in the root exudation patterns or in the hormonal balance of the plant can affect the formation of the mycorrhizal symbiosis. Other soil micro-organisms, however, are known to inhibit AM fungi and/or AM formation (Azcón-Aguilar & Barea 1992; Linderman 1992; Fitter & Garbaye 1994).

INTERACTIONS BETWEEN ARBUSCULAR MYCORRHIZAS AND SOIL MICRO-ORGANISMS INVOLVED IN NUTRIENT CYCLING

A number of soil micro-organisms are able to change the availability of mineral plant nutrients. This could account for the improvement of plant growth by means of synergistic interactions with mycorrhizal fungi. It has also been suggested that a certain selectivity ('specificity') is involved in such interactions. From the perspective of sustainability, the re-establishment of nutrient cycles after any process of soil degradation is a topic of obvious interest, as is also the understanding of such interactions which will allow the subsequent management of these natural microbial resources, either in a low-input agricultural technology (Bethlenfalvay & Linderman 1992), or in the re-establishment of natural vegetation in degraded areas (Skujins & Allen 1986; Miller & Jastrow 1994; Barea & Jeffries 1995). It is, however, a complex task to integrate the information available concerning the interactions between AM and specific groups of soil microbiota components and ecosystem or agro-ecosystem sustainability. The discussions here try to contribute to a better understanding of these inter-relationships. The related microbial groups are as follows.

Rhizobium

Legume species have a great relevance in any sustainability approach for either natural or agro-ecosystems. This derives mainly from the ability of these plant species to form two types of symbiotic associations with soil micro-organisms: N_2-fixing rhizobial nodules and mycorrhizas. The interactions of AM fungi and *Rhizobium* greatly benefit the biological N_2 inputs to soil–plant systems. To achieve this accurately, 'specific' *Rhizobium*/AM fungi combinations must be first selected (Barea *et al.* 1992a, b). Today, the role of AM in improving nodulation (Fig. 4.2a) and N_2 fixation (Fig. 4.2b) is universally recognized and the use of the isotope ^{15}N made it possible to ascertain and quantify the amount of N_2 that has actually fixed, in a particular situation, and the contribution of the AM symbiosis to the process (Barea *et al.* 1992b). Figure 4.3 records representative data on this concern. The 'N_2 transfer' from mycorrhizal and nodulated legumes to non-fixing, but mycorrhizal plants growing nearby, either intercropped (Fig. 4.4) or in natural

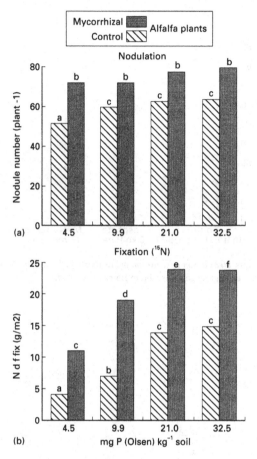

FIG. 4.2. Effect of mycorrhizal inoculation on (a) nodulation and (b) N_2 fixation of alfalfa plants, given different levels of available phosphate. Mean values (five replicates) not sharing a common letter differ significantly at $P = 0.05$ (Tukey test). (From Barea *et al.* 1989.)

ecosystems because of its repercussions in N_2 cycling, is a topic of great relevance to any sustainable approach. Particularly, these complex interactions are of interest for shrubland rehabilitation practices (Herrera *et al.* 1993). Mycorrhizal associations thus play a key role in the re-establishment of nutrient cycles.

There have been many studies aimed at ascertaining whether the AM symbiosis enhances rhizobial activity only through a generalized stimulation of host nutrition, or by means of a more localized effect, either at the precolonization stages or during the development of the tripartite symbiosis, at root or nodule level (Barea *et al.* 1992a). The role of AM fungi in influencing water relations, hormonal balance, photosynthetic rate, or C allocation patterns in plants, could account for these effects

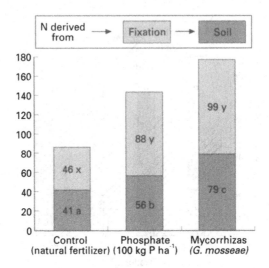

FIG. 4.3. Sources of N for *Hedysarum coronarium* plants inoculated with mycorrhizal fungi, given extra phosphate and corresponding controls. (From Barea *et al.* 1987.)

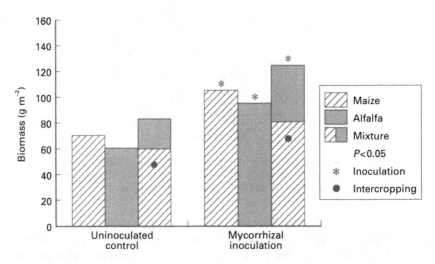

FIG. 4.4. Interactions between mycorrhizal inoculation and intercropping, to benefit N nutrition of maize growing in mixture with alfalfa under field conditions.

(Azcón-Aguilar & Barea 1992). It is interesting that AM can help nodule formation and function under stress conditions (drought, salinity, low-nutrient soil, etc.) which are characteristics of degraded habitats; for example Azcón *et al.* (1988) carried out a study where the legume *Medicago sativa* (+ *Rhizobium meliloti*) was grown under controlled conditions to study the interactions between soluble phosphorus (P) in soil (four levels), or a mycorrhizal inoculum, and the degree of water potential (four levels) in relation to plant development and N_2 fixation. [15]N was added for a qualitative estimate of N_2 fixation. It was found that N_2 fixation, the dry-matter yield, and nodulation increased with the amount of plant available P in the soil, and decreased as the water stress increased, for each P level. The mycorrhizal effect on dry matter, N_2 yield, and on nodulation was little affected by the water status.

Root nodulating N_2-fixing actinomycetes and cyanobacterias

In the root of the so-called actinorrhizal plants, mycorrhizal fungi coexist with actinorrhiza-forming diazotrophic bacteria which belong to the genus *Frankia*. These N_2-fixing non-legume species are of great ecological value (Cervantes & Rodríguez-Barrueco 1992; Puppi *et al.* 1994).

Species belonging to the gymnosperm order Cycadales are nodulated by the cyanobacteria *Nostoc* or *Anabaena*, endophytes coexisting with AM fungi (Trappe 1986).

Nitrogen-fixing free-living bacteria

Azotobacter spp., *Pseudomonas*, *Azospirillum* spp. and other diazotrophic bacteria have been found to enhance mycorrhizal colonization and to improve plant growth. However, these effects seem to be due to hormone production by the bacteria rather than to N_2 fixation. Since the growth substances produced by these micro-organisms can affect rooting patterns, it is logical that they can affect mycorrhiza formation and function (Azcón-Aguilar & Barea 1992; Linderman 1992; Puppi *et al.* 1994). The ^{15}N-based methods have been proposed to ascertain if any bacterial influence on AM response is mediated by the N_2 inputs derived from N_2 fixation (Barea *et al.* 1992a, b). There is evidence supporting the ability of AM symbiosis to improve the establishment of such diazotrophic bacteria in plant rhizosphere (Azcón-Aguilar & Barea 1992).

Phosphate solubilizing micro-organisms (PSM)

A number of *in vitro* experiments have shown that many soil micro-organisms are able to solubilize phosphate ions from sparingly soluble inorganic and organic P compounds. However, the effectiveness of this process in soil is unclear because of: (i) the scarcity of available energy sources for micro-organisms in non-rhizospheric microhabitats; (ii) problems of amensalism in soil; and (iii) difficulties for the translocation of phosphate ions to the root surface, upon any solubilization. However, in the case of mycorrhizal plants, the microbiologically solubilized phosphate would be taken up by AM fungal hyphae. Some experiments support such synergistic microbial interaction. Another explanation for some positive interactions between AM and PSM would be that the PSM might act through plant hormone production. Assays using ^{32}P-based techniques would be useful to ascertain the mechanisms accounting for the above mentioned effects (Azcón-Aguilar & Barea 1992). All in all, these interactions appear of relevance in sustainability approaches, within the context of the biogeochemical cycling of plant nutrients (Jeffries & Barea 1994; Miller & Jastrow 1994); particularly, when considering that among the farming practices that may lead to sustainability, the use of cheaper phosphate sources, such as phosphate-bearing rocks, is commonly recommended. However, to improve plant use of the sparingly soluble natural phosphate sources, like the rock phosphates, it is necessary to enhance the release of P from the natural material. The synergistic activities of AM fungi and PSM have been shown to produce a positive effect on P cycling and the subsequent plant P acquisition, under certain circumstances (Azcón-Aguilar & Barea 1992).

Plant growth promoting rhizobacterial (PGPR)

The attribute 'rhizobacteria' refers to the ability of some soil bacteria to 'aggressively' colonize the root soil interface, where they establish and maintain, for a time, a considerable number of cells. These bacteria, which promote plant growth by means of several mechanisms, are now commonly designated by the acronym PGPR (Kloepper et al. 1991; Linderman 1992; Glick 1995). The significant establishment of these bacteria in the rhizosphere is a prerequisite for them to develop any beneficial activity concerning plant growth improvement. Therefore, it is critical that the free-living bacterial strains, such as N fixers or phosphate solubilizers, get established in the rhizosphere microenvironments in order to induce positive plant responses. The growth promotion activity of PGPR has been shown with effects on seedling emergence, root formation, nutrient cycling and nodulation (Kloepper et al. 1991). However, one of the most remarkable effects of PGPR on plant growth is that deriving from their ability as biological control agents for plant pathogens (Kloepper 1992), as discussed later.

There is evidence that some PGPR, mainly *Pseudomonas* spp. and *Bacillus* spp., interact with AM or AM fungi developing synergistic activities which leads to improved plant growth and nutrition. Besides, both groups of micro-organisms (saprophytes and symbionts) can interact reciprocally, thus improving their establishment in the rhizosphere (Azcón-Aguilar & Barea 1992; Linderman 1992). Figure 4.5(a, b) gives examples of such reciprocal interactions between AM fungi and PGPR at establishing themselves in plant root/rhizosphere. The advantage of these interactions has been particularly associated with plant establishment in degraded habitats (Barea & Jeffries 1995). It has been reported that certain micro-organisms

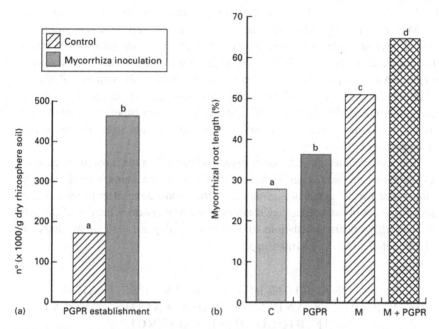

FIG. 4.5. Effect of (a) a PGPR (*Pseudomonas* sp.) on mycorrhiza formation and (b) mycorrhizal inoculation on the number of PGPR associated with maize plants grown in natural field soil. Mean values (10 replicates) not sharing a common letter differ significantly at $P = 0.05$ (Tukey test).

involved in manganese (Mn) and iron (Fe) transformations in soil, interact with AM fungi in helping plants to acquire the appropriate amounts of these elements. However, this is still a controversial topic (Marschner & Dell 1994).

INTERACTIONS BETWEEN ARBUSCULAR MYCORRHIZAS AND SOIL-BORNE PLANT PATHOGENS

In the rhizosphere microhabitats many types of microbe–microbe interactions take place (Stoztky 1972). Among these antagonistic actions develop. Any microorganism which exerts antagonism on harmful, 'soil-borne plant pathogen' organisms is beneficial to the plant in terms of health protection (Linderman 1994). Therefore, the appropriate management of these antagonistic organisms, common components of natural or agro-ecosystems, represents a form of biological control, a key practice in sustainable agriculture.

Both AM fungi and pathogens are 'associated' in the same 'biosubstrate' – that is, the fine feeder roots of their common host. AM associations can reduce, or even suppress, damage caused by soil-borne plant pathogens (Hooker *et al.* 1994;

Linderman 1994). The enhancement of root resistance/tolerance to pathogen attack in AM plants is not applicable to all pathogens, and it is not expressed in all substrates or in all environmental conditions. Nevertheless, there are examples demonstrating that prior colonization of selected AM fungi protected plants against pathogenic fungi, such as *Phytophthora, Gaeumannomyces, Fusarium, Thielaviopsis, Pythium, Rhizoctonia, Sclerotium, Verticillium, Aphanomyces,* or nematodes such as *Rotylenchus, Pratylenchus, Meloidogyne,* etc. The review papers by Hooker *et al.* (1994), Linderman (1994) and Roncadori (this volume) summarize the experiments to support such statements.

The mechanisms that have been suggested to explain the protective action of AM symbiosis include the improvement of plant nutrition, damage compensation, competition for photosynthates, competition for colonization/infection sites, induction of changes in the morphology/anatomy of the root system, induction of changes in mycorrhizosphere populations and activation of plant defence mechanisms (Hooker *et al.* 1994; Linderman 1994).

INTERACTION BETWEEN ARBUSCULAR MYCORRHIZAS AND MICRO-ORGANISMS INVOLVED IN BIOLOGICAL CONTROL

The prophylactic ability of some AM fungi could be exploited in co-operation with other rhizosphere micro-organisms that have demonstrated antagonistic abilities against pathogens, and that are being used as biological control agents (Linderman 1994). Among the micro-organisms catalogued as antagonists to fungal pathogens there are both fungi, such as *Trichoderma* spp. and *Gliocladium* spp., and bacteria, which are actually PGPR, such as *Pseudomonas* spp. and *Bacillus* spp. (Kloepper *et al.* 1991; Linderman 1994; Whipps, this volume).

It has been shown that microbial antagonists of fungal pathogens do not have the same effect against AM fungi. Even more, they can interact positively with them to improve the development of the mycosymbiont and also to facilitate AM formation (Linderman 1994; Vidal *et al.* 1996). Therefore, the accurate management of these interactions, by tailoring appropriate mycorrhizosphere systems relevant to plant growth and health in an integrated approach, is among the main objectives/aims of sustainable agriculture (Linderman 1992, Barea & Jeffries 1995).

INTERACTIONS BETWEEN ARBUSCULAR MYCORRHIZAS AND MICRO-ORGANISMS INVOLVED IN WATER-STABLE AGGREGATE FORMATION

The formation of water-stable aggregates is a crucial process for soil conservation, which is, in turn, a key subject to sustainability (Bethlenfalvay 1992). Mycorrhizal fungi, as colonizers not only of roots and the rhizosphere but also of the bulk soil, appear to play a role in conserving soil as in enhancing plant productivity (Bethlenfalvay & Barea 1994). This property derives from the fact that the AM mycelium is involved in the formation of stable soil aggregates. In fact, the soil mycelium first develops as a skeletal structure which holds soil particles by simple entanglement. Later, binding agents, mostly of microbial origin, co-operate in the process of cementation and stabilization of microaggregates which are finally held together in macroaggregates by an interacting participation of both AM hyphae and soil micro-organisms (Miller & Jastrow 1992; Bethlenfalvay & Schüepp 1994; Tisdall 1994).

CONCLUSIONS

Experimental evidence demonstrates that microbial interactions involving AM fungi appear to play a key role in the soil–plant interfaces. The accurate management, upon a deeper understanding of these interactions, is a crucial, biologically controlled aspect within the context of sustainability of soil–plant systems, since they can benefit plant growth and health, and soil quality.

ACKNOWLEDGEMENTS

The authors thank the CICYT Project AMB 95-0699.

REFERENCES

Azcón, R., El-Atrach, F. & Barea J. M. (1988). Influence of mycorrhiza vs. soluble phosphate on growth and N_2 fixation (^{15}N) in alfalfa under different levels of water potential. *Biology and Fertility of Soils*, 7, 28–31.

Azcón-Aguilar, C. & Bago, B. (1994). Physiological characteristics of the host plant promoting an undisturbed functioning of the mycorrhizal symbiosis. *Impact of Arbuscular Mycorrhizas on Sustainable Agriculture and Natural Ecosystems* (Ed. by S. Gianinazzi & H. Schüepp), pp. 47–60. ALS, Birkhäuser, Basel.

Azcón-Aguilar, C. & Barea, J. M. (1992). Interactions between mycorrhizal fungi and other rhizosphere microorganisms. *Mycorrhizal Functioning: An Integrative Plant–Fungal Process* (Ed. by M. J. Allen), pp. 163–198. Chapman & Hall, New York.

Barea, J. M. & Jeffries, P. (1995). Arbuscular mycorrhizas in sustainable soil plant systems. *Mycorrhiza Structure Function, Molecular Biology and Biotechnology* (Ed. by B. Hock & A. Varma), pp. 521–559. Springer, Heidelberg.

Barea, J. M., Azcón-Aguilar, C. & Azcón, R. (1987). Vesicular–arbuscular mycorrhiza improve both symbiotic N_2 fixation and N uptake from soil as assessed with a ^{15}N technique under field conditions. *New Phytologist*, 106, 717–725.

Barea, J. M., El-Atrach, F. & Azcón, R. (1989). Mycorrhiza and phosphate interactions as affecting plant development, N_2 fixation, N-transfer and N-uptake from soil in legume grass mixtures by using a ^{15}N dilution technique. *Soil Biology and Biochemistry*, 21, 581–589.

Barea, J. M., Azcón, R. & Azcón-Aguilar, C. (1992a). Vesicular–arbuscular mycorrhizal fungi in nitrogen-fixing systems. *Methods in Microbiology* (Ed. by J. R. Norris, D. J. Read & A. K. Varma), pp. 391–416. Academic Press, London.

Barea, J. M., Azcón, R. & Azcón-Aguilar, C. (1992b). The use of ^{15}N to assess the role of VA mycorrhiza in plant N nutrition and its application to evaluate the role of mycorrhiza in restoring mediterranean ecosystems. *Mycorrhizas in Ecosystems* (Ed. by D. J. Read, D. H. Lewis, A. H. Fitter & I. J. Alexander), pp. 190–197. CAB International, Wallingford.

Barea, J. M., Azcón, R. & Azcón-Aguilar, C. (1993). Mycorrhiza and crops. *Advances in Plant Pathology*, Vol. 9. *Mycorrhiza: A Synthesis* (Ed. by I. Tommerup), pp. 167–189. Academic Press, London.

Bethlenfalvay, G. J. (1992). Mycorrhizae and crop productivity. *Mycorrhizae in Sustainable Agriculture* (Ed. by G.J. Bethlenfalvay & R.G. Linderman), pp. 1–25. ASA Special Publication, Madison, WI.

Bethlenfalvay, G. J. & Barea, J. M. (1994). Mycorrhizae in sustainable agriculture. I. Effects on seed yield and soil aggregation. *American Journal of Alternative Agriculture*, 9, 157–161.

Bethlenfalvay, G. J. & Linderman, R. G. (1992). *Mycorrhizae in Sustainable Agriculture.* ASA Special Publication, Madison, WI.

Bethlenfalvay, G. J. & Schüepp, H. (1994). Arbuscular mycorrhizas and agrosystem stability. *Impact of Arbuscular Mycorrhizas on Sustainable Agriculture and Natural Ecosystems* (Ed. by S. Gianinazzi & Schüepp), pp. 117–131. ALS, Birkhäuser, Basel.

Cervantes, E. & Rodríguez-Barrueco, C. (1992). Relationships between the mycorrhizal and actinorhizal symbioses in non-legumes. *Methods in Microbiology* (Ed. by J. R. Norris, D. J. Read & A. K. Varma), pp. 617–432. Academic Press, London.

Fitter, A. H. & Garbaye, J. (1994). Interaction between mycorrhizal fungi and other soil organisms. *Management of Mycorrhizas in Agriculture, Horticulture and Forestry* (Ed. by A. D. Robson, L. K. Abbott & N. Malajczuk), pp. 123–132. Kluwer Academic, Dordrecht.

Francis, R. & Read, D. J. (1994). The contribution of mycorrhizal fungi to the determination of plant community structure. *Management of Mycorrhizas in Agriculture, Horticulture and Forestry* (Ed. by A. D. Robson, L. K. Abbott & N. Malajczuk), pp. 11–25. Kluwer Academic, Dordrecht.

Glick, B. R. (1995). The enhancement of plant growth by free-living bacteria. *Canadian Journal of Microbiology*, 41, 109–117.

Harley, J. L. & Smith, S. E. (1983). *Mycorrhizal Symbiosis.* Academic Press, New York.

Herrera, M. A., Salamanca, C. P. & Barea, J. M. (1993). Inoculation of woody legumes with selected arbuscular mycorrhizal fungi and rhizobia to recover desertified mediterranean ecosystems. *Applied and Environmental Microbiology*, 59, 129–133.

Hooker, J. E., Jaizme-Vega, M. & Atkinson, D. (1994). Bicontrol of plant pathogens using arbuscular mycorrhizal fungi. *Impact of Arbuscular Mycorrhizas on Sustainable Agriculture and Natural Ecosystems* (Ed. by S. Gianinazzi & H. Schüepp), pp. 191–200. ALS, Birkhäuser, Basel.

Jasper, D. A. (1994). Management of mycorrhiza in revegetation. *Management of Mycorrhizas in Agriculture, Horticulture and Forestry* (Ed. by A. D. Robson, L. K. Abbott & N. Malajczuk), pp. 211–219. Kluwer Academic, Dordrecht.

Jeffries, P. & Barea J. M. (1994). Biogeochemical cycling and arbuscular mycorrhizas in the sustainability of plant–soil systems. *Impacts of Arbuscular Mycorrhizas on Sustainable Agriculture and Natural Ecosystems* (Ed. by S. Gianinazzi & H. Schüepp), pp. 101–115. ALS, Birkhäuser, Basel.

Kennedy, A. C. & Smith, K. L. (1995). Soil microbial diversity and the sustainability of agriculture soils. *Plant and Soil*, 170, 75–86.

Kloepper, J. W. (1992). Plant growth-promoting rhizobacteria as biological control agents. *Soil Microbial Ecology. Applications in Agriculture Forestry and Environmental Management* (Ed. by F. Blaine & J. Metting), pp. 255–274. Marcel Dekker, New York.

Kloepper, J. W., Zablotowick, R. M., Tipping, E. M. & Lifshitz, R. (1991). Plant growth promotion mediated by bacterial rhizosphere colonizers. *The Rhizosphere and Plant Growth* (Ed. by D. L. Keister & P. B. Cregan), pp. 315–326. Kluwer, Dordrecht.

Linderman, R. G. (1992). Vesicular–arbuscular mycorrhizae and soil microbial interactions. *Mycorrhizae in Sustainable Agriculture* (Ed. by G. J. Bethlenfalvay & R. G. Linderman), pp. 45–70. ASA Special Publication, Madison, WI.

Linderman, R. G. (1994). Role of VAM fungi in biocontrol. *Mycorrhizae and Plant Health* (Ed. by F. L. Pfleger & R. G. Linderman), pp. 1–26. APS Press, St Paul, MN.

Lynch, J. M. (1990). *The Rhizosphere.* J.H. Wiley and Sons, Chichester.

Marschner, H. & Dell, B. (1994). Nutrient uptake in mycorrhizal symbiosis. *Management of Mycorrhizas in Agriculture, Horticulture and Forestry* (Ed. by A. D. Robson, L. K. Abbott & N. Malajczuk), pp. 89–102. Kluwer Academic, Dordrecht.

Miller, R. M. & Jastrow, J. D. (1992). The role of mycorrhizal fungi in soil conservation. *Mycorrhizae in Sustainable Agriculture* (Ed. by G. J. Bethlenfalvay & R. G. Linderman), pp. 45–70. ASA Special Publication, Madison, WI.

Miller, R. M. & Jastrow, J. D. (1994). Vesicular–arbuscular mycorhizae and biogeochemical cycling. *Mycorrhizae and Plant Health* (Ed. by F. L. Pfleger & R. G. Linderman), pp. 189–212. APS Press, St Paul, MN.

Puppi, G., Azcón, R. & Höflich, G. (1994). Management of positive interactions of arbuscular mycorrhizal fungi with essential groups of soil microorganisms. *Impact of Arbuscular Mycorrhizas on Sustainable Agriculture and Natural Ecosystems* (Ed. by S. Gianinazzi & H. Schüepp), pp. 201–215. ALS, Birkhäuser, Basel.

Read, D. J. (1993). Mycorrhiza in plant communities. *Advances in Plant Pathology*, Vol. 9. *Mycorrhiza: A Synthesis* (Ed. by I. Tommerup), pp. 1–31. Academic Press, London.

Rosendahl, S., Dodd, J. C. & Walker, C. (1994). Taxonomy and phylogeny of the *Glomales*. *Impact of Arbuscular Mycorrhizas on Sustainable Agriculture and Natural Ecosystems* (Ed. by S. Gianinazzi & H. Schüepp), pp. 1–12. ALS, Birkhäuser, Basel.

Skujins, J. & Allen, M. F. (1986). Use of mycorrhizae for land rehabilitation. *MIRCEN Journal*, **2**, 161–176.

Smith, S. E., Gianinazzi-Pearson, V., Koide, R. & Cairney, J. W. G. (1994). Nutrient transport in mycorrhizas: structure, physiology and consequences for efficiency of the symbiosis. *Management of Mycorrhizas in Agriculture, Horticulture and Forestry* (Ed. by A. D. Robson, L. K. Abbott & N. Malajczuk), pp. 103–113. Kluwer Academic, Dordrecht.

Stoztky, G. (1972). Activity, ecology and population dynamic of microorganisms in soil. *Critical Review in Microbiology*, **2**, 59–137.

Thompson, J. P. (1994). What is the potential for management of mycorrhizas in agriculture? *Management of Mycorrhizas in Agriculture, Horticulture and Forestry* (Ed. by A. D. Robson, L. K. Abbott & N. Malajczuk), pp. 191–200. Kluwer Academic, Dordrecht.

Tisdall, J. M. (1994). Possible role of soil microorganisms in aggregation in soils. *Management of Mycorrhizas in Agriculture, Horticulture and Forestry* (Ed. by A. D. Robson, L. K. Abbott & N. Malajczuk), pp. 115–121. Kluwer Academic, Dordrecht.

Trappe, J. M. (1986). Phylogenetic and ecological aspects of mycotrophy in the angiosperms from an evolutionary standpoint. *Ecophysiology and VA Mycorrhizal Plants* (Ed. by G. R. Safir), pp. 5–25. CRC Press, Boca Raton, FL.

Vidal, M. T., Andrade, G., Azcón-Aguilar, C. & Barea, J. M. (1996). Comparative effect of Pseudomonas, strain F113 [biocontrol agent (antifungal)] and its isogenic mutant, strain F113G22 [impaired biocontrol ability] on spore germination and mycelial growth of *Glomus mosseae* under monoxenic conditions. *Mycorrhizas in Integrated Systems: from Genes to Plant Development* (Eds. C. Azcón-Aguilar & J. M. Barea), pp. 673–676. COST 821, Brussels.

5. INTERACTIONS BETWEEN ARBUSCULAR MYCORRHIZAL FUNGI AND FOLIAR PATHOGENS: CONSEQUENCES FOR HOST AND PATHOGEN

H. M. WEST

Department of Life Science, University of Nottingham, Nottingham NG7 2RD, UK

INTRODUCTION

Arbuscular mycorrhizal (AM) fungi form symbiotic associations with the roots of over 70% of land plants and are the most widespread of all plant symbionts (Trappe 1987). This ubiquitous symbiosis is of potential benefit to both plant and fungus, since the latter obtains its carbon from the host and the plant may benefit from fungal-mediated uptake of minerals, particularly phosphorus (Harley & Smith 1983). Other benefits to the host may also occur; for example, improved drought tolerance (Allen & Boosalis 1983) and disease resistance (Sharma *et al.* 1992; West *et al.* 1993a; Newsham *et al.* 1994).

Most evaluations of interactive effects of AM fungi and pathogens have concentrated on events within the rhizosphere (Barea *et al.*, this volume). However, alterations of the plant's nutrient status as a result of fungal colonization may be a key factor in determining relationships between host and other fungal species. There have been relatively few studies of the effects of simultaneous AM fungal and foliar fungal infections on the host. An increasing number of review papers addressing the effects AM fungi (AMF) have on the plant's response to disease can be found in the literature, but there is a conspicuous lack of information pertaining to AMF–foliar pathogen interactions. The reviews to date, which do mention this interaction, point to a common theme of increased disease severity in mycorrhizal compared to non-mycorrhizal plants (Schönbeck 1979; Dehne 1982; Bagyaraj 1984; Sharma *et al.* 1992). Thus, the following discussion concerns data arising from a small number of investigations.

Biotrophic foliar pathogens, unlike AMF, are generally host specific and this group includes the rusts and mildews. The results of infection may range from localized dysfunction to gross whole plant changes (Paul 1992). There is clearly a difference between the two groups of fungi in terms of whether the association is mutualistic or antagonistic, and also in the level of host specificity. This contrast between the two fungal groups is particularly striking because they both live in equal intimacy with their hosts (Law 1988).

79

COSTS AND BENEFITS

The expression 'costs and benefits' is widely used in relation to the AM fungal
symbiosis (Koide & Elliott 1989; Fitter 1991). It implies that there is both a cost
and a benefit of the association to the host plant. Generally, the benefit to the
fungus is not questioned because of its biotrophic nature. However, as the plant
may potentially benefit from the mycorrhizal association in a number of different
ways, Koide and Elliott (1989) and Fitter (1991) suggest that in a cost:benefit
analysis, there should be a common unit of cost – that is, carbon. Thus, growth of
a mycorrhizal plant (positive or negative) is a result of fungal-mediated effects on
the plant's carbon economy (assimilation or expenditure) (Koide & Elliott 1989).
Although plants may benefit in a number of different ways from the symbiosis,
the most widely accepted benefit is that of increased phosphate uptake. Phosphate
diffusion rates in soil are very low and, consequently, phosphate depletion zones
build up around the root. The quantity of phosphate required by a plant for maximum growth often exceeds the amount that can readily diffuse to the absorbing
surface (Nye & Tinker 1977). Mycorrhizal infection helps to overcome this problem,
because hyphae in effect increase the absorbing area of the root surface.

Estimates of the intra-radical mycelial biomass range from 4 to 17% (Hepper
1977) of the root dry weight, with extra-radical hyphae constituting about 1% of
the root fresh weight (Bevege *et al.* 1975). Lipids are abundant within the hyphae,
arbuscules and vesicles (Cox *et al.* 1975), and this abundance, together with
the large fungal biomass, is suggestive of a potentially significant carbon drain
on the host (Peng *et al.* 1993). Indeed, estimates of the amount of the plant's
daily photosynthate which is withdrawn by the fungus, range from 7% (Pang &
Paul 1980) to 20% (Jakobsen & Rosendahl 1990). Despite this apparently high
maintenance cost to the plant, Fitter (1991) points out that construction costs are
lower for production of hyphae than for roots, because mycelium is an order of
magnitude finer. Thus, efficiency of phosphate acquisition is increased in mycorrhizal compared to non-mycorrhizal plants. Since photosynthesis is linked to the
phosphorus (P) status of the plant, and may be sink-limited, any carbon utilized by
the fungus may be compensated for by enhanced carbon fixation. This is not always
the case and growth depressions resulting from mycorrhizal infection have been
reported (Koide 1985; Son & Smith 1988). Peng *et al.* (1993) recently carried out
an analysis of carbon costs in mycorrhizal Volkamer lemon (*Citrus volkameriana*).
Their work differed from other similar analyses of carbon costs because they
eliminated the complicating factor of phosphate nutrition by feeding high rates of
P to the plants. They found that mycorrhizal plants had lower specific growth rates
than non-mycorrhizal hosts, with colonized plants exhibiting a greater root biomass
allocation. This resulted in higher costs of maintenance respiration in mycorrhizal
plants. Not all plant species exhibit increased root biomass allocation in response

to mycorrhizal infection, therefore carbon costs will vary considerably. Thus, in a field situation, the costs of colonization can vary tremendously depending on many factors. Whether the benefits outweigh the costs will depend on, for example, soil P status and plant P demand. When nutrient levels are low, the benefits more than outweigh the costs.

Effects of mycorrhizal infection can thus influence host plant fitness positively, negatively or in an apparently neutral way. Benefits of mycorrhizal colonization in terms of enhanced plant growth and improved nutrition have been unequivocally demonstrated in pot cultures, yet the same benefits are often difficult to demonstrate in the field (Fitter 1985). Despite this, Peat and Fitter (1993), using data derived from the Ecological Flora Database, showed that mycorrhizal perennial species occupy wider habitat ranges than non-mycorrhizal species, indicating an overall benefit of the association.

Fitness and nutrient status are important in determining how well a plant will cope with stress. Infection by foliar biotrophic fungi may be considered as a biotic stress. Indeed, pathogenic infection can impose a considerable carbon drain on the host. In addition to the carbon costs already described for AMF colonization, further defence costs are involved, resulting in less available carbon being exported from the diseased leaf with consequent effects on the carbon balance of the host. For reviews of carbon budgets relating to pathogenic infections see Whipps and Lewis (1981), Farrar (1992) and Scholes (1992).

Since AMF colonization may affect host nutrient status, it is possible that this may in turn influence the relationship between the plant and foliar pathogens. Interactions occurring between the two fungal groups could arguably be dependent on the level of host benefit gained from AMF colonization. Zulu *et al.* (1991) showed that P status of wheat affected both the development of mildew and the effect of the disease on the plant. The outcome of simultaneous infections will undoubtedly result from a balance between costs and benefits.

PLANT–FUNGAL INTERACTIONS
IN INDIVIDUAL HOST PLANTS

Schönbeck (1979) described an experiment in which mycorrhizal and non-mycorrhizal plant species were inoculated with a range of obligate and facultative fungal pathogens affecting aerial plant parts. *Phaseolus* bean was infected with either *Uromyces phaseoli* or *Colletotrichum lindemuthianum*, cucumber with *Erysiphe cichoracearum*, barley with *Cochliobolus sativus* and lettuce with *Botrytis cinerea*. Overall, Schönbeck found that simultaneous infection by mycorrhizal fungi and a pathogen resulted in an increase in the number of visible necroses and spots. Additionally, the rust infection of bean yielded a higher spore count per pustule in mycorrhizal hosts compared with non-mycorrhizal plants, and the incubation time of the rust was reduced. Details of the nutrient status of the growth medium or plants were not given, but the author does discuss improved nutrition in relation to host mycorrhizal status. However, Schönbeck and Dehne (1979) (cited by Dehne 1982) found that when mycorrhizal and non-mycorrhizal plants were grown under conditions allowing for equivalent growth of both treatments, the mycorrhizal plants were still more susceptible to foliar infection.

Meyer and Dehne (1986) studied dual fungal infections of six varieties of lettuce, each infected with *Bremia lactucae* and four varieties of bean, infected with *Uromyces phaseoli*. Their findings corroborated those of previous workers as the level of pathogen infection was increased in mycorrhizal plants. However, despite this observation, mycorrhizal lettuce infected with *B. lactucae* generally had a greater shoot weight than non-mycorrhizal plants, although this depended on the particular AM fungal isolate used. This observation is similar to that of West (1995), who found that the deleterious effects of infection by the rust *Puccinia lagenophorae* on leaf production in *Senecio vulgaris* were not as severe in mycorrhizal compared with non-mycorrhizal groundsel when grown at an adequate level of P nutrition (i.e. sufficiently low P to induce a mycorrhizal response, but not too deleterious in the absence of AMF). Interestingly, this effect was not evident in groundsel grown at high levels of P nutrition. West (1995) did not find any mycorrhizal-mediated difference in disease severity as percentage rust infection of individual leaves was constant, but spore production per pustule was not measured. However, the proportion of leaves showing rust symptoms was dependent on total leaf production. Although when grown at an adequate soil P status, non-mycorrhizal groundsel had fewer leaves than mycorrhizal plants, a higher proportion of those leaves showed rust symptoms. Yet, when groundsel was grown at high soil P fertilization, mycorrhizal plants had fewer leaves than non-mycorrhizal groundsel, but the same proportion of leaves showed rust symptoms. Dehne (1982) stated that differences between comparable mycorrhizal and non-mycorrhizal cucumber in the degree of infection by powdery mildew, were observed only on young leaves. Thus, increased susceptibility was a function of enhanced pathogen development rather than increased frequency of infection. Since in West's (1995) investigation there were no treatment-related differences in leaf number at the time of applying rust spores, the frequency of infection was a function of leaf production, at least at the lower level of P nutrition. In a separate investigation, rust infection of groundsel was influenced by age of leaf. Older leaves at the time of rusting were more resistant to infection than younger leaves although no difference in degree of infection could be related to the mycorrhizal status of the groundsel (West, unpublished data). According to Burdon (1987) ontogenetic changes in the resistance or susceptibility of a host to a pathogen play an important role in controlling epidemics within a natural population.

Differences observed between the groundsel studies and the previous work may be a direct result of comparing wild and cultivated plant species as well as comparing effects derived from using different species and strains of AM fungi. AM fungi differ in their level of effectiveness and plant species vary in their degree of responsiveness (Fitter 1985; Brundrett 1991; Habte & Manjunath 1991). The requirement for P is determined by the plant's growth rate and the minimum concentration of P within the plant tissues which will support maximum growth and reproduction (Koide 1991). When P is present in the soil at very high levels, the plant will not benefit from mycorrhizal colonization. In natural systems, many plants will have a relatively low demand for P since they may be inherently adapted to nutrient deficient soils. Indeed, Fitter (1991) suggested that the demand for

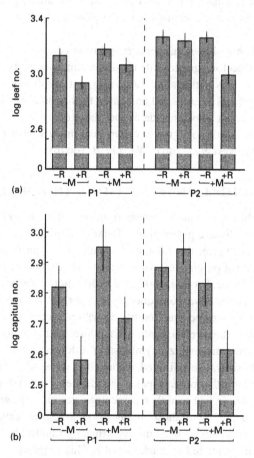

FIG. 5.1. Effects of arbuscular mycorrhizal (AM) and rust fungal treatments on (a) leaf and (b) capitula number of groundsel plants grown at two levels of soil phosphorus. P1, 30 μg g^{-1} P; P2, 120 μg g^{-1} P; R, rust infected, M, mycorrhizal. Values are means with individual standard errors. (From West 1995.)

P may be high only at certain stages within the life-cycle; for example, flowering and seed set. In the study by West (1995), when groundsel was grown at high soil P concentrations the mycorrhizal benefit was lost, but the cost of colonization was still a factor. Rusted, mycorrhizal plants grown at high P levels suffered in terms of leaf and capitula production, with fewer of each component being formed than on rusted, non-mycorrhizal plants. Interestingly, leaf and capitula numbers were similar in rusted, mycorrhizal plants grown at adequate and high P levels (Fig. 5.1a, b). The carbon demand of both the mycorrhizal and rust fungi may have resulted in the growth depressions observed in mycorrhizal, rusted plants at high P. Both the AM fungus and the rust can each utilize up to 20% of the plant's photosynthate. This represents a substantial drain on the host's resources, particularly if the mycorrhizal fungus is not contributing to the plant's fitness as a result of high available P.

Previous work has shown that modern cultivars of crop species have higher growth rates and, therefore, greater P demands than undeveloped ecotypes (Chapin et al. 1989). Thus, it is perhaps not surprising that the outcome of disease on mycorrhizal crop plants is enhanced compared to that on a weed species. However, groundsel is a fast-growing ruderal and, therefore, responds to fertilization and mycorrhizal infection to a greater extent than many slower growing species (e.g. *Vulpia ciliata*, West *et al.* 1993a; West *et al.* 1993b).

PLANT–FUNGAL INTERACTIONS IN POPULATIONS

The level of benefit which a plant sustains from being mycorrhizal may also have an influence on the disease potential of a foliar pathogen simultaneously infecting the plant; for instance, the amount of sporulation is an important factor in determining the rate of progress of a rust epidemic (Plummer *et al.* 1992). Thus, if as a result of mycorrhizal colonization, rust sporulation is enhanced (Schönbeck 1979), then this could have consequences for pathogen survival and spread. Rotem *et al.* (1978) showed that conditions which encourage a higher rate of photosynthesis may enhance sporulation; therefore, in a situation where a plant was benefiting from a mycorrhizal symbiosis (e.g. low nutrient availability), and where photosynthesis was sink-limited, it could be envisaged that rust sporulation might be increased. Furthermore, Gopalan and Manners (1984) showed that spores of *Puccinia striiformis* had a higher percentage germination when formed on healthy leaves compared with those produced on senescent leaves. West (1995) found that senescence was reduced in mycorrhizal groundsel to a greater degree than senescence in plants fed with additional P. This suggested that mycorrhizal infection had an effect on the plant which was additional to, or instead of, any P effect. Whatever the mechanism behind this observation (increased nitrogen uptake or alteration in phytohormone balance), a mycorrhizal-driven delay in senescence could improve the chances of rust spore survival on coming into contact with a new host.

Although under certain circumstances, mycorrhizal colonization may influence rust sporulation and potential infectivity, the influence that AM fungi has on species diversity could be less beneficial to the pathogen. In populations where susceptible plants have been replaced by resistant ones, the effective spread of pathogen inoculum is reduced. This is because the quantity of susceptible tissue is limited and the distance the inoculum has to travel in order to encounter a susceptible host would be increased (Burdon 1987). Indeed, Burdon and Shattock (1980) have shown that these factors influence the occurrence of disease in mixed stands.

Mycorrhizal colonization can influence species diversity within plant communities (Grime *et al.* 1987; Gange *et al.* 1990; Gange, *et al.*, 1993; Newsham *et al.* in press), which in turn will alter the genotypic diversity within the stand. It is arguable that this may affect the intrinsic spread of disease by reducing the chances of a host-specific pathogen from encountering a suitable host. However, Harper (1977) argued that the host-specific pathogen itself may be important in community diversification. This is because other susceptible hosts near to the infected plant will be damaged. Thus, plants which effectively disperse their seeds greater distances are more likely to escape the disease. Seed dispersal in time may also influence potential host survival. West (1995) found that seed set was enhanced in mycorrhizal

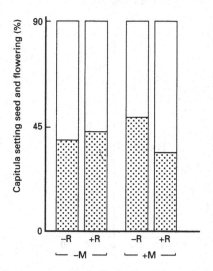

FIG. 5.2. Interactions between mycorrhizal colonization and rust infection affecting the proportion of capitula flowering (▨) and setting seed (□) at the time of harvest. Data are expressed as arcsin (square-root) transformed percentages. (From West 1995.)

rust infected groundsel, compared to rusted non-mycorrhizal plants (Fig. 5.2). It is interesting to speculate that this response to dual infection may strategically benefit the groundsel by encouraging rapid dispersal, thus giving resultant seedlings a competitive advantage. The other facet to the argument is that the rust fungus could also benefit, because the spatial distribution of the seed rain would inevitably result in seedlings germinating near to the parent plant. A more rapid production of the next generation would ensure effective spread of the disease focus. Figure 5.2 also shows that proportionally more capitula were still flowering at the time of harvest in rusted, non-mycorrhizal groundsel compared to the non-rusted, non-mycorrhizal plants. This response to rust infection was only slight and may be explained by the observation of Paul and Ayres (1986b) that rust infection delayed the start of flowering in groundsel.

Relatively little work has been carried out on the effects diseases have on natural plant communities, although the influence of *P. lagenophorae* on groundsel populations has been extensively studied (Paul & Ayres 1986a, b, c). These studies suggest that on the whole, rust infection can lead to reductions in growth and reproduction. Figure 5.3 gives an example of the effects both rust and mycorrhizal fungi have on populations of groundsel. Frequency distributions of plant height were affected by both fungal types. Data are represented as frequency polygons which show that mycorrhizal, rusted populations, produce a greater proportion of plants nearer to the maximum height class than do non-mycorrhizal, non-rusted populations. Rust infection of non-mycorrhizal populations resulted in a greater number of plants nearer to the minimum height class.

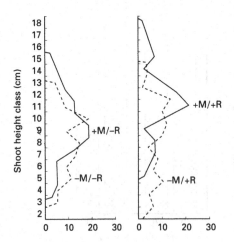

FIG. 5.3. Frequency distributions for height at a planting density of eight plants per pot. M, mycorrhizal; R, rust infected. Mean height: +M/+R = 10.78 ± 0.21; +M/−R = 8.53 ± 0.58; −M/+R = 8.70 ± 0.49; −M/−R = 8.07 ± 0.49. (West, unpublished data.)

CONCLUSION

Interpretation of results obtained to date is difficult because of the paucity of available data, and the complexity of the interactions. The plant and fungal species involved, as well as the soil nutrient status and other environmental variables, will alter the outcome of the tripartite interactions. The importance of these interactions in terms of individual plant fitness will depend on the costs and benefits of the associations. The importance in terms of population fitness and community structure has yet to be established, but clearly the scope for novel research is enormous.

ACKNOWLEDGEMENTS

The study of mycorrhizal/rust interactions in groundsel was supported by the NERC and the work was carried out in Professor P.G. Ayres's laboratory at Lancaster University.

REFERENCES

Allen, M. F. & Boosalis, M. G. (1983). Effects of two species of VA mycorrhizal fungi on drought tolerance of winter wheat. *New Phytologist*, **93**, 67–76.

Bagyaraj, D. J. (1984). Biological interactions with VA mycorrhizal fungi. *VA Mycorrhiza* (Ed. by C. Ll. Powell & D. J. Bagyaraj), pp. 131–153. CRC Press, Boca Raton, FL.

Bevege, D. I., Bowen, G. D. & Skinner, M. F. (1975). Comparative carbohydrate physiology of ecto- and endomycorrhizas. *Endomycorrhizas* (Ed. by F. E. Sanders, B. Mosse & P. B. Tinker), pp. 149–174. Academic Press, London.

Brundrett, M. (1991). Mycorrhizas in natural ecosystems. *Advances in Ecological Research*, **21**, 171–313.

Burdon, J. J. (1987). *Diseases and Plant Population Biology.* Cambridge University Press, Cambridge.

Burdon, J. J. & Shattock, R. C. (1980). Diseases in plant communities. *Applied Biology*, **5**, 145–219.

Chapin, F. S., Groves, R. H. & Evans, L. T. (1989). Physiological determinants of growth rate in response to phosphorus supply in wild and cultivated *Hordeum* species. *Oecologia*, **79**, 96–105.

Cox, G., Sanders, F. E., Tinker, P. B. & Wild, J. A. (1975). Ultrastructural evidence relating to host-endophyte transfer in vesicular–arbuscular mycorrhiza. *Endomycorrhizas* (Ed. by F. E. Sanders, B. Mosse & P. B. Tinker), pp. 297–312. Academic Press, London.

Dehne, H. W. (1992). Interaction between vesicular–arbuscular mycorrhizal fungi and plant pathogens. *Phytopathology*, **72**, 1115–1119.

Farrar, J. F. (1982). Beyond photosynthesis: the translocation and respiration of diseased leaves. *Pests and Pathogens* (Ed. by P. G. Ayres), pp. 107–127. Bios Scientific, Oxford.

Fitter, A. H. (1985). Functioning of vesicular–arbuscular mycorrhizas under field conditions. *New Phytologist*, **99**, 257–265.

Fitter, A. H. (1991). Costs and benefits of mycorrhizas: Implications for functioning under natural conditions. *Experientia*, **47**, 350–355.

Gange, A. C., Brown, V. K. & Farmer, L. M. (1990). A test of mycorrhizal benefit in an early successional plant community. *New Phytologist*, **115**, 85–91.

Gange, A. C., Brown, V. K. & Sinclair, G. S. (1993). Vesicular–arbuscular mycorrhizal fungi: a determinant of plant community structure in early succession. *Functional Ecology*, **7**, 616–622.

Gopalan, R. & Manners, J. G. (1984). Environmental and other factors affecting germination of urediniospores in yellow rust of wheat. *Transactions of the British Mycological Society*, 82, 239–243.

Grime, J. P., Mackey, J. M. L., Hillier, S. H. & Read, D. J. (1987). Floristic diversity in a model system using experimental microcosms. *Nature*, 328, 420–422.

Habte, M. & Manjunath, A. (1991). Categories of vesicular–arbuscular mycorrhizal dependency of host species. *Mycorrhiza*, 1, 3–12.

Harley, J. L. & Smith, S. E. (1983). *Mycorrhizal Symbiosis*. Academic Press, London.

Harper, J. L. (1977). *Population Biology of Plants*. Academic Press, London.

Hepper, C. M. (1977). A colorimetric method for estimating vesicular–arbuscular mycorrhizal infection in roots. *Soil Biology & Biochemistry*, 9, 15–18.

Jakobsen, I. & Rosendahl, L. (1990). Carbon flow into soil and external hyphae from roots of mycorrhizal cucumber plants. *New Phytologist*, 115, 77–83.

Koide, R. (1985). The nature of growth depressions in sunflower caused by vesicular–arbuscular mycorrhizal infection. *New Phytologist*, 99, 449–462.

Koide, R.T. (1991). Nutrient supply, nutrient demand and plant response to mycorrhizal infection. *New Phytologist*, 117, 365–386.

Koide, R. & Elliott, G. (1989). Cost, benefit and efficiency of the vesicular–arbuscular mycorrhizal symbiosis. *Functional Ecology*, 3, 252–255.

Law, R. (1988). Some ecological properties of intimate mutualisms involving plants. *Plant Population Ecology* (Ed. by A. J. Davy, M. J. Hutchings & A. R. Watkinson), pp. 315–341. Blackwell Scientific Publications, Oxford.

Meyer, J. & Dehne, H. W. (1986). The influence of VA mycorrhizae on biotrophic leaf pathogens. *Mycorrhizae: Physiology and Genetics* (Ed. by V. Gianinazzi-Pearson & S. Gianinazzi), pp. 781–786. INRA, Paris.

Newsham, K. K., Fitter, A. H. & Watkinson, A. R. (1994). Root pathogenic and arbuscular mycorrhizal fungi determine fecundity of asymptomatic plants in the field. *Journal of Ecology*, 82, 805–814.

Newsham, K. K., Watkinson, A. R., West, H. M. & Fitter, A. H. (in press). Symbiotic fungi determine plant community structure: changes in a lichen-rich community induced by fungicide application. *Functional Ecology*.

Nye, P. H. & Tinker, P. B. (1977). *Solute Movement in the Soil-Root System*. University of California Press, Berkeley.

Pang, P. C. & Paul, E. A. (1980). Effects of vesicular arbuscular mycorrhiza on ^{14}C and ^{15}N distribution in nodulated fababeans. *Journal of Soil Science*, 60, 241–250.

Paul, N. D. (1992). Partitioning to storage, regrowth and reproduction. *Pests and Pathogens* (Ed. by P. G. Ayres), pp. 129–141. Bios Scientific, Oxford.

Paul, N. D. & Ayres, P. G. (1986a). The impact of a pathogen (*Puccinia lagenophorae*) on populations of groundsel (*Senecio vulgaris*) overwintering in the field. I. Mortality, vegetative growth and the development of size hierarchies. *Journal of Ecology*, 74, 1069–1084.

Paul, N. D. & Ayres, P. G. (1986b). The impact of a pathogen (*Puccinia lagenophorae*) on populations of groundsel (*Senecio vulgaris*) overwintering in the field. II. Reproduction. *Journal of Ecology*, 74, 1085–1094.

Paul, N. D. & Ayres, P. G. (1986c). Interference between healthy and rusted groundsel (*Senecio vulgaris* L.) within mixed populations of different densities and proportions. *New Phytologist*, 104, 257–269.

Peat, H. J. & Fitter, A. H. (1993). The distribution of arbuscular mycorrhizas in the British Flora. *New Phytologist*, 125, 845–854.

Peng, S., Eissenstat, D. M., Graham, J. H., Williams, K. & Hodge, N. (1993). Growth depressions in mycorrhizal citrus at high-phosphorus supply. *Plant Physiology*, 101, 1063–1071.

Plummer, R. M., Hall, R. L. & Watt, T. A. (1992). Effect of leaf age and nitrogen fertilisation on sporulation of crown rust (*Puccinia coronata* var. *lolii*) on perennial ryegrass (*Lolium perenne* L). *Annals of Applied Biology*, 121, 51–56.

Rotem, J., Cohen, Y. & Bashi, E. (1978). Host and environmental influences on sporulation in vivo. *Annual Review of Phytopathology*, **16**, 83–101.

Scholes, J. D. (1992). Photosynthesis: cellular and tissue aspects in diseased leaves. *Pests and Pathogens* (Ed. by P. G. Ayres), pp. 85–106. Bios Scientific, Oxford.

Schönbeck, F. (1979). Endomycorrhiza in relation to plant diseases. *Soil-borne Plant Pathogens* (Ed. by B. Schippers & W. Gams), pp. 271–279. Academic Press, London.

Schönbeck, F. & Dehne, H. W. (1979). Untersuchungen zum Einfluss der endotrophen Mycorrhiza auf Planzenkrankheiten. 4. Pilzliche Sprossparasiten, *Olpidium brassicae*. TMV. *Zeitschrift für Pflanzenkrankheiten und Pflanzenschutz*, **86**, 103–112.

Sharma, A. K., Johri, B. N. & Gianinazzi, S. (1992). Vesicular–arbuscular mycorrhizae in relation to plant disease. *World Journal of Microbiology and Biotechnology*, **8**, 559–563.

Son, C. L. & Smith, S. E. (1988). Mycorrhizal growth responses: interactions between photon irradiance and phosphorus nutrition. *New Phytologist*, **108**, 305–314.

Trappe, J. M. (1987). Phylogenic and ecologic aspects of mycotrophy in the angiosperms from an evolutionary standpoint. *Ecophysiology of VA Mycorrhizal Plants* (Ed. by G. R. Safir), pp. 5–25. CRC Press, Boca Raton, FL.

West, H. M. (1995). Soil phosphate status modifies response of mycorrhizal and non-mycorrhizal *Senecio vulgaris* L. to infection by the rust, *Puccinia lagenophorae* Cooke. *New Phytologist*, **129**, 107–116.

West, H. M., Fitter, A. H. & Watkinson, A. R. (1993a). The influence of three biocides on the fungal associates of the roots of *Vulpia ciliata* ssp. *ambigua* under natural conditions. *Journal of Ecology*, **81**, 345–350.

West, H. M., Fitter, A. H. & Watkinson, A. R. (1993b). Response of *Vulpia ciliata* ssp. *ambigua* to removal of mycorrhizal infection and to phosphate application under natural conditions. *Journal of Ecology*, **81**, 351–358.

Whipps, J. M. & Lewis, D. H. (1981). Patterns of translocation, storage and interconversion of carbohydrate. *Effects of Disease on the Physiology of the Growing Plant* (Ed. by P. G. Ayres), pp. 47–84. Cambridge University Press, Cambridge.

Zulu, J. N., Farrar, J. F. & Whitbread, R. (1991). Effects of phosphate supply on wheat seedlings infected with powdery mildew: carbohydrate metabolism of first leaves. *New Phytologist*, **118**, 553–558.

CONCLUDING REMARKS

N. J. FOKKEMA

The unifying concept in the previous reviews is the beneficial effect of plant-associated micro-organisms for plants in agricultural or natural ecosystems. Often these associations are only beneficial in the presence of stresses: minor or major pathogens or malnutrition. In conventional agriculture, these stresses have been coped with by fungicides and chemical fertilizers. However, the drawbacks of such a large input of agrochemicals are well known and a low-chemical input, sustainable agriculture is the core of the agricultural policy of several nations and the European Union.

RHIZOSPHERE INTERACTIONS

It is relatively easy to produce plant growth promoting rhizobacteria (PGPR) or fungal biocontrol agents compared with that of inoculants of arbuscular mycorrhizal (AM) fungi which can only be grown on living roots. This difference may have controlled the development of experimental research. While mycorrhizal workers have studied differences between mycorrhizal and non-mycorrhizal plants, plant pathologists have studied mechanisms of microbial interactions and population dynamics in the rhizosphere. Large-scale production of AM fungi may never be achieved, and attention has been focused on the effects of naturally occurring AM fungi on host-plant physiology, soil structure and the possibilities of managing the mycorrhizosphere by cultural methods. In addition, in conventional agricultural systems, ploughing is detrimental to AM colonization, but does not affect free-living microbial antagonists. Consequently, PGPRs and fungal biocontrol agents utilized in conventional agriculture will generally operate in systems not promoting colonization by mycorrhizal fungi.

In sustainable agriculture, however, with low input of chemicals and low tillage, natural AM colonization may occur and consequently the beneficial effects of AM colonization can be expressed. So we have the interesting situation that in sustainable agriculture both types of organism not only may play a major role in disease control and plant growth but that sustainable agriculture is also a prerequisite for concerted action of mycorrhizal fungi and free living microbial antagonists (Barea *et al.*).

Bacterial biocontrol agents, affecting various soil-borne pathogens, seem – so far – not to affect mycorrhizal fungi, perhaps because of the biotrophic nature of the latter or spatial separation. It would be interesting to study whether mycorrhizal fungi are insensitive to antibiotics and other mechanisms of microbial interaction which are assumed to play a role in biocontrol. Before introducing bacteria to the soil or root system, the compatibility of biocontrol agents with mycorrhizal fungi should be checked and, if positive, the establishment of biocontrol agents may be enhanced in the mycorrhizosphere. Joint efforts for (more) field studies in this area would be very rewarding.

West has shown that cost–benefits relations with mycorrhizal fungi are not always easy to demonstrate under field conditions. Reasons for possible discrepancies between growth-promoting pot studies and experience under field conditions have not been further analysed. It has been observed, however, that continuous cropping of monocultures may result in the development of less beneficial species in the AM fungal community (Safir 1994).

Mycorrhizas seem to reduce effects of soil-borne diseases but may stimulate infections by both necrotrophic and biotrophic foliar pathogens in some cultivated crops. This indirect interaction might be the result of the effect of the changed nutrient status of the plant on infection. Moreover, under field conditions a denser foliage may create disease promoting microclimatic conditions. Despite this increase in disease severity, damage caused by foliar pathogens is often reduced in mycorrhizal plants. It is unknown whether mycorrhizal fungi can induce host resistance against pathogens or whether they are themselves affected by resistance induced by biocontrol agents or pathogens. The effect of mycorrhizae on foliar diseases could not be noticed in wild plants perhaps because of a different response to phosphate nutrition. Interestingly, none of the mycorrhizal papers considered variations in the density of mycorrhizal colonization, when analysing positive or negative effects.

Research on biological control of soil-borne pathogens has made tremendous progress with the mechanisms involved in microbial interactions by using molecular techniques. Genes responsible for antibiotic production, siderophore production or other traits can be deleted and complemented again. Comparison of loss and restoration of expression with loss and restoration of biocontrol activity has provided insight in to mechanisms of interaction. *In situ* detection of antibiotics and the identification of regulatory genes for antibiotic production in *Pseudomonas fluorescens* allow further studies into the environmental signals triggering antibiotic production. These achievements are equally well applicable to bacterial interactions in the phyllosphere. Insight into the production of antimicrobial metabolites in different situations is one of the prerequisites for improving reliability of biological control.

So far, mechanistic studies in particular have taken advantage of molecular techniques. We should not forget, however, that biocontrol is generally dependent on the density of the antagonist at the proper place. Keel and Défago mention a number of possible reasons for inconsistency in biocontrol. Variability in root colonization is on top of the list, but does not seem to receive a research input that justifies its importance. Such studies may become increasingly complicated by the recognition of the presence of viable but non-culturable bacterial cells in the rhizosphere.

In fungus–fungus interactions, mycoparasitism, particularly by ecologically obligate mycoparasites of sclerotium-forming pathogens, looks promising for biocontrol because the specific interaction may continue for a long period in and around decomposing plant residues. In some non-mycoparasitic associations antibiotics may play an important role. Gliotoxin production by a commercialized isolate of *Gliocladium virens* in biocontrol of a number of soil-borne pathogens has convincingly been demonstrated by using mutants not producing the antibiotic. The enzyme glucose oxidase produced by *Talaromyces flavus* seems to play a key role in controlling *Verticillium dahliae* by the production of hydrogen peroxide. The role of lytic enzymes in biocontrol such as chitinases, β-1,3, glucanases, proteases and lipases is important in mycoparasitic interactions, but Whipps questions their importance in other biocontrol associations. This view may not be shared by plant pathologists trying to improve the production of chitinolytic enzymes by the antagonists *Trichoderma* and *Gliocladium* spp. by genetic manipulation (Lorito *et al.* 1993). Their *in vitro* studies demonstrated that cell wall degrading enzymes and other antifungal compounds may act synergistically (Di Pietro *et al.* 1993). Mechanisms based on nutrient competition are less easy to prove by molecular techniques, but we should be careful that this will not become an argument to ignore this important mechanism. Similarly, some evidence for induced resistance should not exclude nutrient competition during the same interaction. To avoid confusion, the term cross-protection should be considered more as a phenomenon than as a mechanism like induced resistance.

PHYLLOSPHERE INTERACTIONS

Antibiotic production is the major mechanism discussed in the review of bacterial antagonist–fungal pathogen interactions. This is undoubtedly biased by *in vitro* screening of antagonists which will not detect *in vivo* activities such as nutrient competition. Earlier studies by Brodie and Blakeman (1975) and Blakeman and Brodie (1976) demonstrated that competition for exogenous nutrients as well as endogenous nutrient reserves in the fungal spore is a very common phenomenon, preventing spore germination in a similar way as in soil mycostasis. The often observed discrepancies between results of *in vitro* and *in planta* experiments on one hand and results of field experiments on the other hand are probably most extreme in bacteria–fungus interactions because of the differences in water requirements of both type of organisms (Reid 1980). However, in some protected crops and in the control of post-harvest diseases originating from man-made wounds, these water requirements seem to be met. Differences in humidity requirements do not interfere when bacterial antagonists are used for controlling bacterial pathogens. For example, fire blight in pear trees caused by *Erwinia amylovora* has been controlled effectively in several field experiments by *Pseudomonas fluorescens* and *Erwinia herbicola* (Johnson *et al.* 1993).

Very interesting is the discovery of the role of biosurfactants in biocontrol, resulting in adverse microclimatic conditions for spore germination of the pathogen. The ecological importance of this interesting trait for the antagonist population, that would never have been detected during *in vitro* screening, needs further research.

Microbial protection of infection of individual leaves imitates the action of fungicides, but has the disadvantage of preventive applications of the biocontrol agent before the establishment of the pathogen. Moreover, the possible interaction time between antagonist and pathogen is short because under favourable conditions the pathogen rapidly escapes from antagonism by penetrating the leaf.

Interaction with pathogen sporulation allows much longer interaction times (days/weeks) and aims at a retardation of the epidemic (Fokkema 1993). This strategy is followed in biological control of biotrophic pathogens, such as powdery mildews and rusts (Hofstein & Fridlender 1994), and currently investigated for necrotrophic pathogens (Sutton & Peng 1993, Köhl *et al.* 1995). Treatment of dead plant tissue with selected naturally occurring drought-resistant saprophytic fungi prevents colonization by *Botrytis cinerea* and reduces production of air-borne inoculum. It is envisaged that in sustainable agriculture, the amount of crop residue left on the soil surface will increase. This may create a large inoculum source of air-borne pathogens with serious epidemiological consequences (de Nazareno *et al.* 1993). Applications of competing saprophytes as a way of microbial sanitation may reduce this problem.

FUTURE RESEARCH

Exploitation of the beneficial effects of mycorrhizal fungi and antagonistic micro-organisms is handicapped by the often observed inconsistency of the beneficial effects under field conditions. Keel and Défago suggested this is due to: (i) variability in colonization; (ii) variability in production of responsible metabolites; (iii) specificity of the beneficial effect; and (iv) genetic instability of the inoculants.

One may add (v) the presence of naturally occurring beneficial organisms in the control, and (vi) the absence of nutritional or biological stress factors.

For future research it is important to identify the major cause of failures. Adequate representation of beneficial organisms at the proper place and time seems the most crucial factor. However, it is not the most intensively studied issue. There are only a few studies using computer simulation models in predicting colonization and effect of introduced or naturally occurring beneficial micro-organisms in the phyllosphere (Knudsen & Hudler 1987; Dik 1991). With simulation models, we may predict the response of introduced populations to a variety of environmental conditions and this will be helpful in selecting more ecologically competent strains. Moreover, the reliability will be improved when we know the conditions under which our introduced micro-organisms will work or not.

Therefore, future research should concentrate on the ecology of the beneficial micro-organisms, using molecular techniques for identifying specific colonization traits as well as tracing and quantifying introduced micro-organisms. Simulation models should be used for a better understanding of microbial population dynamics in relation to their beneficial role and their environmental impact.

REFERENCES

Blakeman, J. P. & Brodie, I. D. S. (1976). Inhibition of pathogens by epiphytic bacteria on aerial plant surfaces. *Microbiology of Aerial Plant Surfaces* (Ed. by C. H. Dickinson & T. F. Preece), pp. 529–557. Academic Press, London.

Brodie, I. D. S. & Blakeman, J. P. (1975). Competition for carbon compounds by a leaf surface bacterium and conidia of *Botrytis cinerea*. *Physiological Plant Pathology*, 6, 125–135.

De Nazareno, N. R. X., Lips, P. E. & Madden, L. V. (1993). Effect of levels of corn residue on the epidemiology of gray leaf spot of corn in Ohio. *Plant Disease*, 77, 67–70.

Di Pietro, A., Lorito, M., Hayes, C. K., Broadway, R. M. & Harman, G. E. (1993). Endochitinase from *Gliocladium virens*: isolation, characterization, and synergistic antifungal activity in combination with gliotoxin. *Phytopathology*, 83, 308–313.

Dik, A. J. (1991). Interactions among fungicides, pathogens, yeasts and nutrients in the phyllosphere. *Microbial Ecology of Leaves* (Ed. by J. H. Andrews & S. S. Hirano), pp. 412–429. Springer, New York.

Fokkema, N. J. (1993). Opportunities and problems of control of foliar pathogens with micro-organisms. *Pesticide Science*, 37, 411–416.

Hofstein, R. & Fridlender, B. (1994). Development of production, formulation and delivery systems. *Proceedings of the Brighton Crop Protection Conference – Pests and Diseases 1994*, pp. 1273–1280. BCPC, Farnham.

Johnson, K. B., Stockwell, V. O., McLaughlin, R. J., Sugar, D., Loper, J. E. & Roberts, R. G. (1993). Effect of antagonistic bacteria on establishment of honey bee-dispersed *Erwinia amylovora* in pear blossoms and on fire blight control. *Phytopathology*, 83, 995–1002.

Knudsen, G. R. & Hudler, G. W. (1987). Use of a computer simulation model to evaluate a plant disease biocontrol agent. *Ecological Modelling*, 35, 45–62.

Köhl, J., Molhoek, W. M. L., Van der Plas, C. H. & Fokkema, N. J. (1995). Effect of *Ulocladium atrum* and other antagonists on saprophytic colonization of dead lily leaves exposed to field conditions by *Botrytis cinerea*. *Phytopathology*, 85, 393–401.

Lorito, M., Harman, G. E., Hayes, C. K., Broadway, R. M., Tronsmo, A., Woo, S. L. & Di Pietro, A. (1993). Chitinolytic enzymes produced by *Trichoderma harzianum*: Antifungal activity of purified endochitinase and chitobiosidase. *Phytopathology*, 83, 302–307.

Reid, D. S. (1980). Water activity as the criterion of water availability. *Contemporary Microbial Ecology* (Ed. by D. C. Ellwood, J. N. Hedger, M. J. Latham, J. M. Lynch & J. H. Slater), pp. 15–27. Academic Press, London.

Safir, G. R. (1994). Involvement of cropping systems, plant produced compounds and inoculum production in the functioning of VAM fungi. *Mycorrhizae and Plant Health* (Ed. by F. L. Pfleger & R. G. Linderman), pp. 239–259. APS Press, St Paul, MN.

Sutton, J. C. & Peng, G. (1993). Biocontrol of *Botrytis cinerea* in strawberry leaves. *Phytopathology*, 83, 615–621.

PART 2
PLANT–MICROBE–ANIMAL
INTERACTIONS

INTRODUCTORY REMARKS

J. B. WHITTAKER

*Division of Biological Sciences, Institute of Environmental and Biological
Sciences, Lancaster University, Lancaster LA1 4YQ, UK*

Multitrophic interactions in terrestrial systems have been recognized since the pioneering work of Summerhayes and Elton (1923) and others. These remained largely descriptive until attempts were made to quantify them through measurements of energy and nutrient fluxes. But experimental work was largely lacking despite the plea (Varley 1957) to develop ecology as an experimental science. Such quantitative and experimental work as had been done was largely confined to trophic interactions between primary and secondary consumers in agricultural and forestry systems where the main question was one of crop damage, or between trophic levels in the same kingdom such as predator/prey or parasite/host systems. This reflected the prevailing organization of biological research and training into separate disciplines of agriculture, forestry, microbiology, botany or zoology. Just as the development of computer models of interactions was constrained by available computer power (e.g. most models had temporal but not spatial components though it was clear to field ecologists that the latter were crucial), so experimental work in multitrophic interactions was limited by the specialized training and experience of the scientists and it has to be said, often their unwillingness to communicate. The emergence of multidisciplinary departments of biological sciences in the UK in the mid-1960s led first to collaboration between established scientists who saw the merits of exchanging ideas with specialists from other disciplines and ultimately, with the usual lead time associated with such changes, a new generation of individuals capable of crossing some of these boundaries. So, in the 1980s came a rapid increase, reflected in the Society's journals, of papers in which, for example, insects and plants were treated as more than just one being a resource of the other. At this time came also extremely stimulating texts (Crawley 1983) which set the stage for quantitative and experimental studies of ditrophic interactions. As confidence grew, experiments were designed to include three levels of interaction, plant–mycorrhiza–insect, for example (Warnock *et al.* 1982) and in this session we are privileged to hear five scientists who have been more successful than most in exploring these multitrophic interactions at the levels of plant–microbe–animal interactions.

6. INTERACTIONS BETWEEN ARBUSCULAR MYCORRHIZAS AND PLANT PARASITIC NEMATODES IN AGRO-ECOSYSTEMS

R. W. RONCADORI

Department of Plant Pathology, University of Georgia, Athens, GA 30602, USA

INTRODUCTION

Considering the vast number and variety of soil micro-organisms that coexist in the soil, the plant parasitic nematode–arbuscular mycorrhizal fungus relationship is among the most unusual. Both groups are widespread, biotrophic, and each may have a profound but opposite effect upon plant health. Furthermore, they compete for the same substrate – young feeder roots.

Plant parasitic nematodes have been long recognized as a major cause of root diseases in agricultural crops. However, the beneficial symbiotic relationship between arbuscular mycorrhizal (AM) fungi and plants has attracted considerable interest only during the past 25 years. Throughout this period, the role of arbuscular mycorrhizas in facilitating nutrient (primarily phosphorus (P)) and water absorption and translocation as well as other benefits has been reported for a multitude of plant species (Harley & Smith, 1983); yet, neither the physiological nor biochemical basis for symbiosis or nematode parasitism is sufficiently understood to identify to any significant degree the mechanisms involved in interactions. Consequently, the bulk of the literature on this topic is predominantly descriptive. Several reviews on plant parasitic nematode–arbuscular mycorrhiza interactions have been published. Among the most recent are Hussey and Roncadori (1982), Smith (1987), Ingham (1988) and Francl (1993).

The effect of AM on plant parasitic nematodes is likely to become a more important issue due to regulatory removal of a number of effective nematicides from commercial crop management systems in many countries. Other control means, such as cultural practices and biological control, are now being considered as more attractive alternatives, including the potential of AM.

BIOTROPHIC COMPETITORS

The nematode feeding habit is a significant factor affecting any potential interaction with AM fungi (Hussey & Roncadori 1982). Characteristically, the feeding behaviour may be endoparasitic or ectoparasitic with migratory or sedentary forms in each category. A few species may alternate between the endo- or ectoparasitic habit. The most economically important group unquestionably is the sedentary endoparasites, particularly *Meloidogyne* spp. (root-knot nematode), and *Heterodera* and *Globodera* spp. (cyst nematode). This chapter emphasizes research done with the root-knot nematode due to the attention that has been focused on this group. Species typical of *Meloidogyne* have a life-cycle beginning with the egg, developing through four juvenile stages, and maturing into adults. The only infectious stage is the second stage juvenile, which is vermiform-shaped and penetrates the epidermis and cortex immediately behind the root tip and begins to feed near or within the developing stele. Soon afterward the nematode becomes immobile, the musculature deteriorates, and the body swells. The nematode develops into an adult (usually female) and is totally dependent upon several large, extensively modified giant cells around its head to feed, ultimately causing the root to form a gall around the feeding site. In contrast, other species of nematodes that are migratory endoparasites tend to remain vermiform as adults. The entire nematode enters the root and moves through the cortex without establishing a permanent feeding site and in doing so, may cause considerable tissue damage. Ectoparasites also are generally vermiform and remain outside the root feeding on internal cells by means of a long stout stylet. With few exceptions, the latter group seldom causes extensive root damage.

On the other hand, AM fungi have both intra- and extra-radical stages. Structures such as hyphae, arbuscules, spores and vesicles are present primarily in the cortex and less often in the epidermis, but not in the stele. The soil phase may consist of hyphae, vesicles, spores and sporocarps. Based on the spore stage, the fungi have been tentatively placed in the class Zygomycetes family Endogonaceae (Francl 1993).

INTERACTION STUDIES

To provide a basis for discussion of the interaction studies, the definitions introduced by Cook and Evans (1987) are used. Host responses are expressed either as tolerant (little or no suppression of plant growth or yield) or intolerant (significant suppression of plant growth or yield) or are referred to as resistant (retarded nematode development or reproduction) or susceptible (normal nematode development or increased reproduction). In addition, effects on the endophyte may be positive, negative, or neutral in affecting root colonization or sporulation (Hussey & Roncadori 1982). Unless otherwise stipulated, studies have been carried out in fumigated or treated soils in the glasshouse.

The earliest evidence that AM and plant parasitic nematodes interact and affect plant health was from the results of field studies. Prior to these field observations, Fox and Spasoff (1972) reported mutual antagonism between the cyst nematode and an AM fungus on tobacco. It was suggested that the degree of micro-organism antagonism was affected by cultivar and host tolerance to the nematode. Subsequently, field surveys of root-knot nematode on cotton by Bird *et al.* (1974) and root-knot and cyst nematodes on soybean by Schenk and Kinloch (1974) revealed an inverse relationship between incidence of AM and these plant parasitic nematodes. On cotton, a nematicide treatment was associated with increases in arbuscules, vesicles, and sporulation and this effect was more pronounced in a nematode-resistant than a nematode-susceptible cultivar. With soybean, mycorrhizas were more prevalent in fields with fewer nematodes than in highly infested fields and cultivar suceptibility also had an effect. These surveys created interest in studies under controlled conditions and were also the initial indicators of the complexities of tripartite relationships at a time in which little was known about the AM symbiosis. Schenck, *et al.* (1975) carried out a comprehensive study on soybean involving three mycorrhizal fungi, root-knot nematode, and nematode-resistant and nematode-susceptible cultivars. Variability in responses led to the conclusion that the outcome of each interaction may depend upon the specific plant parasitic nematode–mycorrhizal fungus–plant host combination tested. This hypothesis has proven true to the present time.

Since the mid-1970s, approximately 50 co-inoculation or challenge tests have been reported. The earlier studies were usually conducted in low P soils to favour mycorrhizal development and attain maximum plant growth and challenge response. However, under such conditions it was not possible to determine whether tolerance and resistance were due solely to improved host P nutrition or other factors as well (Hussey & Roncadori 1982; Smith 1987). Consequently, later experiments included the use of different P fertilization rates ranging from suboptimal to optimal for plant growth. Because of the extensive number of papers on plant parasitic–arbuscular mycorrhizal fungus interactions, it has been essential to limit the number of reports discussed.

SEDENTARY ENDOPARASITIC NEMATODE–AM STUDIES

The abundance, economic importance, and ease of manipulation of *Meloidogyne* spp. has made them prime candidates for challenge studies with mycorrhizas. Consequently, endophyte interactions with root-knot nematodes have been investigated more frequently than all other plant parasitic nematodes combined. Likewise, the most commonly used endophytes have been in the genera *Glomus* and *Gigaspora* (Hussey & Roncadori 1982; Ingham 1988; Francl 1993). Additional interacting factors affecting challenge studies have included plant cultivar, sequence of microorganism inoculation, soil fertilization (particularly P) rates, initial micro-organism population densities, and the use of high P-tolerant endophytes.

Interactions affecting plant tolerance

Increased plant tolerance to nematode attack has been the most frequently observed benefit of symbiosis (Hussey & Roncadori 1982; Smith 1987; Ingham 1988). In nearly half of the tests, mycorrhizas enabled the plant to compensate for nematode-induced stunting or other symptoms. The degree of compensation varied in each test but occurred along a continuum described in the model by Hussey and Roncadori (1982) in which nematode-induced stunting was offset but growth was usually less than that of mycorrhizal plants not attacked by the nematodes. The earliest challenge tests compared co-inoculated plant responses with that of control plants grown in P-deficient soils (Smith 1987). A wide variety of crops were used, such as carrot, citrus, cotton, grape, oats, onion, peach, peanut, soybean, tomato and tobacco. Tolerance was usually ascribed to improved host P nutrition.

During the 1980s, interaction studies were expanded to include different levels of soil P. Hence, it was possible and essential to make comparisons between co-inoculated plants and appropriate controls that were similar in biomass and not P deficient (Smith 1987). These studies strongly, but not exclusively, linked improved host P status with improved nematode tolerance on crops such as alfalfa, bean, clover, cotton, soybean, tamarillo and tomato. In contrast, Smith, *et al.* (1986b) reported in cotton grown in field microplots that high rates of soil P fertilization reduced growth and yield of plants inoculated with root-knot nematode. The effect was attributed to a Zn deficiency induced by high P fertilization and nematode parasitism and was compensated for by the mycorrhiza. Increased P fertilization in tamarillo actually reduced tolerance to root-knot nematode attack (Cooper & Grandison 1987). However, the effect occurred only in plants co-inoculated at planting but not in plants in which nematode inoculation was delayed by 4 weeks after planting. Since root and shoot P contents were similar in nematode-parasitized and control plants, a P effect was discounted and benefits in mycorrhizal plants were attributed to other factor(s). In onion, P fertilization had little effect on tolerance to root-knot nematode attack (MacGudwin *et al.* 1985).

Plant reaction to nematode attack may also be altered by other factors, such as sequence of inoculation. In cotton, 4 weeks of arbuscular mycorrhizal colonization prior to root-knot nematode inoculation increased tolerance compared with simultaneous inoculation (Smith *et al.* 1986a). Similar results were reported in tamarillo (Cooper & Grandison 1987). Precolonization with AM fungi, however, may not be a realistic sequence of events occurring in the field. Kellam and Schenck (1980) have noted that mycorrhizal colonization of soybean roots may require at least 10 days to 2 weeks in soybean. In our research with cotton, even under the most favourable conditions, root colonization by AM fungi routinely took place between 2 and 4 weeks. On the other hand, most plant parasitic nematodes may penetrate roots in terms of hours to a few days. Certainly, in studies involving simultaneous inoculation with the endophyte and nematode, the latter may have a distinct advantage. Perhaps one of the benefits of nematicides is their capacity to suppress plant parasitic nematodes to very low population densities and thus give mycorrhizas more time to colonize the root system.

Delayed nematode inoculation itself may noticeably suppress symptom expression on plants in the absence of mycorrhizas. In a 4-year microplot interaction study on peach, a significant suppression was caused in growth and yield by *M. incognita* added at planting compared with a 6-week delayed inoculation of non-mycorrhizal and pre-inoculated mycorrhizal plants at a moderate soil P level (Roncadori & Hussey 1986). P fertilization and an AM fungus initially affected tolerance but were no more effective than delayed nematode inoculation after four growing seasons as determined by fruit yield. Hence in root-knot susceptible crops such as peach, it may be difficult to separate the effect from that of pre-inoculation with an endophyte from increasing host tolerance with age without using a proper control – that is, delayed nematode inoculation of non-mycorrhizal as well as mycorrhizal plants. If possible, plant tolerance in interactions should be determined on a yield basis since the early beneficial effects on plant growth may be transient.

Interactions affecting plant resistance

Nematode reproduction

Unfortunately, the outcome of the tripartite relationship on host resistance has been as unpredictable as it has been with host tolerance (Hussey & Roncadori 1982; Smith 1987). AM may suppress, stimulate, or have no effect on nematode reproduction. Nematode reproduction has been reported either as numbers per root system or per gram of root. Each criterion has specific uses. Comparing the effect of AM on the basis of eggs per root system can be useful epidemiologically since mycorrhizal root systems tend to be larger than non-mycorrhizal root systems and support greater egg population densities in the field. Expression of eggs per gram of root or per female is appropriate in determining mycorrhizal-induced physical or physiological differences which may affect the nematode life-cycle. However, expression of data on a whole root system or per gram of root may be used both in an epidemiological capacity or to determine physical or physiological effects induced by the endophyte if treatment comparisons are made between plants with equal-sized root systems. The use of both criteria in the same study is advisable since it would standardize results and allow a comparison of experiments reported by different investigators. In cases in which AM fungi suppress nematode egg production per gram of root, the results may be due to a dilution effect brought on by mycorrhizal root growth stimulation and reflect disease escape rather than alteration in resistance (Smith 1987).

The first report of AM fungi altering plant resistance to nematode attack was published in 1972 when Fox and Spasoff observed a 25–35% reduction in population densities of cyst nematode on mycorrhizal tobacco. Nematode egg production has been the first and most frequently used criterion to evaluate mycorrhizal effects on host resistance. The variation in the way these data have been reported makes direct and reliable comparisons difficult (Hussey & Roncadori 1982; Smith 1987). A summary of challenge studies by Hussey and Roncadori (1982) indicates that root-knot nematode egg production per gram of root in 21 tests was suppressed in eight, increased in five and unaffected in eight. Crops included in the studies were carrot, cotton, grape, oats, peach, peanut, soybean, tobacco and tomato. Smith (1987) reported in eight subsequent co-inoculation studies with AM fungi and root-knot nematode that egg production was suppressed in three cases, increased in one case and unaffected in four trials. Crops used as hosts in these later studies were bean, cotton, onion, soybean and tomato. These challenge tests differed from earlier experiments through the evaluation of different P fertilization concentrations on root-knot nematode reproduction both with and without mycorrhizas. The effect of improved host P status on root-knot nematode egg production was variable. With cotton inoculated at planting, high P fertilization of non-mycorrhizal plants stimulated egg production per plant by 75% over low P fertilized plants and was associated with a greater number of eggs per ovipositing female rather than root system size differences (Smith *et al.* 1986a). An arbuscular fungus added at planting had no effect on egg production per plant under similar conditions. A 28-day delay in root-knot nematode inoculation after mycorrhizal development suppressed egg production per plant by 54%, but increased P fertilization had no effect on eggs per female. The nematode population density was higher, however, due to larger root systems.

On bean, results varied in root-knot nematode egg production at four different soil P fertilization levels in a challenge study (Smith 1987). At the lowest added P rate (0 mg kg^{-1} soil) eggs per plant were greater on mycorrhizal plants due to larger root systems. At two intermediate P fertilization rates (50 and 100 mg P kg^{-1} soil), nematode reproduction was markedly suppressed by the mycorrhiza even though the plants were over 20% larger than the non-mycorrhizal plants. At these levels, AM root colonization was 56–59%. At the highest P level (200 mg g^{-1} soil), root colonization was only 7% and there was little difference in egg population densities compared with non-mycorrhizal plants.

Penetration and developmental stages

Mycorrhizal effects on nematode reproduction have led to studies on other stages in the nematode life-cycle to better understand the mechanisms involved in tolerance and resistance. Generally, both mycorrhizal development and higher P fertility rates have tended to increase root-knot nematode penetration and infection (Smith 1987). However, Sikora (1978) reported a 64% decrease in root-knot nematode second stage juveniles in mycorrhizal tomato roots compared with non-mycorrhizal roots 72 h after inoculation and juveniles in mycorrhizal roots developed more slowly than in non-mycorrhizal roots up to 16 days later. Nematode development has usually been increased by P fertilization and suppressed on mycorrhizal plants (Smith 1987).

Host resistance should not be determined by observing juvenile penetration of the root system over a short time period. Herman *et al.* (1991) reported that on some soybean resistant and susceptible genotypes, root-knot nematode second stage juveniles penetrated roots of both groups abundantly but often emerged from the roots of resistant genotypes shortly after entry. In fact, the rate of emergence with one genotype was 72% 5 days after inoculation and the juveniles had decreased ability to attack other roots.

A study by Cooper and Grandison (1986) illustrated the interacting effects of P fertilization rates (0, 8, 30, 120 mg of added P kg^{-1} soil), AM and inoculation sequence (nematodes added at planting or 28 days later) on root-knot nematode infection and development on tomato and white clover expressed on a per plant and per gram root basis. Greater P fertility rates increased the total nematode population densities as well as adult females on non-mycorrhizal plants of both host species regardless of the inoculation sequence. Mycorrhizal effects, even though they increased plant growth at all P levels, were less consistent. At the two lower P fertilization rates, total nematode population densities were greater on mycorrhizal plants than on non-mycorrhizal plants, possibly due to improved P status and larger root systems of the former. At the two higher P rates on tomato, fewer nematodes infected mycorrhizal plants than non-mycorrhizal plants, but this trend occurred on clover only at the highest P fertility rate with delayed nematode inoculation. With clover, both total nematodes per gram root and number of adult females were less on mycorrhizal plants than non-mycorrhizal plants at all fertility levels, regardless of inoculation sequence. However, on tomato the mycorrhizas suppressed nematode infection per gram root only at the two higher P levels during co-inoculation at planting, but were effective at all P levels in the delayed nematode inoculation. Adult females per gram of tomato root were reduced by the mycorrhizas and the effect was not changed by P level or inoculation sequence. This mycorrhizal suppression of root-knot nematode development was attributed to factor(s) unrelated to P since the P content of roots and shoots were similar or greater in mycorrhizal plants than in non-mycorrhizal plants.

A growth chamber study demonstrating the interacting effects of an AM fungus, P fertilization and nematode inoculation sequence on cotton was reported by Smith *et al.* (1986a). Co-inoculation at planting resulted in more total nematodes and ovipositing females on P-fertilized, non-mycorrhizal plants compared with mycorrhizal plants or non-mycorrhizal, and non-fertilized plants. In contrast, total nematodes and number of ovipositing females were lower on plants pre-inoculated with the arbuscular fungus 28 days before nematode inoculation than on P-fertilized or non-fertilized plants.

Interactions affecting the fungal symbiont

The effects of sedentary endoparasitic nematodes on AM fungus activity have been extremely inconsistent as determined by sporulation and root colonization. Endophyte activities may be altered by increased P fertilization, nematode-resistance or susceptibility of the cultivar, sequence of inoculation, host mycorrhizal dependency and fungal species. Hussey and Roncadori (1982) summarized the literature on antagonism of sedentary endoparasites *Meloidogyne* spp. and *Heterodera glycines* toward activities of different AM fungi. In 17 tests, four reported antagonism toward the endophyte, two demonstrated stimulation of the endophyte, and there was no effect in 11 studies. Later investigations tend to confirm this variability (Ingham 1988; Francl 1993).

The unpredictable response of AM fungi to nematode attack on the host is illustrated by an investigation by Carling *et al.* (1989) on soybean. In a challenge test involving *M. incognita*, *Gigaspora margarita* and *Glomus etunicatum* at P fertilization levels of 0, 25, 50 and 150 mg P g^{-1} soil, root colonization of both endophytes was only slightly affected. However, sporulation of *G. margarita* was stimulated by the nematode by 340% and 163% at the two lowest P levels and by 179% at the highest fertilization rate. On the other hand, sporulation of *G. etunicatum* was suppressed by 50% and 37% at the 0 and 25 mg P fertilization rates, respectively. Root colonization and sporulation of both endophytes decreased markedly with increased P fertilization between 50 and 150 mg rates.

MIGRATORY ENDOPARASITIC NEMATODE–AM INTERACTION STUDIES

The extensive occurrence of necrosis in cortical and epidermal tissues distinguish this group of plant parasitic nematodes from the sedentary endoparasites. Likewise, spatially their activities in the cortex may place them in direct competition with AM fungi for a common location (Hussey & Roncadori 1978). The obligate biotrophism of the endophyte seems to be in direct contrast and vulnerable to such a destructive feeding habit.

Migratory endoparasite–AM interactions have been given little attention by researchers. Only a few studies have been published dealing with a diversity of hosts, such as citrus, plum and cotton (Hussey & Roncadori 1982; Smith 1987; Francl 1993). Again, test results have revealed considerable variation in host tolerance and resistance. Increased tolerance in *Citrus limon* to attack by the burrowing nematode, *Radopholus similis*, was induced by *G. etunicatum* (O'Bannon & Nemec 1979), but not in a test with *R. citrophilus* challenged by *Glomus intraradices* (Smith & Kaplan 1988). There was no mycorrhizal effect on tolerance of navy bean (Elliott *et al.* 1984), cotton (Hussey & Roncadori 1978), and plum (Camprubi *et al.* 1993) challenged with the lesion nematode *Pratylenchus* spp. since the nematodes did not suppress growth at the inoculum rates used. In cotton, the population density of *P. brachyurus* per gram root was significantly suppressed by the mycorrhiza, but was stimulated on navy bean. There were no antagonistic mycorrhizal effects toward nematode reproduction in the other tests.

Activities of the AM fungi were inconsistently affected by this group of nematodes. Root colonization and sporulation of *G. etunicatum* on *C. limon* were suppressed by *R. similis* (O'Bannon & Nemec 1979) but root colonization in the same crop by *G. intraradices* was not affected by *R. citrophilus* (Smith & Kaplan 1988). Lesion nematode (*Pratylenchus* spp.) parasitism reduced root colonization on navy bean (Elliott *et al.* 1984) but had no negative effect on endophyte activities on cotton or plum.

MECHANISMS OF TOLERANCE OR RESISTANCE INVOLVED IN INTERACTIONS

Understanding of mycorrhizal mechanisms involved in tolerance or resistance to plant parasitic nematode attack is rudimentary and reported conclusions may be conflicting. Since these various hypotheses have been discussed in previous reviews (Hussey & Roncadori 1982; Smith 1987; Ingham 1988; Francl 1993), treatment here is brief. Effects of each group of micro-organisms on the other is apparently mediated through host changes rather than directly. However, AM fungsal parasitism of cyst nematode can occur, but is considered an exception since the endophyte has access to nutrients from the plant host (Francl 1993).

Improved plant nutrition and growth

Greater tolerance of mycorrhizal plants to nematode attack often has been attributed to improved P nutrition (Smith 1987). Enhanced nutrient and water absorption can increase root growth and function to compensate for loss and dysfunction. Extremely rapid root growth could dilute the nematode population density and thus promote disease escape of some roots (Hussey & Roncadori 1978). Furthermore, since plant parasitic nematodes are not known to directly attack the extra-radical hyphae of arbuscular fungi, it is possible that the more extensive network of hyphae in the soil reduces the need for extensive root branching and, hence, fewer infection sites that would be available to the nematode.

Effects of nutrients other than P

AM also may improve absorption and translocation of Ca, Cu, Mn, S and Zn as
well as P (Harley & Smith 1983). Plants attacked by nematodes may show de-
ficiencies of certain micronutrients such as Zn. In a field microplot study on cotton,
AM increased tolerance to root-knot nematode. However, high P fertilization of
non-mycorrhizal plants was ineffective and, instead, created a nutrient imbalance
and Zn deficiency and became a predisposing factor to disease (Smith *et al.* 1986b).
In tamarillo, tolerance to root-knot nematode attack was improved by a mixture
of endophytes tolerant to high soil P, but could not be attained by increased P
fertilization (Cooper & Grandison 1987).

Competition for space

The hypothesis that competition for space may play a significant role in nematode–
endophyte interactions has not yet been adequately documented. It can be speculated
that this mechanism would be most important to migratory endoparasites which
can cause extensive damage and necrosis in the cortex and epidermis, thus ren-
dering the tissues unsuitable for symbiotic development (Hussey & Roncadori
1978). However, the observation of fewer nematodes in mycorrhizal than non-
mycorrhizal roots may simply not support this hypothesis, although split-root system
studies have revealed that nematode populations are reduced only when both micro-
organisms inhabit the same half-root system (Strobel *et al.* 1982; Tylka *et al.* 1991).
Smith (1987) discounts the relative importance of this hypothesis noting that
nematode activities may not be affected in root systems in which only 40–60%
fungal root colonization rates have been reported (Saleh & Sikora 1984) and that
initial nematode feeding sites in the zone of elongation adjacent to the root tip
rarely become mycorrhizal. Use of the root organ culture technique (Bécard &
Piché 1990), which permits direct observation of mycorrhizal development might
be adapted to directly observe nematode and fungal behaviour consistently over
a period of time and, thus, determine the effects of cohabitation and sequence
of colonization in interactions. This approach would represent a considerable
improvement over past studies carried on in pots where environmental conditions
are not subject to rigid control.

Changes in root exudation and root physiology

The influences of P nutrition on root membrane permeability and exudates affecting
penetration and colonization by AM fungi have been documented (Ratnayake *et
al.* 1978). Increases in sugar and amino-acid concentrations in the rhizosphere and
rhizoplane in P-deficient plants favoured endophyte activities. Such changes in

mycorrhizal root exudates might have a direct effect on nematode attraction or an indirect effect by altering populations of rhizosphere or rhizoplane micro-organisms when antagonists are present.

Mycorrhizal-induced biochemical changes within root tissue have been reported and could suppress root infection and development of nematodes (Smith 1987; Francl 1993). Increased phytoalexin production in soybean was associated with mycorrhizal development but was detected at low levels (Morandi *et al.* 1984). Other potentially important inhibiting compounds have been detected in mycorrhizal roots of tomato and banana (Francl 1993). This is an area that clearly requires further investigation.

Competition for host photosynthate

This hypothesis seems to have merit since there is competition for a common host resource by the endophyte and nematode as well as the root system. General estimates of carbon cost exacted by the fungus vary around 10% (Fitter 1991) or 4–12% (Smith 1987) and under certain conditions of high sucrose concentration the endophyte is considered a strong physiological sucrose sink (Amijee *et al.* 1993). Similarly, the nematode functions as a carbohydrate sink (Bird & Loveys 1975). Should the fungal symbiont have primary access to photosynthate, it is logical to assume that nematode growth and reproduction would be adversely affected.

FUTURE CONSIDERATIONS

Control measures for reducing crop losses to plant parasitic nematodes have been seriously hampered by removal of highly effective and frequently used nematicides from use due to environmental and safety concerns. Alternatives, such as biocontrol, are now being more seriously considered. With disease-control constraints now a reality, additional research on mycorrhizae–plant parasitic nematodes relationships is warranted. Considering the complexity and variability inherent in the tripartite relationships, some suggested study approaches by Smith (1987), such as varied initial nematode and endophyte population densities, inoculation sequence and appropriate statistical models, might be considered in research planning. It is essential that fundamental information gathered from studies under controlled conditions be eventually transferred to the field, or at the least, to specialized management systems, such as nurseries. With AM inoculum production a serious, persistent obstacle, we perhaps should consider cultural practices that would encourage activities of native endophyte populations in the field rather than excessive reliance on introduction of a highly efficient but imported fungal symbiont. In view of the possibility of negative interactions, we would be remiss not to consider the possibility that arbuscular mycorrhizal activities might not need to be promoted in certain instances. Adopting uniform criteria to assess plant and micro-organism growth and reproduction would eliminate many of the inconsistencies reported in the past. The greatest challenge though will be to develop research methods that allow progress beyond the descriptive stage and provide a better understanding of the physical and physiological bases of the interactions.

REFERENCES

Amijee, F., Stribley, D. P. & Tinker, P. B. (1993). The development of endomycorrhizal root systems. VII. Effects of soil phosphorus and fungal colonization on the concentration of soluble carbohydrates in roots. *New Phytologist*, **123**, 297–306.

Bécard, G. & Piché, Y. (1990). Physiological factors determining vesicular–arbuscular mycorrhizal formation in host and nonhost Ri T-DNA transformed roots. *Canadian Journal of Botany*, **68**, 1260–1264.

Bird, A. F. & Loveys, B. R. (1975). The incorporation of photosynthates by *Meloidogyne incognita*. *Journal of Nematology*, **7**, 111–113.

Bird, G. W., Rich, J. R. & Glover, S. U. (1974). Increased endomycorrhizae of cotton roots in soil treated with nematicides. *Phytopathology*, **64**, 48-51.

Camprubi, A., Pinochet, J., Calvet, C. & Estaun, V. (1993). Effects of the root-lesion nematode *Pratylenchus vulnus* and the vesicular–arbuscular mycorrhizal fungus *Glomus mosseae* on the growth of three plum root stocks. *Plant and Soil*, **153**, 223–229.

Carling, D. E., Roncadori, R. W. & Hussey, R. S. (1989). Interactions of vesicular–arbuscular mycorrhizal fungi, root-knot nematode, and phosphorus fertilization on soybean. *Plant Disease*, **73**, 730–733.

Cook, R. & Evans, K. (1987). Resistance and tolerance. *Principles and Practice of Nematode Control in Crops* (Ed. by R. H. Brown & B. R. Kerry), pp. 179–231. Academic Press, Orlando. FL.

Cooper, K. M. & Grandison, G. S. (1986). Interaction of vesicular–arbuscular mycorrhizal fungi with root-knot nematode on cultivars of tomato and white clover susceptible to *Meloidogyne hapla*. *Annals of Applied Biology*, **108**, 1–11.

Cooper, K. M. & Grandison, G. S. (1987). Effect of vesicular–arbuscular mycorrhizal fungi on infection of tamarillo (*Cyphomandra betacea*) by *Meloidogyne incognita* in fumigated soil. *Plant Disease*, **71**, 1101–1106.

Elliott, A. P., Bird, G. W. & Safir, G. R. (1984). Joint influence of *Pratylenchus penetrans* (Nematoda) and *Glomus fasciculatum* (Phycomyceta) on the ontogeny of *Phaseolus vulgaris*. *Nematropica*, **14**, 111–119.

Fitter, A. H. (1991). Costs and benefits of mycorrhizae: Implications for functioning under natural conditions. *Experientia*, **47**, 350–355.

Fox, J. A. & Spasoff, L. (1972). Interaction of *Heterodera solanacearum* and *Endogone gigantea* on tobacco. *Journal of Nematology*, **4**, 224–225.

Francl, L. J. (1993). Interactions of nematodes with mycorrhizae and mycorrhizal fungi. *Nematode interactions* (Ed. by M. W. Khan), pp. 203–216. Chapman & Hall, London.

Harley, J. L. & Smith, S. E. (1983). *Mycorrhizal Symbiosis*. Academic Press, London.

Herman, M., Hussey, R. S. & Boerma, H. R. (1991). Penetration and development of *Meloidogyne incognita* on roots of resistant soybean genotypes. *Journal of Nematology*, **23**, 155–161.

Hussey, R. S. & Roncadori, R. W. (1978). Interaction of *Pratylenchus brachyurus* and *Gigaspora margarita* on cotton. *Journal of Nematology*, **10**, 16–20.

Hussey, R. S. & Roncadori, R. W. (1982). Vesicular–arbuscular mycorrhizae may limit nematode activity and improve plant growth. *Plant Disease*, **66**, 9–14.

Ingham, R. E. (1988). Interactions between nematodes and vesicular–arbuscular mycorrhizae. *Agriculture, Ecosystems and Environment*, **24**, 169–182.

Kellam, M. L. & Schenck, N. C. (1980). Interactions between a vesicular–arbuscular mycorrhizal fungus and root-knot nematode on soybean. *Phytopathology*, **70**, 293–296.

MacGudwin, A. E., Bird, G. W. & Safir, G. R. (1985). Influence of *Glomus fasciculatum* on *Meloidogyne hapla* infecting *Allium cepa*. *Journal of Nematology*, **17**, 389–395.

Morandi, D., Bailey, J. A. & Gianinazzi-Pearson, V. (1984). Isoflavenoid accumulation in soybean roots infected with vesicular-arbuscular mycorrhizae fungi. *Physiological Plant Pathology*, **24**, 357–364.

O'Bannon, J. H. & Nemec, S. (1979). The response of seedlings to a symbiont, *Glomus etunicatus*, and a pathogen, *Radopholus similis*. *Journal of Nematology*, **11**, 270–275.

Ratnayake, M., Leonard, R. T. & Menge, J. A. (1978). Root exudation in relation to supply of phosphorus and its possible relevance to mycorrhizal formation. *New Phytologist*, **82**, 543–552.

Roncadori, R. W. & Hussey, R. S. (1986). Effect of *Gigaspora margarita, Meloidogyne incognita*, and phosphorus fertility on peach growth and yield. *Phytopathology*, **76**, 1108.

Saleh, H. & Sikora, R. A. (1984). Relationship between *Glomus fasciculatus* root colonization of cotton and its effects on *Meloidogyne incognita*. *Nematologica*, **30**, 230–237.

Schenck, N. C. & Kinloch, R. A. (1974). Pathogenic fungi, parasitic nematodes, and endomycorrhizal fungi associated with soybean roots in Florida. *Plant Disease Reporter*, **58**, 169–173.

Schenck, N. C., Kinloch, R. A. & Dickson, D. W. (1975). Interaction of endomycorrhizal fungi and root-knot nematode on soybean. *Endomycorrhizas* (Ed. by F. E. Sanders, B. Mosseae & P. B. Tinker), pp. 607–617. Academic Press, New York.

Sikora, R. A. (1978). Effect of the endotrophic mycorrhizal fungus, *Glomus mosseae*, on the host-parasite relationship of *Meloidogyne incognita* in tomato. *Journal of Plant Disease Protection*, **185**, 197–202.

Smith, G. S. (1987). Interactions of nematodes with mycorrhizal fungi. *Vistas on Nematology* (Ed. by J. A. Veech & D. W. Dickson), pp. 292–300. Society of Nematologist. Hyattsville, MD.

Smith, G. S. & Kaplan, D. T. (1988). Influence of mycorrhizal fungus, phosphorus, and burrowing nematode interactions on growth of rough lemon citrus seedlings. *Journal of Nematology*, **20**, 539–544.

Smith, G. S., Hussey, R. S. & Roncadori, R. W. (1986a). Penetration and postinfection development of *Meloidogyne incognita* on cotton as affected by *Glomus intraradices* and phosphorus. *Journal of Nematology*, **18**, 429–435.

Smith, G. S., Roncadori, R. W. & Hussey, R. S. (1986b). Interaction of endomycorrhizal fungi, superphosphate, and *Meloidogyne incognita* on cotton in microplot and field studies. *Journal of Nematology*, **18**, 429–435.

Strobel, N. S., Hussey, R. S. & Roncadori, R. W. (1982). Interactions of vesicular–arbuscular mycorrhizal fungi, *Meloidogyne incognita*, and soil fertility on peach. *Phytopathology*, **72**, 690–694.

Tylka, G. L., Hussey, R. S. & Roncadori, R. W. (1991). Interactions of vesicular–arbuscular mycorrhizal fungi, phosphorus, and *Heterodera glycines* on soybean. *Journal of Nematology*, **23**, 122–133.

7. INTERACTIONS BETWEEN INSECTS AND MYCORRHIZAL FUNGI

A. C. GANGE AND E. BOWER

School of Biological Sciences, Royal Holloway University of London, Egham Hill, Egham, Surrey TW20 0EX, UK

INTRODUCTION

A mycorrhiza can be defined as the mutualistic association between plant roots and one or more soil fungi. Between 5000 and 6000 species of fungi enter into this relationship (Molina *et al.* 1992), with representatives from the Ascomycetes, Basidiomycetes and Zygomycetes. These associations are extremely widespread; a form of mycorrhiza occurs in all ecosystems of the world (Brundrett 1991) and it has been estimated that between 70 and 80% of all land plants form a mycorrhiza of some kind (Malloch *et al.* 1980; Trappe 1987), the majority of which are arbuscular mycorrhizas (AM).

Despite the fact that in all the ecosystems listed by Brundrett (1991), the vast majority (often over 90%) of plants are arbuscular mycorrhizal, fungal abundance in the soil and consequent infection of plants varies at all spatial scales (e.g. Brundrett & Kendrick 1988; Miller & Allen 1992). AM fungi rely on passive movement of propagules through soil or the action of wind and rain for disperal (Allen 1987, 1988). Subterranean invertebrates are notable dispersal agents (Rabatin & Stinner 1988) and the first aim of this review is to examine whether the spatial distributions of some mycorrhizal fungi may be linked to the spatial distributions of subterranean insects.

There is a voluminous literature detailing the beneficial effect that mycorrhizal infection can have on plant growth. Many reviews discuss how mycorrhizal plants have enhanced growth and yield, improved water relations and increased disease resistance compared with conspecific individuals which are fungus-free (Harley & Smith 1983; Allen 1991; West, this volume). However, a feature of the literature is that the vast majority of these experiments have taken place in controlled conditions. Indeed, very few field experiments have demonstrated a benefit to the mycorrhizal plant (McGonigle 1988) and it appears that mycorrhizas may often be ineffective under field conditions (Fitter 1985). However, it is possible that mycophagous insects, particularly Collembola, which exist in field soils may feed on the fungi and reduce their effectiveness (McGonigle & Fitter 1988). We will therefore review the evidence for mycorrhizal feeding by Collembola and will assess the consequences for the functioning of the mycorrhiza and plant growth.

Fossil evidence suggests that some of the earliest land plants from the Devonian period may have possessed AM fungi (Pirozynski 1981), although more persuasive evidence has been found in roots from the Triassic (Stubblefield *et al.* 1987). However, despite the facts that some mycorrhizas are apparently ubiquitous in nature and geologically ancient, a number of plant species in a wide range of plant families never form an association or do so only rarely (Tester *et al.* 1987; Brundrett 1991). It is still a puzzle why this should be so, for if we assume that a mycorrhiza is the ancient condition, then this feature has been independently lost in many plant families.

The earliest recorded evidence of arthropod feeding damage to plants also comes from the Devonian (Stephenson & Scott 1992). Again, the evidence is scanty, and much clearer examples come from the Carboniferous period onwards. However, it is reasonable to assume that throughout most of geological time, most plants have been subject to mycorrhizal infection and insect attack, and therefore the potential for host-plant mediated interactions between these trophic levels has been great.

There are over 900 000 named species of insects in the present biota, although estimates of the actual number vary between 1.5 million and 30 million (Stork 1993). Approximately half of these species feed on living plant material (Gullan & Cranston 1994). Chewing, sucking, mining or galling insects feeding on above- or below-ground plant parts can have a dramatic effect on the vegetative and reproductive performance of individual plants, populations or communities (Brown & Gange 1990, 1992; Gange 1990). Faced with such attack, one of the best means of defence for a plant is to erect a barrier of chemicals, which are deterrent or toxic to insects. It has often been suggested that there has been a coevolutionary 'arms race' between plants and herbivores, as new chemicals are produced and subsequently overcome by insects (Harborne 1988). However, in view of the fact that mycorrhizal fungi can affect the biochemical nature of plants, through altering levels of carbon (C), nitrogen (N) or phosphorus (P) (Bonfante-Fasolo & Scannerini 1992), it is surprising that their role in insect–plant defence mechanisms has received little attention. A third aim of this review will therefore be to consider the interactions between mycorrhizas and phytophagous insects, from the point of view of fungus and insect. In particular, we will enquire whether mycorrhizas can substitute for, or aid in the production of a chemical defence battery, and whether the mycorrhizal habit has been maintained in response to insect herbivory.

MYCORRHIZAL PROPAGULE DISPERSAL BY INSECTS

All AM fungi produce spores below ground and a range of vertebrate and invertebrates are involved in their dispersal (Fitter & Sanders 1992). As 83% of ecto- and ectendomycorrhizal fungi fruit above ground (Molina *et al.* 1992), wind is an efficient dispersal mechanism and it is not surprising that dispersal of these fungi by insects is of less importance.

The most important invertebrate AM fungal dispersal agents are earthworms (Rabatin & Stinner 1991; Gange 1993). However, a range of insect orders has also been implicated, although in many cases only spore presence in guts has been recorded, rather than transmission, as demonstrated by the successful infection of plants with soil containing insect faeces. Spores have been found inside the guts of Orthoptera, such as grasshoppers and crickets (Hanson & Ueckert 1970; Ponder 1980; Rabatin & Stinner 1985; Warner *et al.* 1987) and root-feeding larvae of Lepidoptera, Diptera and Coleoptera (Rabatin & Stinner 1988, 1989, 1991). Given that Rabatin and Stinner (1988) have reported successful infection of plant roots with soil containing woodlouse (Isopoda) and millipede (Diplopoda) faeces, it is reasonable to assume that all the insects mentioned above are capable of dispersing viable spores of AM fungi.

Gange (unpublished data) sampled 100 larvae of the garden chafer *Phyllopertha horticola* (Coleoptera: Scarabaeidae), from the tee area on a golf course. Although mycorrhizal fungi are rare in such situations, because of fungicide application (Gange 1994), 37% of larvae contained spores. The mean value of spores per larva was 2.04 ± 0.2 (range 0–6) and given that chafer densities in this tee averaged 329 m^{-2}, with a maximum of 784 m^{-2} (Gange *et al.* 1991), this represents a large turnover of AM propagules. Furthermore, faeces of chafers, when mixed with sterilized sand, successfully initiated infection (mean $5.4 \pm 1.1\%$) in roots of *Plantago lanceolata* seedlings.

Virtually all root-feeding larvae have highly aggregated spatial distributions, resulting from a number of biotic and abiotic factors (Brown & Gange 1990). In the case of *P. horticola* (above) larval distribution fitted a negative binomial because larvae aggregated towards their preferred host plant, *Agrostis capillaris* (Gange *et al.* 1991). *A. capillaris* is highly mycorrhizal (Harley & Harley 1987) and it is likely that larvae ingested AM spores while feeding on roots. Other species of scarabaeid larvae also aggregate towards a preferred host (Zhen-Rong *et al.* 1986; Allsopp & Bull 1989) and we therefore suggest that one reason for the heterogeneous nature of mycorrhizas in soil is the heterogeneous nature of their dispersal agents.

Rabatin and Stinner (1988, 1991) present data which show that 13% of sampled predatory ground-dwelling carabid beetles contained AM fungal spores, derived either by ingestion of spores while the beetle is in the soil or through predation on other invertebrates. *Glomus intraradices* spores were fed to black cutworm (*Agrotis ipsilon*) larvae, which in turn were fed to carabids. Photographic evidence showed that many intact spores occurred in beetle faeces, after passage through two trophic levels, but spore viability was not tested. Another example of multitrophic spore dispersal was provided by Lakshman and Raghavendra (1990) who found that AM fungal spores were present in the nests of predatory Hymenoptera and in the dung balls produced by scarabaeid beetles. Material taken from wasps and beetles was added to sterilized soil and successfully initiated mycorrhizal infections in sunflower.

Ants (Hymenoptera: Formicidae) are also important agents of AM spore dispersal (McIlveen & Cole 1976; McGinley *et al.* 1994). In a study of harvester ant (*Pogonomyrmex occidentalis*) mounds in semiarid shrub steppe habitats in Wyoming, Friese and Allen (1993) found mats of densely packed clipped roots in the mounds with spore densities 2000–5000 times higher than surrounding soil. The mounds are abandoned after 12–20 years, resulting in an environment where mycorrhizal inoculum occurs in patches. These patches are excellent sites for the establishment of AM plants resulting in patchy distributions of colonizing mycorrhizal plants (Friese & Allen 1993).

MYCORRHIZAS AND FUNGUS-FEEDING INSECTS

Roots are the principal energy source for the majority of below-ground consumers, but a significant proportion of the carbon allocated below ground may be directed to mycorrhizal fungi. Reports have suggested that as much as 40–50% of total plant photosynthate may be directed to ectomycorrhizal (ECM) fungi (Fogel & Hunt 1979), although these figures assume that all roots are active, which is probably not true. More realistic figures of 10–20% have been recorded for AM fungi (Koch & Johnson 1984; Jakobsen & Rosendahl 1990), and although some of this feeds the higher respiration rate in infected roots, it represents a significant energy source for subterranean mycophagous insects.

Ectomycorrhizal (ECM) feeding by insects

The report by Zak (1965) is the only record of sucking insects feeding on mycorrhizal hyphae. This describes how two species of aphid feed directly on the ECM mycelium associated with *Pseudotsuga menziesii*, *Tsuga heterophylla* and *Picea sitchensis*.

In field soils, the most abundant mycophagous insects are Collembola, which can reach densities of 6×10^5 m^{-2} (Curry 1994). Several papers have demonstrated that Collembola show distinct feeding preferences when presented with ECM fungi in laboratory experiments, despite the fact that field-collected animals have a wide range of foodstuffs in their guts (Anderson & Healey 1972). However, results of all laboratory feeding trials should be treated with caution, as Leonard (1984) has shown that the preferences recorded depend on the nature of the substrate used to culture the fungi. Shaw (1985, 1988) tested several ECM fungi for feeding by the collembolan *Onychiurus armatus*. Preferences were shown for the ECM-forming fungus *Lactarius rufus*, while other ECM fungi were less preferred (e.g. *Suillus luteus*) or avoided (e.g. *Hebeloma crustuliniforme*). Schultz (1991) studied the ECM feeding preferences of two Collembola, one (*Proisotoma minuta*) extracted from a woodland soil where ECM fungi are common and a second (*Folsomia candida*) from a soybean field where one would expect AM fungi. Both species showed preferences, although the results of Shaw (1988) with *O. armatus* were not repeated with either of these two species. This suggests a number of conclusions: that culture conditions may be the overriding factor in these laboratory experiments, but also, with this proviso, certain fungi are consistently eaten while others are avoided and that this is not necessarily consistent between Collembola species.

A recent paper (Hiol Hiol *et al.* 1994) has extended Schultz's (1991) work with *P. minuta* to a pot trial. In three out of four fungi tested, addition of the Collembola resulted in a significantly lower percentage of colonized lateral roots of *Pinus taeda*. The experiment was run for only 10 weeks, and no differences in plant biomass were observed between fungal or Collembola treatments. There is a need to examine a situation such as this over a much longer growth period, in order to determine the importance of Collembola in ECM functioning and plant growth.

It is likely that Collembola grazing on hyphae does not just directly affect plant growth, but also indirectly through the mobilization of N (see below for a discussion of this point with AM fungi). Verhoef and de Goede (1985) present some interesting data which shows that Collembola presence in soil can affect the nutrient concentration of that soil. In a field experiment, removal of Collembola by non-chemical means resulted in a decrease in the N concentration, so that plots with Collembola had 2.3 times the concentration of N. A subsequent laboratory experiment showed that the response was non-linear; N concentration peaked at intermediate densities of Collembola. These results suggest that low densities of Collembola result in mineralization of N and P, which counteracts the loss of the mycorrhiza, while at high densities, this mineralization is insufficient to offset the mycorrhizal loss. Clearly, now that laboratory feeding trials with Collembola have established feeding preferences, we need to extend these to field situations, in order to understand the role of Collembola in below-ground nutrient cycling.

Arbuscular mycorrhizal (AM) feeding by insects

The impossibility of axenic culturing of AM fungi has meant that the majority of AM fungi/Collembola interactions have taken place in pot trials or field plots. One exception is provided by Moore *et al.* (1985), in which spores of four AM fungi were germinated on aqueous agar and elongated germ tubes offered to four species of Collembola, (*F. candida*, *P. minuta* and two species of *Onychiurus* (not *armatus*)). Feeding preferences were observed, in that *F. candida* alone ate hyphae of *Glomus fasciculatum*, while all species fed on hyphae of *Gigaspora rosea*. The data suggest that spore shape was not the determining factor in feeding, but that preference may be related to chemical production. Toxic metabolite production has been suggested as the reason for feeding preferences of *P. minuta* on non-mycorrhizal fungi (Lartey *et al.* 1989). Production of metabolites by mycorrhizal fungi has rarely been documented, although Caron *et al.* (1986) have reported the production of antibiotics or other inhibitory compounds by *Glomus intraradices*.

The effects of Collembola on the establishment of mycorrhizal infection were studied by Kaiser and Lussenhop (1991). Addition of Collembola to pots of soybean (*Glycine max*) at the time of mycorrhizal addition resulted in a lower infection level of plants than if the insects were added 15 days after planting. However, there were no corresponding effects on plant growth.

Collembola grazing may reduce the functioning of a mycorrhiza, with serious consequences for the host plant. Warnock *et al.* (1982) grew seedlings of leek (*Allium porrum*) with or without mycorrhiza (*G. fasciculatum*) (= *G. fasciculatus*)), with or without soil leachings (an aqueous suspension of the soil microflora, without AM fungi) and with or without the Collembolan *F. candida*, in a factorial design. Both mycorrhizal infection and leachate increased plant growth, but the addition of *F. candida* to infected plants meant that these grew little better than uninfected individuals. The authors concluded that grazing on the external hyphae rendered the infection ineffective.

Finlay (1985) extended this work and in one experiment, Collembolan densities were studied after inoculation into mycorrhizal or non-mycorrhizal pots of leek and red clover (*Trifolium pratense*). Collembola numbers increased in the presence of mycorrhizas, and evidence suggested that this was because they were feeding on the hyphae. Furthermore, Collembolan feeding was noticeable in its effects on plant yield, as in four out of five trials, there was a significant increase in plant biomass between mycorrhizal and non-mycorrhizal plants when Collembola were absent, but no differences when they were present (Finlay 1985).

An interesting result from Finlay's (1985) experiments was that there appeared to be a non-linear relation between Collembola density and shoot P content, with low densities of Collembola increasing P uptake, while high densities decreased it. Another non-linear effect with Collembola grazing on mycorrhizas associated with *Geranium robertianum* was recorded by Harris and Boerner (1990). In this experiment, low densities of Collembola stimulated plant growth (but not P uptake), while high densities inhibited it. A number of explanations have been provided for these results. Collembola may disperse fungal inoculum (Finlay 1985), so that at low densities, plants become infected more readily, or it may be that infection of roots is above the optimum required for maximum benefit, and grazing reduces the infection levels to this optimum (Fitter & Sanders 1992). Meanwhile, Finlay (1985) and Harris and Boerner (1990) suggest that plant growth may be stimulated by nutrients which are mineralized or mobilized by the Collembola, through feeding on the fungi. A similar non-linear response to grazing intensity has also been reported in a decomposer system (Hanlon 1981) and it is interesting that the results with mycorrhizal pot trials suggest a similar mechanism to that in Verhoef and de Goede's (1985) field trial discussed above. Furthermore, Boerner and Harris (1991) have shown that N availability is increased by Collembolan grazing and that this could shift the balance of competition between the mycorrhizal *G. robertianum* and the non-mycorrhizal *Brassica nigra*. However, a note of caution must be sounded with these laboratory trials, regarding Collembolan density. It is difficult to compare densities of Collembola used in laboratory experiments (where abundance is expressed in numbers dm^{-3}) with those of field abundance (where numbers are per m^2). Despite this, a quick calculation reveals that these pot trials have often taken place at very low densities of insects, compared with field situations. Thus, Warnock *et al.* (1982) appear to have used 3000–20 000 Collembola m^{-2}, while Finlay (1985) used 1000–10 000 m^{-2}, Harris and Boerner (1990) 1000–3000 m^{-2} and Boerner and Harris (1991) 3000 m^{-2}. These figures are very low compared with numbers in grassland soils given by Curry (1994). Over 34 studies, the average density was 57 000 m^{-2}, with some pastures having densities up to 100 000 m^{-2}. Furthermore, Takeda (1973) and Tamura (1976) recorded densities of 100 000 m^{-2} for *Folsomia octoculata* in pine forest soils. In order to understand these potentially important non-linear effects, and thus below-ground nutrient cycling by Collembola, we must perform controlled experiments which use a range of realistic insect densities, or manipulate field situations.

Finlay (1985) sterilized field plots with methyl bromide and 10 months later seeded these with *Trifolium pratense* and the AM fungus *Glomus occultum*. Some plots were then left as control, others received the soil insecticide chlorfenvinphos and others the systemic fungicide benomyl. Insecticide-treated plants showed greater shoot mass, shoot P content and shoot P accumulation rates than controls. These data suggest that the mycorrhiza functioned more efficiently when the Collembolan populations were reduced. Estimated (by us) Collembola densities in this experiment were 5000 m^{-2} in untreated plots and 1000 m^{-2} in insecticide plots.

In an extension of this work, McGonigle and Fitter (1988) applied chlorfenvinphos and benomyl in a 2×2 factorial design to a semi-natural grassland, containing a wide variety of forbs and grasses. Unfortunately, the results of the full factorial design, with any interactions, were not presented, but insecticide application reduced Collembolan density from 57 000 m^{-2} to 2000 m^{-2} and resulted in increased shoot dry mass and P inflow of *Holcus lanatus*. The authors concluded that the simplest explanation was that the removal of Collembolan grazers enabled plants to benefit from the mycorrhiza.

As Rabatin and Stinner (1991) point out, the interpretation of these field experiments is complicated, because benomyl and chlorfenvinphos are toxic to earthworms (Edwards *et al.* 1968; Edwards 1983). Benomyl is also phytotoxic to some plants

(Reyes 1975) and the Collembolan *F. candida* is highly sensitive to this fungicide (Tomlin 1977). In addition, it is likely that the application of chlorfenvinphos by McGonigle and Fitter (1988) would have significantly reduced the numbers of larger root-feeding insects, which one would expect to be common in such a grassland (Brown & Gange 1990). The reduction of these insects by a similar insecticide (chlorpyrifos) resulted in significant increases in the growth of perennial grasses, such as *H. lanatus* in a developing grassland community (Brown & Gange 1992). Despite the problems, these field experiments are vitally important as they provide evidence that soil invertebrates are important in mycorrhizal functioning, and are really the best we can do, as there are no insecticides specific to fungivores, and no fungicides specific to mycorrhizas.

In an attempt to alleviate some of these problems Gange and Brown (1992) applied chlorpyrifos (insecticide) and iprodione (fungicide) in a factorial design to a ruderal plant community. An extensive series of tests revealed little in the way of toxic effects of these two chemicals on non-target organisms (Gange *et al.* 1992; Brown, Gange & Montalvo, unpublished data). In addition, larger root-feeding insects were very rare in this community, because of the soil disturbance caused by ploughing (Brown & Gange 1992). The insecticide caused a 400-fold reduction in Collembola numbers and there was a significant interaction between the pesticides, because the benefit from mycorrhizal infection was clearly seen when insects were absent, but not when they were present. These data suggest that feeding by Collembola is one of the major reasons for mycorrhizal ineffectiveness in field situations. There is now an urgent need to study the roles of Collembola and other root-feeding insects in mycorrhizal functioning in 'controlled' field or microcosm experiments. Furthermore, these experiments should also consider the interactions with soil nutrients such as phosphorus, as field trials have so far tended to take place in P-deficient soils (Finlay 1985; Gange & Brown 1992); neither have these experiments considered the times in the life of a plant, such as seedling and flowering, when P is most in demand (Fitter & Sanders 1992).

MYCORRHIZAS AND ROOT-FEEDING INSECTS

As noted above, it is impossible to differentiate between the effects of soil-dwelling mycophagous and rhizophagous insects when applying insecticide to field plots. One way of solving this problem is to examine the interactions of mycorrhizal fungi and root-feeding insects in pot trials, similar to those involving Collembola. It is surprising that this experiment, while being relatively easy to execute, has hardly ever been attempted.

One exception is provided by Gange *et al.* (1994). In this study, first instar larvae of the rhizophagous black vine weevil (*Otiorhynchus sulcatus*) were added to plants of *Taraxacum officinale* (dandelion) at the rate of 0, 150 m^{-2} and

600 m^{-2}. Plants were grown in sterilized soil and half were inoculated with the mycorrhizal fungus *Glomus mosseae*. It was found that survival of larvae was halved on the mycorrhizal plants and growth rate was also significantly reduced. There were corresponding effects seen on plant biomass, so that 150 larvae m^{-2} had no effect on mycorrhizal plants, but significantly reduced both shoot and root dry mass in non-mycorrhizal plants. Therefore, it appeared that the presence of the mycorrhiza effectively raised the threshold density of larvae required to elicit damage in this plant.

In a current experiment, Gange and Hamblin (unpublished data) established strawberry (*Fragaria* × *ananassa*) plants (cv 'Red Gauntlet') with infection by either *G. mosseae* or *G. intraradices* or both fungi. There was a non-mycorrhizal control. Half of the plants then received an infestation of 10 *O. sulcatus* eggs (equivalent to 750 eggs m^{-2}). It was found that either fungus as a single infection significantly reduced insect survival (*G. mosseae*: $F_{1,36} = 6.5$, $P < 0.05$; *G. intraradices*: $F_{1,36} = 9.6$, $P < 0.001$) (Fig. 7.1), but when both fungi were present, survival did not differ from insects on non-mycorrhizal (control) plants, leading to a significant interaction ($F_{1,36} = 28.4$, $P < 0.001$). Corresponding effects were also seen in plant biomass.

The mechanisms for mycorrhizal-induced resistance of roots to rhizophagous insects have yet to be studied. It is known that *G. intraradices* can increase flavonoid levels in roots of alfalfa (*Medicago sativa*) (Volpin *et al.* 1994), lower total amino acid levels in rose (*Rosa hybrida*) roots (Augé *et al.* 1992) and increase the amount of wall-bound phenolic acids in onion (*Allium cepa*) roots (Grandmaison *et al.* 1993). Meanwhile, *G. mosseae* infection of onion can induce the production of eight polypeptides, not found in uninfected tissue (Garcia-Garrido *et al.* 1993) and ECM fungal colonization can change the levels of hundreds of polypeptides found

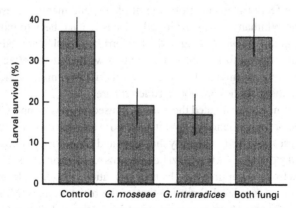

Fig. 7.1. Effect of infection of strawberry roots by two mycorrhizal fungi on survival of vine weevil larvae. Bars represent means with one standard error.

in uninfected roots (Hilbert & Martin 1988). Indeed, the terms ectomycor-
rhizins and endomycorrhizins have been used for these specific classes of proteins
(Bonfante-Fasolo & Scannerini 1992). Clearly mycorrhizal fungi can significantly
change root chemistry and thereby affect rhizophagous insects, and this could be
another productive area for research.

Brundrett (1991) lists the types of chemical commonly associated with the
roots of so-called non-mycorrhizal plant families. He suggests that plant families
in which the majority of species are non-mycorrhizal tend to accumulate chemicals
which are considered evolutionarily advanced, such as alkaloids, glucosinolates
and saponins. Brundrett (1991) hypothesizes that these chemicals are responsible
for the absence of mycorrhizas in the families in which they accumulate. However,
an equally plausible hypothesis would be that some plant families have 'used' the
mycorrhizal association as a protectant against rhizophage attack, while others
have dispensed with the fungus, relying on their own secondary metabolites. Clearly,
there are costs and benefits of either evolutionary pathway, but the major cost, that
of rhizophage herbivore pressure, is unknown as root-feeding insect faunas associated with specific plants are so poorly documented (Brown & Gange 1990).
However, given that root-feeding specialization appears to occur at the level of
the plant family (Brown & Gange 1990), we may speculate that one reason for
the retention of mycorrhizal fungi in many plant families has been the need for
protection against root-feeding insects.

MYCORRHIZAS AND FOLIAR-FEEDING INSECTS

Foliar-feeding insects and mycorrhizas both require energy from plants, and therefore there are likely to be interactions between these organisms, mediated through
the host (Gehring & Whitham 1994). There are a number of studies which have
examined the effect of manual defoliation of plants, and many of these are listed
by Gehring and Whitham (1994). In virtually all cases, it has been found that severe
foliage loss impairs the functioning of the mycorrhiza, but that at moderate levels
of removal, the mycorrhiza enables the plant to withstand the effects of tissue
removal. However, as simulated herbivory rarely, if ever, mimics the real thing
(Gange 1990), these studies are not considered here.

Two studies have investigated the effects of insect herbivory on ECM colonization. The first of these (Gehring & Whitham 1991) showed that pinyon pine trees
(*Pinus edulis*) susceptible to attack by the stem- and cone-boring moth *Dioryctria
albovitella* had 33% less ECM colonization than insect-resistant trees. When insects
were excluded from susceptible trees by the application of insecticide, mycorrhizal
infection increased to a level similar to that of the resistant trees. More recently,
Del Vecchio *et al.* (1993) have examined the effects of the scale insect *Matsucoccus
acalyptus* also on the mycorrhizas of *P. edulis*. Scale removal was performed by

hand, and it was found that scale-susceptible trees had 28% lower ECM infection than resistant trees, but that these levels were similar when scale insects were removed. Currently, there are no published studies on the effect of insect herbivory on AM infection levels of plants. However, Gange and Brown (unpublished data) performed an experiment which involved growing plants of *Plantago lanceolata* with or without mycorrhizal infection by *G. mosseae* in a controlled environment room at 15°C and a light regime of 16:8 L:D. Half the plants were subjected to a herbivory 'event' (feeding by one-third instar larva of the Lepidopteran *Arctia caja*) for 1 week in every 3 weeks. Harvests were performed at 3-weekly intervals. Mycorrhizal infection data of the plants are presented in Fig. 7.2. It can be seen that once plants were 15 weeks old (i.e. attacked plants had suffered a total of 5 weeks herbivory in this time), there was a significant reduction in mycorrhizal infection caused by herbivory ($F_{1,18} = 6.8$, $P < 0.05$). This effect was maintained until the end of the experiment, when herbivory resulted in a 33% reduction in AM infection. The most likely explanation for these results is that foliage consumption by insects results in a reduction of the C-source capacity of the plant, so that there is not enough available to meet the demands of the fungus (Gehring & Whitham 1994). Therefore, it is highly likely that the lack of growth responses to mycorrhizal infection so often observed in field trials (McGonigle 1988) has been due to the failure to control for foliar insect attack. Foliar-feeding insects are likely to be one of the main reasons for the apparent ineffectiveness of mycorrhizas under field conditions (Fitter 1985).

The first study to examine the reverse interaction, that is the effects of mycorrhizal infection on the performance of a foliar herbivore, was that of Rabin and Pacovsky (1985). These authors found that mycorrhizal infection of four soybean cultivars (*Glycine max*) by *Glomus fasciculatum* resulted in lower larval growth

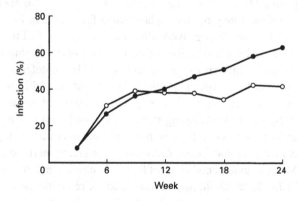

Fɪɢ. 7.2. Effect of foliar herbivory on the development of mycorrhizal infection in pot-grown *Plantago lanceolata*. (●) No herbivory; (○) attack by *Arctia caja* larvae, 1 week in every 3 weeks.

rates and pupal weights and higher larval mortality in two generalist Lepidopteran species, *Spodoptera frugiperda* and *Heliothis zea*. Although a number of plant biochemical parameters were measured, larval growth reductions did not appear to be related to leaf N, amino acid, carbohydrate, micronutrient or phenolic contents.

Gange and West (1994) have recently performed a field experiment, in which *P. lanceolata* was grown with and without the application of the fungicide iprodione. A set of 10 plants in each treatment was harvested every 2 weeks and examined for mycorrhizal infection, plant size and foliar insect attack. It was found that fungicide application successfully reduced mycorrhizal infection of plants and that this led to an increase in the proportions of leaves damaged by chewing insects. A feeding bioassay at the end of the experiment with the generalist insect *A. caja* larvae found that larval growth and food consumption was greater on the plants treated with fungicide. Interestingly, the growth and performance of a sucking aphid *Myzus persicae* showed the reverse trend, with insects growing larger and being more fecund on untreated plants. Chemical analysis of leaf material revealed that mycorrhizal infection increased the C/N ratio in leaves, mainly through a decrease in total nitrogen content. Levels of the C-based defences, aucubin and catalpol, which are deterrents to generalist insects (Bowers & Puttick 1988) were higher in mycorrhizal plants and are a likely explanation for the patterns of feeding damage observed in the main experiment.

A current experiment has examined the effect of mycorrhizal infection on the performance of a galling insect. Plants of creeping thistle (*Cirsium arvense*) were grown in pots, with iprodione added to pots to reduce AM infection. Plants were then exposed to ovipositing females of the thistle gall fly (*Urophora cardui*). Plants were maintained throughout one season and insect performance recorded at the end of the summer, when galls were mature. Mycorrhizal plants had smaller galls, which contained fewer chambers, with fewer live and smaller larvae, than did non-mycorrhizal plants. The C/N ratio was also higher in mycorrhizal plants and one reason for these results may be that a chemical defence system is enhanced by mycorrhizal infection which negatively affects the performance of the insect.

The C/N balance hypothesis (Bryant *et al.* 1983) proposes that the relative availability of C and N influences the production of C-based defence compounds in plants. Jones and Last (1991) combine this hypothesis with the effects of ECM colonization on plant physiology. They develop a series of scenarios in which the defence response of ECM sapling trees is compared with that of uninfected individuals, under varying conditions of high, low and intermediate light and soil nutrients. It is predicted that ECM plants will have greater herbivore resistance than non-ECM plants under conditions of low soil nutrients and low or high light. Meanwhile, ECM plants should have lower herbivore resistance under conditions of high soil nutrients and low or high light. This is because the nature of the mycorrhizal-induced change in C/N ratios is likely to be affected by the extent of fungal infection; for example, if fungal biomass is large, or conditions for photo-synthesis less than optimal (e.g. high nutrients, low light), then the fungus represents a large C drain and it can even become parasitic, reducing plant growth (e.g. Bethlenfalvay *et al.* 1982). So far, the data of Gange and West (1994) and Gange and Nice (unpublished data) with AM fungi are consistent with the hypothesis of Jones and Last (1991). Both these experiments demonstrated an increase in the C/N ratio in mycorrhizal plants and both were performed in con-ditions of low soil nutrients and high light.

Finally, we come to the question of whether the non-mycorrhizal habit may have evolved in response to insect herbivory. In order to examine this question, we have taken the data of Ward and Spalding (1993), which lists the 30 plant families in the UK with the highest numbers of insects feeding on them. These data also include the proportions of the insects associated with each plant family which only feed on that family (so called specialist feeders). From this table, we have excluded all plant families which are not arbuscular mycorrhizal and which are not members of the Dicotyledons. For the remaining families, we have calculated the proportion of plant species examined in each family which appear to form mycorrhizas, using the data given by Harley and Harley (1987). Using the combined data of each paper, we have calculated an index of 'herbivore load' for each plant family, by dividing the total number of associated insects by the total number of plant species. Some very interesting trends emerge when these data sets are plotted together (Fig. 7.3). First, there is a suggestion that plant families which have a high proportion of mycorrhizal species also have a high proportion of specialist insects (Fig. 7.3a). The data set as plotted yields a Spearman Rank Correlation of $r_s = 0.187$, $P = 0.270$, but if the two outliers (Plantaginaceae and Primulaceae) are removed, then $r_s = 0.765$, $P = 0.003$. An interpretation of this relation, based on the experimental evidence above, is that mycorrhizal infection can enhance plant defence levels, resulting in defence against generalist insects, and hence the pro-portion of associated specialists is high. It is an accepted fact that specialist insects are well able to cope with the chemical defences of their hosts (Harborne 1988). A further interesting fact is that there is a significant positive relation between the proportion of mycorrhizal species in a plant family and the herbivore load on that family (Fig. 7.3b) (Spearman Rank $r_s = 0.564$, $P = 0.022$). One interpretation of these data is that plant families which have lost mycorrhizal infection may have done so because the herbivore pressure on them, and thus selection pressure to produce chemical defences, has been lower. It is therefore possible (although speculative) that the loss of mycorrhizal infection in certain plant families has been because infection did not confer selective advantages in terms of herbivore resistance. It is accepted that many factors, including family size, geographic range and geological history, determine the number of insects associated with plant families (Ward & Spalding 1993). Indeed, one reason for the Primulaceae being

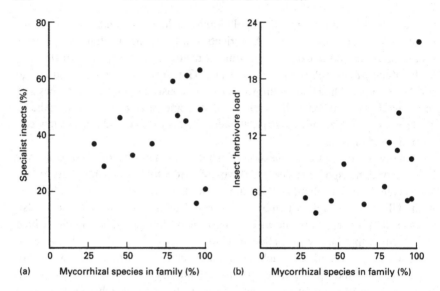

FIG. 7.3. Relation between the proportion of mycorrhizal species in a plant family and (a) the proportion of insects specialist on that family, and (b) the associated insect 'herbivore load'.

an outlier, with a very low proportion of specialist insects, is that a high (33%) proportion of this family is introduced to the British Isles (Stace 1991). However, we suggest that a hitherto unconsidered factor may be the possession of mycorrhizas, and it may be extremely instructive to continue these analyses to the level of the plant species, in order to establish if certain plant families have tended to retain the mycorrhizal habit because of insect herbivore pressure. Of course, one is left asking 'which came first, the mycorrhiza or the insect?' If the mycorrhiza did, then the possession of these widespread fungal associations may be one reason why the majority of insects are specialist feeders (Ward & Spalding 1993).

ACKNOWLEDGEMENTS

Some of the experimental work described was supported by NERC, the British Ecological Society and the British Entomological and Natural History Society. Their financial support is gratefully acknowledged. The Agricultural Genetics Co. have been very helpful with technical advice and AM inoculum supplies.

REFERENCES

Allen, M. F. (1987). Re-establishment of mycorrhizas on Mount St Helens: Migration vectors. *Transactions of the British Mycological Society*, **88**, 413–417.

Allen, M. F. (1988). Re-establishment of mycorrhizas following severe disturbance: Comparative patch dynamics of a shrub desert and a subalpine volcano. *Proceedings of the Royal Society of Edinburgh,* **94B,** 63–71.

Allen, M. F. (1991). *The Ecology of Mycorrhizae.* Cambridge University Press, New York.

Allsopp, P. G. & Bull, R. M. (1989). Spatial patterns and sequential sampling plans for melolonthine larvae (Coleoptera: Scarabaeidae) in southern Queensland sugarcane. *Bulletin of Entomological Research,* **79,** 251–258.

Anderson, J. M. & Healey, I. N. (1972). Seasonal and interspecific variation in major components of the gut contents of some woodland Collembola. *Journal of Animal Ecology,* **41,** 359–368.

Augé, R. M., Foster, J. G., Loescher, W. H. & Stodola, A. J. W. (1992). Symplastic molality of free amino acids and sugars in *Rosa* roots with regard to mycorrhizae and drought. *Symbiosis,* **12,** 1–17.

Bethlenfalvay, G. J., Brown, M. S. & Pacovsky, R. S. (1982). Parasitic and mutualistic associations between a mycorrhizal fungus and soybean: Development of the host plant. *Phytopathology,* **72,** 889–893.

Boerner, R. E. J. & Harris, K. K. (1991). Effects of collembola (Arthropoda) and relative germination date on competition between mycorrhizal *Panicum virgatum* (Poaceae) and non-mycorrhizal *Brassica nigra* (Brassicaceae). *Plant and Soil,* **136,** 121–129.

Bonfante-Fasolo, P. & Scannerini, S. (1992). The cellular basis of plant–fungus interchanges in mycorrhizal associations. *Mycorrhizal Functioning* (Ed. by M. F. Allen), pp. 65–101. Chapman & Hall, New York.

Bowers, M. D. & Puttick, G. M. (1988). Response of generalist and specialist insects to qualitative allelochemical variation. *Journal of Chemical Ecology,* **14,** 319–334.

Brown, V. K. & Gange, A. C. (1990). Insect herbivory below ground. *Advances in Ecological Research,* **20,** 1–58.

Brown, V. K. & Gange, A. C. (1992). Secondary plant succession: how is it modified by insect herbivory? *Vegetatio,* **101,** 3–13.

Brundrett, M. C. (1991). Mycorrhizas in natural ecosystems. *Advances in Ecological Research,* **21,** 171–313.

Brundrett, M. C. & Kendrick, B. (1988). The mycorrhizal status, root anatomy, and phenology of plants in a sugar maple forest. *Canadian Journal of Botany,* **66,** 1153–1173.

Bryant, J. P., Chapin, F. S. & Klein, D. R. (1983). Carbon/nutrient balance of boreal plants in relation to vertebrate herbivory. *Oikos,* **40,** 357–368.

Caron, M., Fortin, J. A. & Richard, C. (1986). Effect of phosphorus concentration and *Glomus intraradices* on fusarium crown and root rot of tomatoes. *Phytopathology,* **76,** 942–946.

Curry, J. P. (1994). *Grassland Invertebrates.* Chapman & Hall, London.

Del Vecchio, T. A., Gehring, C. A., Cobb, N. S. & Whitham, T. G. (1993). Negative effects of scale insect herbivory on the ectomycorrhizae of juvenile pinyon pine. *Ecology,* **74,** 2297–2302.

Edwards, C. A. (1983). Earthworm ecology in cultivated soils. *Earthworm Ecology: From Darwin to Vermiculture* (Ed. by J. E. Satchell), pp. 123–137. Chapman & Hall, New York.

Edwards, C. A., Thompson, A. R. & Benyon, K. I. (1968). Some effects of chlorfenvinphos, an organophosphorus insecticide, on earthworm populations and soil animals. *Review of Ecology and Biology of Soils,* **5,** 199–224.

Finlay, R. D. (1985). Interactions between soil micro-arthropods and endomycorrhizal associations of higher plants. *Ecological Interactions in Soil* (Ed. by A. H. Fitter, D. Atkinson, D. J. Read & M. B. Usher), pp. 319–331. Blackwell Scientific Publications, Oxford.

Fitter, A. H. (1985). Functioning of vesicular–arbuscular mycorrhizas under field conditions. *New Phytologist,* **99,** 257–265.

Fitter, A. H. & Sanders, I. R. (1992). Interactions with the soil fauna. *Mycorrhizal Functioning* (Ed. by M. F. Allen), pp. 333–354. Chapman & Hall, New York.

Fogel, R. & Hunt, G. (1979). Fungal and arboreal biomass in a western Oregon Douglas fir ecosystem. *Canadian Journal of Forest Research,* **9,** 245–256.

Friese, C. F. & Allen, M. F. (1993). The interaction of harvester ants and VA mycorrhizal fungi in a patchy environment: The effects of mound structure on fungal dispersion and establishment. *Functional Ecology*, **7**, 13–20.

Gange, A. C. (1990). Effects of insect herbivory on herbaceous plants. *Pests, Pathogens and Plant Communities* (Ed. by J. J. Burdon & S. R. Leather), pp. 49–62. Blackwell Scientific Publications, Oxford.

Gange, A. C. (1993). Translocation of mycorrhizal fungi by earthworms during early plant succession. *Soil Biology and Biochemistry*, **25**, 1021–1026.

Gange, A. C. (1994). Subterranean insects and fungi: Hidden costs and benefits to the greenkeeper. *Science and Golf II* (Ed. by A. J. Cochran & M. R. Farrally), pp. 461–466. E. & F.N. Spon, London.

Gange, A. C. & Brown, V. K. (1992). Interactions between soil-dwelling insects and mycorrhizas during early plant succession. *Mycorrhizas in Ecosystems* (Ed. by D. J. Read, D. H. Lewis, A. H. Fitter & I. J. Alexander), pp. 177–182. CAB International, Wallingford.

Gange, A. C. & West, H. M. (1994). Interactions between arbuscular mycorrhizal fungi and foliar-feeding insects in *Plantago lanceolata* L. *New Phytologist*, **128**, 79–87.

Gange, A. C., Brown, V. K., Barlow, G. S., Whitehouse, D. M. & Moreton, R. J. (1991). Spatial distribution of garden chafer larvae in a golf tee. *Journal of the Sports Turf Research Institute*, **67**, 8–13.

Gange, A. C., Brown, V. K. & Farmer, L. M. (1992). Effects of pesticides on the germination of weed seeds: Implications for manipulative experiments. *Journal of Applied Ecology*, **29**, 303–310.

Gange, A. C., Brown, V. K. & Sinclair, G. S. (1994). Reduction of black vine weevil growth by vesicular–arbuscular mycorrhizal infection. *Entomologia Experimentalis et Applicata*, **70**, 115–119.

Garcia-Garrido, J. M., Toro, N. & Ocampo, J. A. (1993). Presence of specific polypeptides in onion roots colonized by *Glomus mosseae*. *Mycorrhiza*, **2**, 175–177.

Gehring, C. A. & Whitham, T. G. (1991). Herbivore-driven mycorrhizal mutualism in insect-susceptible pinyon pine. *Nature*, **353**, 556–557.

Gehring, C. A. & Whitham, T. G. (1994). Interactions between aboveground herbivores and the mycorrhizal mutualists of plants. *Trends in Ecology and Evolution*, **9**, 251–255.

Grandmaison, J., Olah, G. M., Van Calsteren, M-R. & Furlan, V. (1993). Characterization and localization of plant phenolics likely involved in the pathogen resistance expressed by endomycorrhizal roots. *Mycorrhiza*, **3**, 155–164.

Gullan, P. J. & Cranston, P. S. (1994). *The Insects: An Outline of Entomology*. Chapman & Hall, London.

Hanlon, R. D. G. (1981). Influence of grazing by Collembola on the activity of senescent fungal colonies grown on media of different nutrient concentration. *Oikos*, **36**, 362–367.

Hanson, R. M. & Ueckert, D. N. (1970). Dietary similarity of some primary consumers. *Ecology*, **51**, 640–648.

Harborne, J. B. (1988). *Introduction to Ecological Biochemistry*. Academic Press, London.

Harley, J. L. & Harley, E. L. (1987). A check-list of mycorrhizas in the British flora. *New Phytologist*, (Suppl.) **105**, 1–102.

Harley, J. L. & Smith, S. E. (1983). *Mycorrhizal Symbiosis*. Academic Press, New York.

Harris, K. K. & Boerner, R. E. J. (1990). Effects of belowground grazing by collembola on growth, mycorrhizal infection, and P uptake of *Geranium robertianum*. *Plant and Soil*, **129**, 203–210.

Hilbert, J. L. & Martin, F. (1988). Regulation of gene expression in Ectomycorrhizae I. Protein changes and the presence of ectomycorrhiza specific polypeptides in the *Pisolithus–Eucalyptus* symbiosis. *New Phytologist*, **110**, 339–346.

Hiol Hiol, F., Dixon, R. K. & Curl, E. A. (1994). The feeding reference of mycophagous Collembola varies with the fungal symbiont. *Mycorrhiza*, **5**, 99–103.

Jakobsen, I. & Rosendahl, L. (1990). Carbon flow into soil and external hyphae from roots of mycorrhizal cucumber plants. *New Phytologist,* **115,** 77–83.

Jones, C. G. & Last, F. T. (1991). Ectomycorrhizae and trees: implications for aboveground herbivory. *Microbial Mediation of Plant–Herbivore Interactions* (Ed. by P. Barbosa, V. A. Krischik & C. G. Jones), pp. 65–103. John Wiley & Sons, Chichester.

Kaiser, P. A. & Lussenhop, J. (1991). Collembolan effects on establishment of vesicular-arbuscular mycorrhizae in soybean (*Glycine max*). *Soil Biology and Biochemistry,* **23,** 307–308.

Koch, K. E. & Johnson, C. R. (1984). Photosynthate partitioning in split-root seedlings with mycorrhizal and non-mycorrhizal root systems. *Plant Physiology,* **75,** 26–30.

Lakshman, H. C. & Raghavendra, S. (1990). Spore dispersal of endogonaceae by worms, wasps, dung rollers and Indian domestic fowls. *Current Trends in Mycorrhizal Research* (Ed. by B. L. Jalali & H. Chand), pp. 40–41. Tata Energy Research Institute & Haryana Agricultural University, Hisar.

Lartey, R. T., Curl, R. A., Peterson, C. M. & Harper, J. D. (1989). Mycophagous grazing and food preference of *Proisotoma minuta* (Collembola: Isotomidae) and *Onychiurus encarpatus* (Collembola: Onychiuridae). *Environmental Entomology,* **18,** 334–337.

Leonard, M. J. (1984). Observations on the influence of culture conditions on the fungal feeding preferences of *Folsomia candida* (Collembola: Isotomidae). *Pedobiologia,* **26,** 361–367.

Malloch, D. W., Pirozynski, K. A. & Raven, P. H. (1980). Ecological and evolutionary significance of mycorrhizal symbiosis in vascular plants (a review). *Proceedings of the National Academy of Science, USA,* **77,** 2113–2118.

McGinley, M. A., Dhillion, S. S. & Neumann, J. C. (1994). Environmental heterogeneity and seedling establishment: ant–plant–microbe interactions. *Functional Ecology,* **8,** 607–615.

McGonigle, T. P. (1988). A numerical analysis of published field trials with vesicular-arbuscular mycorrhizal fungi. *Functional Ecology,* **2,** 473–478.

McGonigle, T. P. & Fitter, A. H. (1988). Ecological consequences of arthropod grazing on VA mycorrhizal fungi. *Proceedings of the Royal Society of Edinburgh,* **94B,** 25–32.

McIlveen, W. D. & Cole, H. (1976). Spore dispersal of Endogonaceae by worms, ants wasps and birds. *Canadian Journal of Botany,* **54,** 1486–1489.

Miller, S. L. & Allen, E. B. (1992). Mycorrhizae, nutrient translocation, and interactions between plants. *Mycorrhizal Functioning* (Ed. by M. F. Allen), pp. 301–332. Chapman & Hall, New York.

Molina, R., Massicotte, H. & Trappe, J. M. (1992). Specificity phenomena in mycorrhizal symbioses: Community–ecological consequences and practical implications. *Mycorrhizal Functioning* (Ed. by M. F. Allen), pp. 357–423. Chapman & Hall, New York.

Moore, J. C., St John, T. V. & Coleman, D. C. (1985). Ingestion of vesicular–arbuscular mycorrhizal hyphae and spores by soil microarthropods. *Ecology,* **66,** 1979–1981.

Pirozynski, K. A. (1981). Interactions between fungi and plants through the ages. *Canadian Journal of Botany,* **59,** 1824–1827.

Ponder, F. (1980). Rabbits and grasshoppers: Vectors of endomycorrhizal fungi on new coal mine spoil. *Forestry Service Research Note,* NC-250, 1p. Forest Service, Washington.

Rabatin, S. C. & Stinner, B. R. (1985). Arthropods as consumers of vesicular–arbuscular fungi. *Mycologia,* **77,** 320–322.

Rabatin, S. C. & Stinner, B. R. (1988). Indirect effects of interactions between VAM fungi and soil-inhabiting invertebrates on plant processes. *Agriculture, Ecosystems and Environment,* **24,** 135–146.

Rabatin, S. C. & Stinner, B. R. (1989). The significance of vesicular–arbuscular mycorrhizal fungal-soil macroinvertebrate interactions in agroecosystems. *Agriculture, Ecosystems and Environment,* **27,** 195–204.

Rabatin, S. C. & Stinner, B. R. (1991). Vesicular–arbuscular mycorrhizae, plant, and invertebrate interactions in soil. *Microbial Mediation of Plant–Herbivore Interactions* (Ed. by P. Barbosa, V. A. Krischik & C. G. Jones), pp. 141–168. John Wiley & Sons, Chichester.

Rabin, L. B. & Pacovsky, R. S. (1985). Reduced larva growth of two Lepidoptera (Noctuidae) on excised leaves of soybean infected with a mycorrhizal fungus. *Journal of Economic Entomology*, **78**, 1358–1363.

Reyes, A. A. (1975). Phytotoxicity of benomyl to crucifers. *Phytopathology*, **65**, 535–539.

Schultz, P. A. (1991). Grazing preferences of two collembolan species, *Folsomia candida* and *Proisotoma minuta*, for ectomycorrhizal fungi. *Pedobiologia*, **35**, 313–325.

Shaw, P. J. A. (1985). Grazing preferences of *Onychiurus armatus* (Insecta: Collembola) for mycorrhizal and saprophytic fungi of pine plantations. *Ecological Interactions in Soil* (Ed. by A. H. Fitter, D. Atkinson, D. J. Read & M. B. Usher), pp. 333–337. Blackwell Scientific Publications, Oxford.

Shaw, P. J. A. (1988). A consistent hierarchy in the feeding preferences of the Collembola *Onychiurus armatus*. *Pedobiologia*, **31**, 179–187.

Stace, C. A. (1991). *New Flora of the British Isles*. Cambridge University Press, Cambridge.

Stephenson, J. & Scott, A. C. (1992). The geological history of insect-related plant damage. *Terra Nova*, **4**, 542–552.

Stubblefield, S. P., Taylor, T. N. & Trappe, J. M. (1987). Vesicular–arbuscular mycorrhizae from the Triassic of Antarctica. *American Journal of Botany*, **74**, 1904–1911.

Stork, N. E. (1993). How many species are there? *Biodiversity and Conservation*, **2**, 215–232.

Takeda, H. (1973). A preliminary study on collembolan populations in a pine forest. *Researches on Population Ecology*, **15**, 76–89.

Tamura, H. (1976). Population studies on *Folsomia octoculata* (Collembola: Isotomidae) in a subalpine coniferous forest. *Revue Ecologie et Biologie du Sol*, **13**, 69–91.

Tester, M., Smith, S. E. & Smith, F. A. (1987). The phenomenon of 'non mycorrhizal' plants. *Canadian Journal of Botany*, **65**, 419–431.

Tomlin, A. D. (1977). Toxicity of soil applications of the fungicide benomyl, and two analogues, to three species of Collembola. *Canadian Entomologist*, **109**, 1619–1620.

Trappe, J. M. (1987). Phylogenetic and ecologic aspects of mycotrophy in the Angiosperms from an evolutionary standpoint. *Ecophysiology of VA mycorrhizal plants* (Ed. by G. R. Safir), pp. 5–25. CRC Press, Boca Raton, FL.

Verhoef, H. A. & de Goede, R. G. M. (1985). Effects of collembolan grazing on nitrogen dynamics in a coniferous forest. *Ecological Interactions in Soil* (Ed. by A. H. Fitter, D. Atkinson, D. J. Read & M. B. Usher), pp. 367–376. Blackwell Scientific Publications, Oxford.

Volpin, H., Elkind, Y., Okon, Y. & Kapulnik, Y. (1994). A vesicular arbuscular mycorrhizal fungus (*Glomus intraradix*) induces a defense response in alfalfa roots. *Plant Physiology*, **104**, 683–689.

Ward, L. K. & Spalding, D. F. (1993). Phytophagous British insects and mites and their food plant families: total numbers and polyphagy. *Biological Journal of the Linnean Society*, **49**, 257–276.

Warner, N. J., Allen, M. F. & MacMahon, J. M. (1987). Dispersal agents of vesicular–arbuscular mycorrhizal fungi in a disturbed arid ecosystem. *Mycologia*, **79**, 721–730.

Warnock, A. J., Fitter, A. H. & Usher, M. B. (1982). The influence of a springtail *Folsomia candida* (Insecta, Collembola) on the mycorrhizal association of leek (*Allium porrum*) and the vesicular–arbuscular mycorrhizal endophyte *Glomus fasciculatus*. *New Phytologist*, **90**, 285–292.

Zak, B. (1965). Aphids feeding on mycorrhizae of Douglas-Fir. *Forest Science*, **11**, 410–411.

Zhen-Rong, W., Zhan-Ou, C., Dong-Sheng, Z. & Ji-Kang, X. (1986). Underground distribution pattern of white grubs and sampling method in peanut and soybean fields. *Acta Entomologica Sinica*, **29**, 395–400.

8. INDIRECT INTERACTIONS BETWEEN INSECT HERBIVORES AND PATHOGENIC FUNGI ON LEAVES

P. E. HATCHER AND P. G. AYRES

Division of Biological Sciences, Institute of Environmental and Biological Sciences, Lancaster University, Lancaster LA1 4YQ, UK

INTRODUCTION

Insects and fungi are the two most numerous groups of living organisms in the world (Hawksworth 1991). Small wonder that they should often interact, particularly since a sizeable proportion of each group feeds on the tissues of living plants. There are direct interactions where, for example, fungi are dispersed by insect vectors, or utilize insect feeding wounds to invade plant tissues, or insects feed on fungal mycelium or spores. There are also indirect interactions where the fungus or insect alters the plant so that the other organism is affected indirectly, even if the organisms are not coincident in time or space (Hatcher 1995).

Direct insect–fungus interactions have been well covered by Carter (1973), Agrios (1980) and Wilding *et al.* (1989), but indirect interactions have been largely ignored. Therefore, we concentrate on the latter interactions and, because of limitations of space and our own interests, concentrate on interactions involving folivorous insects, pathogenic leaf-infecting fungi, and non-woody plants. Aspects of these and other fungus–insect–plant interactions have recently been reviewed by Barbosa (1991), de Nooij *et al.* (1992) and Hatcher (1995).

Our own interest has been in the *Gastrophysa–Uromyces–Rumex* system. *Gastrophysa viridula* is a small chrysomelid beetle that overwinters as an adult and emerges in the UK in mid-April. It undergoes up to three generations before hibernating in the soil in October. *Uromyces rumicis* is a macrocyclic heteroecious non-systemic biotrophic rust fungus which has *Rumex* spp. as primary hosts. Although present throughout the year, severe infections normally occur between August and October, during which time the beetle and fungus can coincide. Both beetle and fungus are sufficiently damaging to their hosts alone to have been considered as biological control agents (Foster 1989).

We describe first the effect of insect–fungus combinations on their host plant, and then discuss how these effects are shaped by inhibitory and facilitatory interactions between the insect and fungus.

EFFECTS ON HOST PLANTS

The effect of insect and fungus combinations on their hosts has been studied both from crop pest and weed biological control perspectives. Reviewing studies which measured the effect of insect and fungus (alone and combined) on host parameters, Hatcher (1995) placed the effects of interactions into four categories: synergistic (causes a reduction in a plant variable significantly greater than that obtained from adding damage from insect and fungus alone), additive (causes a reduction in a plant variable equivalent to that obtained from adding damage from insect and fungus alone), equivalent (causes a reduction in a plant variable equivalent to the damage obtained from either insect or fungus (usually the agent causing the greater damage)) and inhibitory (causes a reduction in a plant variable significantly less than that caused by the weaker of the two agents alone). He recognized that, in elucidating the nature of these tripartite interactions, it is important to determine how damage caused by insect and fungus combined compares with that caused by insect and fungus alone. Obviously, from the plant's perspective (or for weed and pest control) the total amount of damage sustained is most important.

We measured the effects of insect (*Gastrophysa viridula*) and fungus (*Uromyces rumicis*) alone, and combined in sequence, on growth of *Rumex crispus* and *R. obtusifolius* for 3.5 months during early summer in the field (Hatcher *et al.* 1994b). A direct comparison of the effects of insect–fungus combinations with those caused by insect and fungus alone was not straightforward, as in one treatment ('Beetle–Rust') *G. viridula* was added to the plant first, followed by infection with *U. rumicis*, and in the other treatment ('Rust–Beetle') the sequence was reversed; therefore, beetle and rust were present on the plants for different lengths of time. Thus, the damage caused by these treatments was compared with values predicted from a model assuming (i) that the effects of fungus and beetle were additive, and (ii) that the amount of damage from any one organism was a linear function of the length of time that organism was present on the plant (Hatcher *et al.* 1994b). Most parameters of 'Beetle–Rust' and 'Rust–Beetle' treatments for *R. obtusifolius*, and some of the 'Rust–Beetle' treatment for *R. crispus* were accurately predicted (Fig. 8.1), suggesting that these treatments produced additive effects. However, the model significantly overestimated damage for the 'Beetle–Rust' treatment on *R. crispus* (Fig. 8.1), suggesting that in this case the damage was only equivalent. Possible reasons for this effect are discussed subsequently.

Some insect–fungus combinations can damage the plant synergistically. Most of these pairings involve necrotrophic stalk and root rots, in which infection is enhanced by the direct effects of stem and root-boring insects (Hatcher 1995). However, the rust *Uromyces appendiculatus* and aphid *Aphis craccivora* also interacted synergistically in reducing the height of *Vigna sesquipedalis* plants (Chang & Thrower 1981). The only inhibitory interaction noted was between the weevil

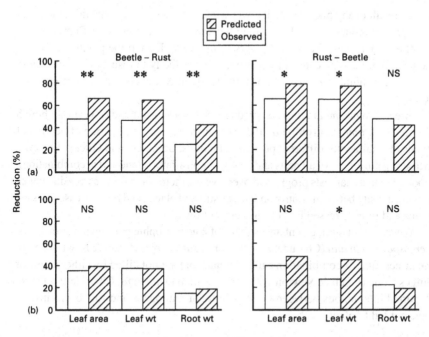

FIG. 8.1. Comparison of observed (□) and predicted (▨) damage to (a) *Rumex crispus* and (b) *R. obtusifolius* from combined herbivory and rust infection. Beetle–Rust, grazing by *Gastrophysa viridula* with subsequent infection by *Uromyces rumicis*; Rust–Beetle, infection with subsequent grazing. Percentage reduction in variables compared to control, undamaged plants. See text for calculation of predicted damage. Comparisons using the G-statistic: *, $P < 0.05$; **, $P < 0.01$; NS; $P > 0.05$. (From Hatcher *et al.* 1994b.)

Perapion antiquum and the necrotrophic fungus *Phomopsis emicis* (Shivas & Scott 1993). Whereas the fungus alone reduced weight of *Emex australis* fruit by 45%, and weevil damage increased fruit weight by 40%, the insect–fungus combination reduced fruit weight by only 5% (Shivas & Scott 1993).

Most investigations have measured plant damage over one season only, in spite of some host species being perennial, with the potential for longer-term damage. We have gone some way towards determining these longer-term effects. A combination of *G. viridula* and *U. rumicis* applied to *R. crispus* and *R. obtusifolius* seedlings planted in August had an additive effect on leaf area during that autumn, and a similar effect on root weight at the start of regrowth in February the following year (Hatcher 1996). Root weight was reduced by 82% and 72% in *R. crispus* and *R. obtusifolius* respectively, compared with healthy controls in February. Root quality was also affected: the insect–fungus combination led to a 92% increase in the concentration of alcohol-soluble carbohydrate in *R. crispus* roots, and a 40% reduction in starch in *R. obtusifolius* roots (Hatcher 1996). Although this treatment did not result in any plant mortality over winter, studies suggest that this amount of damage will reduce the ability of the plants to regrow (Monaco & Cumbo 1972) and flower that year (Weaver & Cavers 1979, 1980). Even if the plants did flower, the degree of damage caused by this treatment is likely to result in a reduction in seed weight, number or size (Inman 1971; Maun & Cavers 1971; Bentley *et al.* 1980).

Insect–fungus interactions can produce effects over a number of years. Although Godfrey and Yeargan (1989) could detect few differences in the rate of decline of stands of alfalfa with different pest combinations, Summers and Gilchrist (1991), measuring the effects of all pests and pathogens on alfalfa (alone and in combination), reported that, as harvests progressed over 5 years, there was a progressive difference in stand density between control and insect-stressed plants, and both disease-stressed plants and those stressed by both insects and fungi.

Where insect damage enhances entry of a necrotrophic pathogen, for example *Cercospora rodmanii* (Charudattan 1986) or *Fusarium* spp. (Leath & Newton 1969), the insect–fungus combination can kill plants that are not killed by either insect or fungus alone (Leath & Newton 1969; Charudattan *et al.* 1978; Charudattan 1986). Reported interactions between biotrophic foliar pathogens and folivorous insects have not led to plant death.

INHIBITORY INTERACTIONS

Effects of fungi on insects

Plant pathogenic fungi can not only induce the production of allelochemicals, but can also alter the nutritional quality of the plant. Hatcher (1995) reviewed fungally induced chemical changes that could adversely affect folivorous insects. These included increases in leaf toughness and lignification, increases in leaf fibre content, and reductions in carbohydrate and nitrogen concentration. Nitrogen quality can also be affected, with increases in nitrate (Piening 1972) and changes in total free and protein amino-acid concentration and composition (Reddy & Rao 1976) being reported.

G. viridula was inhibited during all its life stages when it was reared upon *Rumex* leaves infected with *U. rumicis* (Hatcher *et al.* 1994c). During the last instar, larvae took up to 15% longer to complete feeding, and had a relative growth rate up to 25% lower than beetles feeding on uninfected leaves, although they consumed up to 63% more food.

The fecundity of *G. viridula* feeding on infected leaves was significantly reduced – these adults laid approximately 55% fewer eggs, due to a combination of shorter life span and reduced rate of egg laying (Fig. 8.2). Eggs laid by beetles feeding on rust-infected *R. crispus* leaves also had a reduced viability (Fig. 8.2).

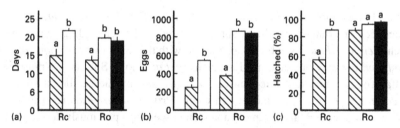

FIG. 8.2. Effect of infection of *Rumex crispus* (Rc) and *R. obtusifolius* (Ro) with *Uromyces rumicis* on *Gastrophysa viridula*: (a) duration of egg laying, (b) number of eggs laid by individuals, (c) percentage of eggs that hatched. Beetles reared on infected leaves (▨), uninfected leaves (□), or on uninfected leaves as an adult but infected leaves as a larva (■). $n = 15$ except infected *R. crispus* ($n = 9$), and uninfected leaves/infected leaves as a larva ($n = 13$). Bar represents one SE. The same letter above columns within species indicates no significant difference ($P > 0.05$), Tukey–HSD test from a one-way ANOVA. (From Hatcher *et al.* 1994c.)

These effects were due to the food consumed as an adult, and not effects carried over from larval development, as beetles fed on rusted *R. obtusifolius* as larvae and transferred to uninfected leaves as adults had fecundity similar to adults which had been reared from egg on uninfected *R. obtusifolius* (Fig. 8.2).

These effects were consistent with the chemical changes occurring in *Rumex* leaves infected with *U. rumicis*, especially a reduction in leaf nitrogen concentration of up to 50% (Hatcher *et al.* 1995a). This fall in nitrogen concentration in infected leaves, in conjunction with a small rise in carbohydrate concentration, increased the ratio of carbohydrate: nitrogen from 2.5 and 4.0 in the healthy leaves of both species to maxima of 6.8 and 7.7 in infected *R. crispus* and *R. obtusifolius* leaves respectively (Hatcher *et al.* 1995b).

Karban *et al.* (1987) reported that cotton infected with *Verticillium dahliae* supported populations of spider mite *Tetranychus urticae* only half the size of those supported on uninfected plants. While growth and reproduction of aphids was inhibited by the biotrophs *Uromyces viciae-fabae* and *Erysiphe graminis* (and by the necrotrophs *Phytophthora erythroseptica* and *Fusarium* sp.) (Leath & Byers 1977; Pratt *et al.* 1982; Ellsbury *et al.* 1985; Sipos & Sági 1987; Pesel & Poehling 1988; Prüter & Zebitz 1991), the only comparable data for a leaf chewing insect involve the vascular wilt fungus *Verticillium albo-atrum* (Kingsley *et al.* 1983). Rate of consumption by larvae of *Spodoptera eridania* (Lepidoptera) increased, and efficiency of conversion of digested food to biomass was significantly reduced, when fed on alfalfa infected with *V. albo-atrum*, although leaf water and organic nitrogen in infected leaves remained unchanged (Kingsley *et al.* 1983).

The few tripartite studies carried out on trees have demonstrated that fungal infection may have an effect on insect herbivores even when insect and fungus are considerably temporally separated, and the insect feeds on the next generation of leaves. For example, the pupal weight of the moth *Epirrita autumnata*, larvae of which feed on young leaves in the spring, was reduced when feeding on mountain birch, *Betula pubescens*, which was infected with the foliar rust *Melampsoridium betulinum* the previous autumn (Lappalainen *et al.* 1995). Similarly, Krause and Raffa (1992) found that infection of *Larix decidua* by the needle cast fungus *Mycosphaerella laricinia* reduced larch sawfly, *Pristiphora erichsonii* consumption rates 1 year later and larvae rapidly abandoned seedlings previously defoliated by *M. laricinia*.

Effects of insects on fungi

Feeding by some insects and other invertebrates has been demonstrated to have an inhibitory effect on the subsequent fungal infection of their host plants; for example, herbivory on *R. crispus* and *R. obtusifolius* by *G. viridula* induced an 80% reduction in *U. rumicis* pustule density within and immediately around the feeding site within 1 day of feeding damage (Hatcher *et al.* 1994a). It also induced resistance throughout the undamaged portion of the leaf, and a less effective and slower resistance in

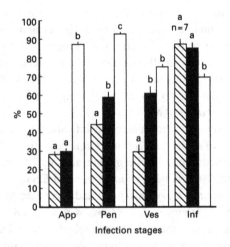

FIG. 8.3. Percentage of *Uromyces rumicis* sporelings reaching one infection stage that passed to the next 3 days after inoculation on to *Rumex obtusifolius* leaves, either with simulated insect herbivory (▨), damaged by *Gastrophysa viridula* grazing (■), or undamaged (□). Infection stages: App, sporelings having a germ tube that also had an appressorium; Pen, sporelings having an appressorium that also had a penetration hypha; Ves, sporelings having a penetration hypha that also had at least one substomatal vesicle; Inf, sporelings having at least one substomatal vesicle that also had intercellular hyphae. *n* = 10 unless stated, bar indicates one SE. The same letter above columns indicates no significant difference (*P* > 0.05) between treatments for that infection stage, Tukey–HSD test from a one-way ANOVA. (From Hatcher *et al.* 1995a.)

undamaged leaves – that is, there was systemic induced resistance. A reduction of 74% and 86% in pustule density was also found on artificially holed leaves of *R. crispus* and *R. obtusifolius* respectively, when exposed to natural *U. rumicis* infection over a 1-month period in the field (Hatcher *et al.* 1994a). Herbivory by *G. viridula* and artificial damage mainly affected the early stages of uredinial development (Fig. 8.3); once intercellular hyphae had started to develop, there was no difference between the rate of development of uredinia on artificially damaged and undamaged leaves (Hatcher *et al.* 1995a). Similarly, for another biotroph, the degree of infection of barley by *Erysiphe graminis* decreased when plants were infested with the aphid *Macrosiphum avenae* (Sipos & Sági 1987).

Defoliation of soybean by larvae of *Pseudoplusia includens* (Lepidoptera) led to a reduction in natural *Diaporthe phaseolorum* infection, including an 84% reduction in total canker area and a 58% reduction in density of perithecia (Padgett *et al.* 1994), compared to undefoliated plants. Cotton plants with spider mite, *Tetranychus urticae*, infestations were less likely than controls to develop wilting due to *Verticillium dahliae* infection (Karban *et al.* 1987).

In all the examples above it is not clear what the mechanism of these inhibitory effects is. We have suggested that alterations in leaf surface topography or volatile production may inhibit appressorium formation by *U. rumicis* on damaged *Rumex* leaves (Hatcher *et al.* 1995a), but this remains untested.

Natural fungal infections can be grazed by invertebrates sufficiently severely to warrant investigation of the role of mycophagy on fungal epidemiology; for example, Powell (1971) found 137 species of insect, 19 species of mite and four species of spider associated with the pine stem rust *Cronartium comandrae*. These damaged up to 62% of rust cankers during 1 year, feeding and matting together aeciospores and destroying large parts of the aecial and spermatogonial layers of the cankers, and 57% of *Puccinia sessilis* pustules on *Allium ursinum* sampled in the field were grazed by slugs and snails (Ramsell & Paul 1990).

No study has examined the effects of grazing by insects on the populations of fungi, although vertebrate grazing has been implicated in reducing disease levels in grassland systems (Skipp & Lambert 1984; Bowers & Sacchi 1991; Wennström & Ericson 1991). Bowers and Sacchi (1991) suggested that the main effect of grazing was to keep plant (or tissue) abundance below that at which fungal infection became epidemic. However, the degree of correlation between plant density and level of fungal infection is very variable (Burdon & Chilvers 1982; Kranz 1990), and it is likely that many factors, for example more favourable conditions for foliage infection in denser, more humid canopies (Skipp & Lambert 1984), or the failure of seedlings to establish in ungrazed localities (Wennström & Ericson 1991) are important.

Plant allelochemical responses

Insect feeding and pathogenic infection affect both primary and secondary plant metabolism – in the case of secondary metabolism often affecting enzymes or products having a recognized role in defence responses against *either* insects *or* fungi, and in some cases affecting enzymes whose products are known to have deleterious effects on *both* insects *and* fungi. Examples of each of the major types of defence response are listed in Table 8.1. The possibility that they may confer cross-protection

TABLE 8.1. Types of major plant resistance response induced both by insect herbivores and pathogens.

Type I
Hypersensitive cell death, a burst of oxidative metabolism, and membrane depolarization. Associated lignification of cell walls.
A hypersensitive response (HR), with production of H_2O_2 and increased lipoxygenase activity occurs in *Solanum tuberosum* infected by *Phytophthora infestans* (Bostock *et al.* 1992). An HR, and increased peroxidase activity, is induced in *S. dulcamara* by gall mite *Aceria cladophthirus* (Westphal, Dreger & Bronner 1991).
Cell walls of cucumber are protected by lignin after challenge by *Colletotrichum lagenarium* (Stein *et al.* 1993). There is no direct evidence of lignification induced by herbivory, but in *Salix* leaves, toughness, usually associated with lignification, damages the mandibles of *Plagiodera versicolora* (Raupp 1985).

Type II
Synthesis of allelochemicals, including phytoalexins.
Glyceollin, an isoflavonoid phytoalexin, helps to protect soybean against *Phytophthora megasperma glycinea* (Classen & Ward 1985) and has anti-feedant effects on herbivores such as the beetle *Epilachna varivestis* (Kogan & Fischer 1991). It is unknown whether insect herbivores induce glyceollin.

Type III
Synthesis of defence-related proteins.
Includes enzymes for metabolic pathways leading to I and II above. In addition, herbivores may induce a family of proteinase inhibitor (PI) proteins, and pathogens induce a family of pathogenesis-related (PR) proteins, such as chitinase and β-1, 3-glucanases. There could be cross-protection. Herbivores can also induce the latter enzymes, while PR6 has proteinase activity, and McGurl *et al.* (1995) have recently shown soil micro-organisms can induce trypsin inhibitors.

Type IV
Systemic signals (to protect distant tissues).
Chemical (or, some claim, electrical or hydraulic) signals induce Type II and III responses in unwounded tissues and healthy leaves. Systemin, ethylene, jasmonic acid and methyl jasmonate may all be involved in signalling both pathogen- and pest-induced wounding (Enyedi *et al.* 1992). Signal transmission after pathogen infection also involves salicylic acid, though probably not as the long-distance messenger (Ryals *et al.* 1994).

has rarely been examined, but, as noted under Type IV, insects and pathogens may induce common signals. Indeed, Farmer and Ryan (1992) have proposed for tomato a scheme by which systemic signals (possibly in the form of systemin, an 18 amino-acid polypeptide) arising from herbivore attack, and local signals (probably oligouronides) resulting from pathogen attack may bind to different plasma membrane receptors before activating the same membrane-bound lipase. This enzyme liberates from the membrane linolenic acid, the substrate of lipoxygenase and precursor of jasmonic acid, which, in turn, activates the gene(s) regulating proteinase inhibitors.

Given this degree of commonality, why is it that inhibitory interactions between insects and fungi have not been recorded more often? Have they simply been overlooked in the field? This is quite possible because most biologists have been happy to busy themselves with simpler, two-way interactions. There may also be a subtle imbalance, such that herbivory has greater effects on biotrophic fungi, which are less visible and more difficult to quantify than necrotrophs, while herbivores are likely to be affected in less immediately obvious ways by biotrophs than by necrotrophs. Another explanation may be that, as Baldwin (1994) has suggested, although it has been accepted that a plant's response to pathogen invasion can be specific both to the attacker and the elicited response, anti-herbivore responses have been assumed to be more general. Anti-herbivore studies have not employed methods that would detect specificity in either the signalling of the damage response or in their consequences for the attacker.

Is it possible that mechanisms effective in the laboratory are not effective in the field? Disagreement between laboratory and field results is all too familiar and there may be many reasons for this. Often the growth conditions of the plant are critical. Thus, for example, amounts of pathogenesis-related (PR) proteins produced by plants are strongly dependent upon growth conditions (Jung *et al.* 1995).

It is also possible that the idea of extensive commonality is illusory. In one plant there may be different responses to different pathogens and different insects (Felton *et al.* 1994; Stout *et al.* 1994); for example, while caterpillars feeding on tomato induced polyphenoloxidase and lipoxygenase (LOX) activity, leafminer feeding induced only peroxidase (POD) activity and mite feeding induced both POD and LOX activity (Stout *et al.* 1994). Thus, although pathogens and herbivores may activate the same area of metabolism, and in some cases the same pathways, the overlap may not be extensive. There are sound biochemical arguments against identical responses to herbivores and pathogens. Phenylalanine ammonia lyase (PAL), the key enzyme in the biosynthetic pathway of phenols, exists in multiple isomeric forms, which are differentially transcriptionally induced by distinct stimuli (Baldwin 1994). Similarly, peroxidase and catalase also exist as multiple isomers, these forms often having different locations and roles within the cell.

FACILITATORY INTERACTIONS

Effects of fungi on insects

All reported instances of invertebrates selectively grazing on leaves infected with foliar fungi (e.g. Barbe 1964; Lewis 1979; Ramsell & Paul 1990, and as above) have been described as examples of facultative mycophagy (Hatcher 1995). This is because these invertebrates tend to remove spores or pustules with only the minimum leaf material, and avoid uninfected areas of leaves. However, the nutritional quality, especially concentrations of nitrogen and carbohydrates, of whole leaves infected by foliar pathogens may be increased (Farrar & Lewis 1987; Paul 1989; Hatcher 1995), not just in the area around the pustule. This increase in leaf nitrogen can be considerable, for example up to 30% in *Tussilago farfara* infected with *Puccinia poarum* (Ramsell & Paul 1990) and between 28 and 50% in *Allium porrum* infected with *P. allii* (Roberts & Walters 1989) and could have significant, albeit untested, effects on herbivorous insects.

Several studies have noted that infection by necrotrophic pathogens facilitated insect development; for example, the European corn borer, *Ostrinia nubilalis* (Lepidoptera), had a faster larval development when reared on *Fusarium graminearum*-infected corn (Chiang & Wilcoxson 1961). Infection of maize with *Colletotrichum graminicola* alone (Carruthers *et al.* 1986), or in combination with other fungi (Jarvis *et al.* 1990), resulted in a 25% decrease in development time, allowing *O. nubilalis* to proceed through a second generation in some seasons. Carruthers *et al.* (1986) suggested that this effect may be due to the predigestion of complex carbohydrates by enzymes released from the fungus. This may also be important in *Botrytis*–insect interactions; both aphids (Zebitz & Kehlenbeck 1991) and Lepidoptera (Savopoulou-Soultani & Tzanakakis 1988) had increased development rates when reared on *Botrytis*-infected tissue.

Effects of insects on fungi

Most reported facilitatory effects of insects on fungi are due to insect-mediated transport and entry of fungi into their host plants (Carter 1973; Agrios 1980). It is often difficult to disentangle indirect insect facilitation from these direct effects.

One good example, however, is the enhancement of fungal infections, including those of necrotrophs *Septoria nodorum* (Dik *et al.* 1991) and *Botrytis fabae* (Last 1960) on leaves with honeydew from aphids. These pathogens may use nutrients from the phyllosphere for germination or superficial mycelial growth before penetrating the leaf (Dik & van Pelt 1992), and may act as scavengers, removing much of the deposited honeydew and thus reducing its adverse effect on leaf photosynthesis and ageing (Rabbinge *et al.* 1981). Honeydew could also affect biotrophic pathogens, because a variety of nutrients, including low concentrations of glucose and amino acids, stimulated uredospore germination and germ-tube growth of *Uromyces viciae-fabae* (Parker & Blakeman 1984).

The infection of plants by some fungi, for example *Fusarium* spp., is enhanced by plant stress, including that caused by aphids (Leath & Byers 1977) and homopteran leaf hoppers (Moellenbeck *et al.* 1992). These fungi also tend not to enter insect-caused wounds directly, but rather close by (Stutz *et al.* 1985).

IMPLICATIONS FOR COMMUNITY STRUCTURE

This review has demonstrated the fragmentary nature of the studies on the fungus–insect–plant interactions defined here. Most of the studies have been conducted with a particular agenda: crop pest control, weed biological control, studies of the effects of altered food plants on insects and so on, and thus have examined only some of the possible interactions. They reveal little about the role of such interactions in communities. One reason for the amount of space devoted here to the *Uromyces–Gastrophysa–Rumex* system is, authors' bias aside, that we believe it remains the only system in which these interactions have been studied from the perspective of all of the participating organisms. It also represents a real, and not artificial, system.

We have studied this system by investigating the effects of different interactions in controlled, laboratory systems, then testing the predictions of these experiments in the field. One of the next steps must be to investigate the importance of such interactions in community functioning. This will not be a trivial undertaking. Already, the step from laboratory to field has produced some surprises. For example, how can an additive effect on the host result from an insect–fungus interaction that is inhibitory to both parties (Hatcher *et al.* 1994b; Hatcher 1996)? We believe that the beetle and fungus may be spatially separated on the plant. *U. rumicis* is poor at infecting young *Rumex* leaves (Hatcher *et al.* 1994a; Hatcher *et al.* 1995a), while these leaves are preferred by *G. viridula* for oviposition (Hatcher *et al.* 1994c). The between-leaf effects are much weaker than the within-leaf effects: *G. viridula* was unaffected by feeding on healthy leaves from plants infected with *U. rumicis* (Hatcher *et al.* 1994c), and the systemic effect of *G. viridula* herbivory on *U. rumicis* infection was relatively weak (Hatcher *et al.* 1994a). The outcome of this for the plant, however, is unfortunate as it tends to maximize the amount of damage caused.

Some plant fungal infections may also change the local distribution of insects. Along with insects orienting towards (Batra & Batra 1985; Roy 1993) or away from (Jennersten & Kwak 1991) fungally altered plant nectar sources, fungal infection may alter host choice in phytophagous insects. Fewer aphids, *Acyrthosiphon pisum*,

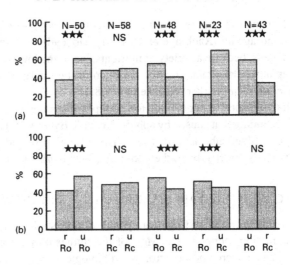

FIG. 8.4. Effect of foliar fungal infection on intra- and inter-specific oviposition choice in *Gastrophysa viridula*. Percentage of total number of eggs laid on different choices in (a) laboratory and (b) field choice tests. In laboratory tests, individual females were offered unexcised leaves for 1 day. In field tests, one gravid female was placed in the intersection of every four plants in a 10×10 checkerboard grid of each choice, the number of eggs laid on non-edge plants ($n = 32$ for each choice) was recorded after 8 days. u, uninfected; r, infected with *Uromyces rumicis*; Rc, *Rumex crispus*; Ro, *R. obtusifolius*. ***, $P < 0.001$; NS, $P > 0.05$, chi square test. (From Hatcher *et al.* 1994c and Hatcher & Whittaker unpublished data.)

colonized clover plants infected with *Phytophthora* root disease than healthy plants in greenhouse (Pratt *et al.* 1982) and field (Ellsbury *et al.* 1985) choice tests, and *Gastrophysa viridula* laid a significantly greater proportion of eggs on healthy as opposed to *Uromyces rumicis*-infected *Rumex obtusifolius* plants in laboratory and field choice tests (Hatcher *et al.* 1994c; Hatcher & Whittaker, unpublished data; Fig. 8.4). Conversely, the grasshopper *Melanoplus differentialis* preferred feeding on *Helianthus annuus* leaves infected with the rust *Puccinia helianthii* to healthy leaves (Lewis 1979, 1984). Differential infection of *Rumex* species with *Uromyces rumicis* may also affect interspecific oviposition choice by *Gastrophysa viridula*. Although in a choice between healthy *R. crispus* and *R. obtusifolius*, *G. viridula* laid a greater proportion of eggs on *R. obtusifolius*, if one species was infected with *U. rumicis* then egg laying was shifted away from that species (Hatcher & Whittaker unpublished data; Fig. 8.4).

These experiments suggest that insect herbivory may be a cost inherent in the development of plant fungal resistance. *R. crispus* and *R. obtusifolius* often occur in grassland together (Cavers & Harper 1964, 1967) and in a community also containing significant herbivory from *G. viridula*, individuals of one *Rumex* species that developed resistance to *U. rumicis* could attract greater herbivory from *G.*

viridula. This would be both from adult feeding and from larvae hatching from the greater number of eggs laid upon the uninfected plants.

Recently, Nakamura *et al.* (1995) have also suggested a cost of pathogen resistance because of greater herbivory. An apparent trade-off existed between susceptibility of *Brassica rapa* to *Alternaria* sp. and seed predation by cecidomyiid gall midges, as plants with the highest fungal infection scores had the lowest seed predation rate (Nakamura *et al.* 1995).

Several papers and books (including Burdon & Leather 1990; Linhart 1991; Ayres 1992; Fritz & Simms 1992; Marquis & Alexander 1992) have purported to consider varying aspects of plant responses, including resistance, community structure and plant evolution, to herbivores *and* pathogens. With a few notable exceptions, these have actually considered mainly effects of herbivores *or* pathogens. While comparisons of herbivore–plant and fungus–plant interactions are intrinsically interesting, we believe they have a low predictive power for more complex fungus–insect–plant studies. This is especially true as our studies are beginning to demonstrate that the outcome of these tripartite interactions for the participants may often not be predictable from a study of pair-wise interactions. Therefore, echoing the above authors, we suggest that greater understanding is required of both the mechanisms underpinning these tripartite interactions, and their effects on populations of the participants, before grander constructs can be attempted.

ACKNOWLEDGEMENTS

We thank the Natural Environment Research Council for funding our work on tripartite interactions at Lancaster, and Nigel Paul and John Whittaker for many useful discussions and comments.

REFERENCES

Agrios, G. N. (1980). Insect involvement in the transmission of fungal pathogens. *Vectors of Plant Pathogens* (Ed. by K. F. Harris & K. Maramorosch), pp. 293–324. Academic Press, New York.

Ayres, P. G. (Ed.) (1992). *Pests and Pathogens: Plant Responses to Foliar Attack*. BIOS, Oxford.

Baldwin, I. T. (1994). Chemical changes rapidly induced by folivory. *Insect–Plant Interactions*, Vol. V (Ed. by E. A. Bernays), pp. 1–23. CRC Press, Boca Raton, FL.

Barbe, G. D. (1964). Relation of the European earwig to snapdragon rust. *Phytopathology*, **54**, 369–371.

Barbosa, P. (1991). Plant pathogens and nonvector herbivores. *Microbial Mediation of Plant–Herbivore Interactions* (Ed. by P. Barbosa, V. A. Krischik & C. G. Jones), pp. 341–382. John Wiley & Sons, New York.

Batra, L. R. & Batra, S. W. T. (1985). Floral mimicry induced by mummy-berry fungus exploits host's pollinators as vectors. *Science*, **228**, 1011–1013.

Bentley, S., Whittaker, J. B. & Malloch, A. J. C. (1980). Field experiments on the effects of grazing by a chrysomelid beetle (*Gastrophysa viridula*) on seed production and quality in *Rumex obtusifolius* and *Rumex crispus*. *Journal of Ecology*, **68**, 671–674.

Bostock, R. M., Yamamoto, H., Choi, D., Ricker, K. E. & Ward, B. L. (1992). Rapid stimulation of

5-lipoxygenase activity in potato by the fungal elicitor arachidonic acid. *Plant Physiology*, **100**, 1448–1456.

Bowers, M. A. & Sacchi, C. F. (**1991**). Fungal mediation of a plant–herbivore interaction in an early successional plant community. *Ecology*, **72**, 1032–1037.

Burdon, J. J. & Chilvers, G. A. (**1982**). Host density as a factor in plant disease ecology. *Annual Review of Phytopathology*, **20**, 143–166.

Burdon, J. J. & Leather, S. R. (Eds) (**1990**). *Pests, Pathogens and Plant Communities*. Blackwell Scientific Publications, Oxford.

Carruthers, R. I., Bergstrom, G. C. & Haynes, P. A. (**1986**). Accelerated development of the European corn borer, *Ostrinia nubilalis* (Lepidoptera: Pyralidae), induced by interactions with *Colletotrichum graminicola* (Melanconiales: Melanconiaceae), the causal fungus of maize anthracnose. *Annals of the Entomological Society of America*, **79**, 385–389.

Carter, W. (**1973**). *Insects in Relation to Plant Disease*. John Wiley & Sons, New York.

Cavers, P. B. & Harper, J. L. (**1964**). *Rumex obtusifolius* L. and *R. crispus* L. *Journal of Ecology*, **52**, 737–766.

Cavers, P. B. & Harper, J. L. (**1967**). Studies in the dynamics of plant populations. I. The fate of seed and transplants introduced into various habitats. *Journal of Ecology*, **55**, 59–71.

Chang, L.-H. & Thrower, L. B. (**1981**). The effect of *Uromyces appendiculatus* and *Aphis craccivora* on the yield of *Vigna sesquipedalis*. *Phytopathologische Zeitschrift*, **101**, 143–152.

Charudattan, R. (**1986**). Integrated control of waterhyacinth (*Eichhornia crassipes*) with a pathogen, insects, and herbicides. *Weed Science*, **34** (Suppl. 1), 26–30.

Charudattan, R., Perkins, B. D. & Littell, R. C. (**1978**). Effects of fungi and bacteria on the decline of arthropod-damaged waterhyacinth (*Eichornia crassipes*) in Florida. *Weed Science*, **26**, 101–107.

Chiang, H. C. & Wilcoxson, R. D. (**1961**). Interactions of the European corn borer and stalk rot in corn. *Journal of Economic Entomology*, **54**, 850–852.

Classen, D. & Ward, E. W. B. (**1985**). Temperature-induced susceptibility of soybeans to *Phytophthora megasperma* f. sp. *glycinea*: production and activity of elicitors of glyceollin. *Physiological Plant Pathology*, **26**, 289–296.

de Nooij, M. P., Biere, A. & Linders, E. G. A. (**1992**). Interaction of pests and pathogens through host predisposition. *Pests and Pathogens: Plant Responses to Foliar Attack* (Ed. by P. G. Ayres), pp. 143–160. BIOS, Oxford.

Dik, A. J. & van Pelt, J. A. (**1992**). Interaction between phyllosphere yeasts, aphid honeydew and fungicide effectiveness in wheat under field conditions. *Plant Pathology*, **41**, 661–675.

Dik, A. J., Fokkema, N. J. & van Pelt, J. A. (**1991**). Interference of nutrients with fungicide activity against *Septoria nodorum* on wheat leaves. *Plant Pathology*, **40**, 25–37.

Ellsbury, M. M., Pratt, R. G. & Knight, W. E. (**1985**). Effects of single and combined infection of arrowleaf clover with bean yellow mosaic virus and a *Phytophthora* sp. on reproduction and colonization by pea aphids (Homoptera: Aphididae). *Environmental Entomology*, **14**, 356–359.

Enyedi, A. J., Yalpani, N., Silverman, P. & Raskin, I. (**1992**). Signal molecules in systemic plant resistance to pathogens and pests. *Cell*, **70**, 879–886.

Farmer, E. E. & Ryan, C. A. (**1992**). Octadecanoid precursors of jasmonic acid activate the synthesis of wound-inducible proteinase inhibitors. *Plant Cell*, **4**, 129–134.

Farrar, J. F. & Lewis, D. H. (**1987**). Nutrient relations in biotrophic infections. *Fungal Infection of Plants* (Ed. by G. F. Pegg & P. G. Ayres), pp. 92–132. Cambridge University Press, Cambridge.

Felton, G. W., Summers, C. B. & Mueller, A. J. (**1994**). Oxidative responses in soybean foliage to herbivory by bean leaf beetle and three-cornered alfalfa hopper. *Journal of Chemical Ecology*, **20**, 639–650.

Foster, L. (**1989**). The biology and non-chemical control of dock species *Rumex obtusifolius* and *R. crispus*. *Biological Agriculture and Horticulture*, **6**, 11–25.

Fritz, R. S. & Simms, E. L. (Eds) (**1992**). *Plant Resistance to Herbivores and Pathogens*. University of Chicago Press, Chicago.

Godfrey, L. D. & Yeargan, K. V. (1989). Effects of clover root curculio, alfalfa weevil (Coleoptera: Curculionidae), and soil-borne fungi on alfalfa stand density and longevity in Kentucky. *Journal of Economic Entomology*, **82**, 1749–1756.

Hatcher, P. E. (1995). Three-way interactions between plant pathogenic fungi, herbivorous insects and their host plants. *Biological Reviews*, **70**, 639–694.

Hatcher, P. E. (1996). The effect of insect–fungus interactions on the autumn growth and over-wintering of *Rumex cripsus* and *R. obtusifolius* seedlings. *Journal of Ecology*, **84**, 101–109.

Hatcher, P. E., Ayres, P. G. & Paul, N. D. (1995a). The effect of natural and simulated insect herbivory, and leaf age, on the process of infection of *Rumex crispus* L. and *R. obtusifolius* L. by *Uromyces rumicis* (Schum.) Wint. *New Phytologist*, **130**, 239–249.

Hatcher, P. E., Paul, N. D., Ayres, P. G. & Whittaker, J. B. (1994a). Interactions between *Rumex* spp., herbivores and a rust fungus: *Gastrophysa viridula* grazing reduces subsequent infection by *Uromyces rumicis*. *Functional Ecology*, **8**, 265–272.

Hatcher, P. E., Paul, N. D., Ayres, P. G. & Whittaker, J. B. (1994b). The effect of an insect herbivore and a rust fungus individually, and combined in sequence, on the growth of two *Rumex* species. *New Phytologist*, **128**, 71–78.

Hatcher, P. E., Paul, N. D., Ayres, P. G. & Whittaker, J. B. (1994c). The effect of a foliar disease (rust) on the development of *Gastrophysa viridula* (Coleoptera: Chrysomelidae). *Ecological Entomology*, **19**, 349–360.

Hatcher, P. E., Paul, N. D., Ayres, P. G. & Whittaker, J. B. (1995b). Interactions between *Rumex* spp., herbivores and a rust fungus: the effect of *Uromyces rumicis* infection on leaf nutritional quality. *Functional Ecology*, **9**, 97–105.

Hawksworth, D. L. (1991). The fungal dimension of biodiversity: magnitude, significance, and conservation. *Mycological Research*, **95**, 641–655.

Inman, R. E. (1971). A preliminary evaluation of *Rumex* rust as a biological control agent for curly dock. *Phytopathology*, **61**, 102–107.

Jarvis, J. L., Ziegler, K. E., Webber, D. F. & Guthrie, W. D. (1990). Effect of second-generation European corn borers and stalk rot fungi on popcorn hybrids. *Maydica*, **35**, 259–265.

Jennersten, O. & Kwak, M. M. (1991). Competition for bumblebee visitation between *Melampyrum pratense* and *Viscaria vulgaris* with healthy and *Ustilago*-infected flowers. *Oecologia*, **86**, 88–98.

Jung, J.-L., Maurel, S., Fritig, B. & Hahne, G. (1995). Different pathogenesis-related-proteins are expressed in sunflower (*Helianthus annuus* L.) in response to physical, chemical and stress factors. *Journal of Plant Physiology*, **145**, 153–160.

Karban, R., Adamchak, R. & Schnathorst, W. C. (1987). Induced resistance and interspecific competition between spider mites and a vascular wilt fungus. *Science*, **235**, 678–680.

Kingsley, P., Scriber, J. M., Grau, C. R. & Delwiche, P. A. (1983). Feeding and growth performance of *Spodoptera eridania* (Noctuidae: Lepidoptera) on 'vernal' alfalfa, as influenced by *Verticillium* wilt. *Protection Ecology*, **5**, 127–134.

Kogan, M. & Fischer, D. C. (1991). Inducible defenses in soybean against herbivorous insects. *Phytochemical Induction by Herbivores* (Ed. by D. W. Tallamy & M. J. Raupp), pp. 347–378. John Wiley & Sons, New York.

Kranz, J. (1990). Fungal diseases in multispecies plant communities. *New Phytologist*, **116**, 383–405.

Krause, S. C. & Raffa, K. F. (1992). Comparison of insect, fungal, and mechanically induced defoliation of larch: effects on plant productivity and subsequent host susceptibility. *Oecologia*, **90**, 411–416.

Lappalainen, J., Helander, M. L. & Palokangas, P. (1995). The performance of the autumnal moth is lower on trees infected by birch rust. *Mycological Research*, **99**, 994–996.

Last, F. T. (1960). Longevity of conidia of *Botrytis fabae* Sardiña. *Transactions of the British Mycological Society*, **43**, 673–680.

Leath, K. T. & Byers, R. A. (1977). Interaction of *Fusarium* root rot with pea aphid and potato leafhopper feeding on forage legumes. *Phytopathology*, **67**, 226–229.

Leath, K. T. & Newton, R. C. (1969). Interaction of a fungus gnat, *Bradysia* sp. (Sciaridae) with *Fusarium* spp. on alfalfa and red clover. *Phytopathology*, **59**, 257–258.

Lewis, A. C. (1979). Feeding preference for diseased and wilted sunflower in the grasshopper, *Melanoplus differentialis*. *Entomologia Experimentalis et Applicata*, **26**, 202–207.

Lewis, A. C. (1984). Plant quality and grasshopper feeding: effects of sunflower condition on preference and performance in *Melanoplus differentialis*. *Ecology*, **65**, 836–843.

Linhart, Y. B. (1991). Disease, parasitism and herbivory: multidimensional challenges in plant evolution. *Trends in Ecology and Evolution*, **6**, 392–396.

Marquis, R. J. & Alexander, H. M. (1992). Evolution of resistance and virulence in plant–herbivore and plant–pathogen interactions. *Trends in Ecology and Evolution*, **7**, 126–129.

Maun, M. A. & Cavers, P. B. (1971). Seed production and dormancy in *Rumex crispus*. I. The effects of removal of cauline leaves at anthesis. *Canadian Journal of Botany*, **49**, 1123–1130.

McGurl, B., Mukherjee, S., Kahn, M. & Ryan, C. A. (1995). Characterization of two proteinase inhibitor (ATI) cDNAs from alfalfa leaves (*Medicago sativa* var. Vernema): the expression of ATI genes in response to wounding and soil microorganisms. *Plant Molecular Biology*, **27**, 995–1001.

Moellenbeck, D. J., Quisenberry, S. S. & Colyer, P. D. (1992). *Fusarium* crown-rot development in alfalfa stressed by threecornered alfalfa hopper (Homoptera: Membracidae) feeding. *Journal of Economic Entomology*, **85**, 1442–1449.

Monaco, T. J. & Cumbo, E. L. (1972). Growth and development of curly dock and broadleaf dock. *Weed Science*, **20**, 64–67.

Nakamura, R. R., Mitchell-Olds, T., Manasse, R. S. & Lello, D. (1995). Seed predation, pathogen infection and life-history traits in *Brassica rapa*. *Oecologia*, **102**, 324–328.

Padgett, G. B., Russin, J. S., Snow, J. P., Boethel, D. J. & Berggren, G. T. (1994). Interactions among the soybean looper (Lepidoptera: Noctuidae), threecornered alfalfa hopper (Homoptera: Membracidae), stem canker, and red crown rot in soybean. *Journal of Entomological Science*, **29**, 110–119.

Parker, A. & Blakeman, J. P. (1984). Nutritional factors affecting the behaviour of *Uromyces viciae-fabae* uredospores on broad bean leaves. *Plant Pathology*, **33**, 71–80.

Paul, N. D. (1989). Effects of fungal pathogens on nitrogen, phosphorus and sulphur relations of individual plants and populations. *Nitrogen, Phosphorus and Sulphur Utilization by Fungi* (Ed. by L. Boddy, R. Marchant & D. J. Read), pp. 155–180. Cambridge University Press, Cambridge.

Pesel, E. & Poehling, H.-M. (1988). Zum Einfluss von abiotischen (Wassermangel) und biotischen (echter Mehltau, *Erysiphe graminis* f. sp. *tritici*) Stressfaktoren auf die Populationsentwicklung der getreideblattläuse *Metopolophium dirhodum* Walk. und *Sitobion avenae* F. *Mitteilungen der Deutschen Gesellschaft für Allgemeine und Angewandte Entomologie*, **6**, 531–536 (In German with English summary).

Piening, L. J. (1972). Effects of leaf rust on nitrate in rye. *Canadian Journal of Plant Science*, **52**, 842–843.

Powell, J. M. (1971). The arthropod fauna collected from the comandra blister rust, *Cronartium comandrae*, on lodgepole pine in Alberta. *Canadian Entomologist*, **103**, 908–918.

Pratt, R. G., Ellsbury, M. M., Barnett, O. W. & Knight, W. E. (1982). Interactions of bean yellow mosaic virus and an aphid vector with *Phytophthora* root diseases in arrowleaf clover. *Phytopathology*, **72**, 1189–1192.

Prüter, C. & Zebitz, C. P. W. (1991). Effects of *Aphis fabae* and *Uromyces viciae-favae* on the growth of a susceptible and an aphid resistant cultivar of *Vicia faba*. *Annals of Applied Biology*, **119**, 215–226.

Rabbinge, R., Drees, E. M., van der Graaf, M., Verberne, F. L. M. & Wesselo, A. (1981). Damage effects of cereal aphids in wheat. *Netherlands Journal of Plant Pathology*, **87**, 217–232.

Ramsell, J. & Paul, N. D. (1990). Preferential grazing by molluscs of plants infected by rust fungi. *Oikos*, **58**, 145–150.

Raupp, M. J. (1985). Effects of leaf toughness on mandibular wear of the leaf beetle, *Plagiodera versicolora. Ecological Entomology*, **10**, 73–79.

Reddy, M. N. & Rao, A. S. (1976). Changes in the composition of free and protein amino acids in groundnut leaves induced by infection with *Puccinia arachidis* Speg. *Acta Phytopathologica, Academiae Scientiarum Hungaricae*, **11**, 167–172.

Roberts, A. M. & Walters, D. R. (1989). Shoot:root interrelationships in leeks infected with the rust, *Puccinia allii* Rud.: growth and nutrient relations. *New Phytologist*, **111**, 223–228.

Roy, B. A. (1993). Floral mimicry by a plant pathogen. *Nature*, **362**, 56–58.

Ryals, J., Uknes, S. & Ward, E. (1994). Systemic acquired resistance. *Plant Physiology*, **104**, 1109–1112.

Savopoulou-Soultani, M. & Tzanakakis, M. E. (1988). Development of *Lobesia botrana* (Lepidoptera: Tortricidae) on grapes and apples infected with the fungus *Botrytis cinerea. Environmental Entomology*, **17**, 1–6.

Shivas, R. G. & Scott, J. K. (1993). Effect of the stem blight pathogen, *Phomopsis emicis*, and the weevil, *Perapion antiquum*, on the weed *Emex australis. Annals of Applied Biology*, **122**, 617–622.

Sipos, L. & Sági, G. (1987). Tavaszi árpán károsító gabonalevéltetű (*Macrosiphum avenae*) és lisztharmatgomba (*Erysiphe graminis* f. sp. *hordei*) kölcsönhatásának vizsgálata üvegházban. *Növénytermelés*, **36**, 31–34 (In Hungarian with English summary).

Skipp, R. A. & Lambert, M. G. (1984). Damage to white clover foliage in grazed pastures caused by fungi and other organisms. *New Zealand Journal of Agricultural Research*, **27**, 313–320.

Stein, B. D., Klomparens, K. L. & Hammerschmidt, R. (1993). Histochemistry and ultrastructure of the induced resistance response of cucumber plants to *Colletotrichum lagenarium. Journal of Phytopathology*, **137**, 177–188.

Stout, M. J., Workman, J. & Duffey, S. S. (1994). Differential induction of tomato foliar proteins by arthropod herbivores. *Journal of Chemical Ecology*, **20**, 2575–2594.

Stutz, J. C., Leath, K. T. & Kendall, W. A. (1985). Wound-related modifications of penetration, development, and root rot by *Fusarium roseum* in forage legumes. *Phytopathology*, **75**, 920–924.

Summers, C. G. & Gilchrist, D. G. (1991). Temporal changes in forage alfalfa associated with insect and disease stress. *Journal of Economic Entomology*, **84**, 1353–1363.

Weaver, S. E. & Cavers, P. B. (1979). The effects of date of emergence and emergence order on seedling survival rates in *Rumex crispus* and *R. obtusifolius. Canadian Journal of Botany*, **57**, 730–738.

Weaver, S. E. & Cavers, P. B. (1980). Reproductive effort of two perennial weed species in different habitats. *Journal of Applied Ecology*, **17**, 505–513.

Wennström, A. & Ericson, L. (1991). Variation in disease incidence in grazed and ungrazed sites for the system *Pulsatilla pratensis–Puccinia pulsatillae. Oikos*, **60**, 35–39.

Westphal, E., Dreger, F. & Bronner, R. (1991). Induced resistance in *Solanum dulcamara* triggered by the gall mite *Aceria cladophthirus. Experimental and Applied Acarology*, **12**, 111–118.

Wilding, N., Collins, N. M., Hammond, P. M. & Webber, J. F. (Eds) (1989). *Insect–Fungus Interactions.* Academic Press, London.

Zebitz, C. P. W. & Kehlenbeck, H. (1991). Performance of *Aphis fabae* on chocolate spot disease-infected faba bean plants. *Phytoparasitica*, **19**, 113–119.

9. FUNGAL ENDOPHYTES, HERBIVORES AND THE STRUCTURE OF GRASSLAND COMMUNITIES

K. CLAY

Department of Biology, Indiana University, Bloomington, Indiana 47405, USA

INTRODUCTION

Plant species typically engage in a simultaneous suite of interspecific interactions involving competition, antagonism and mutualism. It is well documented that herbivores can influence the structure and dynamics of plant communities by differentially inflicting damage upon interacting species (Crawley 1983, Louda *et al.* 1990). Microbial parasites may also influence plant interactions in a similar manner (Paul & Ayres 1986; Kelley & Clay 1987). Less well studied, but perhaps equally important, are the interactions between microbial parasites of plants, in this case fungi, and herbivores of those same plants. The outcome of these interactions may in turn influence the outcome of competition and the dynamics of plant communities. Herbivores may predispose plants to microbial infections by creating wounds through which colonization occurs (Carter 1973); they may serve as vectors of infective propagules (Power 1991); and they may be differentially attracted to or repulsed by infected plants (Clay 1988a; Ramsell & Paul 1990). The objective of this chapter is to consider how the multitrophic interaction between grasses, fungal endophytes and herbivores may influence the dynamics of vegetation in grassland communities. A companion paper (Prestidge, this volume) will focus specifically on agricultural communities.

First I briefly review the basic biology of fungal endophytes of grasses and then consider how endophyte infection affects plant–herbivore and plant–plant interactions. Then I explore how these multitrophic interactions may affect the structure and dynamics of grassland communities. While my focus is on natural communities, most published information comes from two important agricultural grasses, *Festuca arundinacea* and *Lolium perenne*. Grass/endophyte interactions are highly relevant to our understanding of the structure and dynamics of many terrestrial communities given the widespread distribution and ecological importance of grasses. Fungal endophytes represent a model system for understanding how symbiotic micro-organisms can have important effects cascading upwards through the entire community.

FUNGAL ENDOPHYTES OF GRASSES

Many grasses are infected by endophytic fungi in the family Clavicipitaceae (Ascomycotina). A large number of hosts and fungal taxa have been described (White *et al.* 1993). Three features of these fungi are of special interest. First, the fungi are typically systemic and perennial in host plants. Partial and complete losses of infection can occur (DeBattista *et al.* 1990; Fowler & Clay 1995) but the same plants, and clones of the same plant, are usually infected year after year (Diehl 1950; Harberd 1961; Clay 1990a). Intercellular hyphae occur in above-ground leaves, stems, and reproductive organs of their hosts, and can easily be visualized microscopically (Bacon *et al.* 1986, Clay 1988a). The endophytes produce no haustoria and apparently subsist on plant materials secreted into the apoplast.

Second, endophytes can spread contagiously from plant to plant; they can spread vertically from infected plants to their seed progeny; or they can do both (Clay 1988b, White 1988). The fungal fruiting body (stroma) develop coincident with flowering and cause various deformations of leaves and inflorescences, often resulting in host sterility (Clay 1990a). The reproductive potential of the host population is reduced as a result of endophyte infection, although host plants are vegetatively vigorous. The sexual fungus *Epichloë* has also spawned a complex of asexual forms classified in the genus *Acremonium* (White *et al.* 1993). Unlike *Epichloë*, *Acremonium* does not sporulate in nature, infections are asymptomatic, and transmission occurs maternally through the seed of infected plants (Sampson 1933; Clay 1994). The close phylogenetic relationship of *Epichloë* and *Acremonium* has been confirmed by molecular techniques (Schardl *et al.* 1991) and is emphasized by the co-occurrence in one population of asymptomatic plants bearing infected seeds, sterilized plants bearing fungal fruiting bodies and plants with a mixture of stroma-bearing and seed-bearing inflorescences (White & Chambless 1991, Leuchtmann & Clay 1993).

Third, there is a strong correlation between fungal taxonomy and host taxonomy and life history. Seed-transmitted endophytes in the genus *Acremonium* or *Epichloë* are limited to cool-season grasses in the subfamily Pooideae, which dominate temperate areas (Clay 1990a). Seed transmission also occurs in the related fungus *Atkinsonella hypoxylon*, which infects *Danthonia* grasses (subfamily Arundinoideae) in eastern North America (Clay 1994). Seed-transmitted endophytes are significantly more common in clump grasses and annual grasses than in rhizomatous grasses, while the reverse is true for endophytes that spread contagiously (Clay 1988b). Warm-season grasses with more tropical distributions are infected primarily by the clavicipitaceous genus *Balansia* where only contagious transmission by spores is known to occur. The asymptomatic, seed-transmitted endophytes more clearly reveal the dynamics of multitrophic level interactions because host infection frequencies are directly dependent on the relative fitness of infected vs. uninfected plants.

FREQUENCY OF INFECTION IN COMMUNITIES

The significance of endophyte infection in grasses is a function of both the frequency of infection within host species and the frequency of host species within a multispecies community. Relatively few data are available on the prevalence of endophyte-infected species in different communities. A summary of the results of six studies is presented in Table 9.1. In midwestern US forest communities, Swiss forests and British grasslands, many of the grass hosts were infected at essentially 100% frequencies (Clay & Leuchtmann 1989; Leuchtmann 1992; White & Baldwin 1992). Anecdotal observations have indicated that four to ten infected grass species can often be found co-occurring at a single site in southeastern US grasslands and in montane habitats in North Carolina (Clay, personal observation). In his monograph, Diehl (1950) also refers to the common co-occurrence of multiple endophyte hosts in the same site. Thus, endophyte-infected grasses occur in many communities and are often abundant. Reports of individual hosts indicate that clavicipitaceous endophytes of grasses have a world-wide distribution (Diehl 1950; Latch *et al.* 1987). However, very few natural grassland communities have been censused and this represents an important area for future research.

More data are available on frequencies of infection within single host species. The frequency of infection of host populations by stroma-forming fungi is easily enumerated given the visible manifestations of infection but asymptomatic infections of grasses by endophytes have not been examined until recently. Published accounts of infection frequencies in samples of selected US grasses (and sedges) are presented in Table 9.2. With few exceptions, the species are native and are not found in managed or agricultural communities. This list represents only a small fraction of known endophyte hosts in the US and does not include reports of infection frequencies in other parts of the world (see Large 1952; Bradshaw 1959; Lewis

TABLE 9.1. Estimates of the frequency of endophyte-infected grass species in multispecies communities.

Community-type/location	Grass species examined	Species infected (%)	Reference
Deciduous forest/midwestern USA	26	16 (62)	1
Beech forest/Switzerland	14	8 (57)	2
Limestone grassland/Switzerland	12	2 (17)	2
Interior Alaska/USA	14	1 (7)	3
Successional fields/southeastern UK	12	5 (42)	4
Grasslands/west central UK	44	10 (23)	5

References: 1, Clay & Leuchtmann 1989; 2, Leuchtmann 1992; 3, Halisky *et al.* 1990; 4, Clay & Brown, unpublished data; 5, White & Baldwin 1992.

TABLE 9.2. Frequency of infection in grasses in the USA. Evidence for infection comes from microscopic examination of hyphae in leaves (H), seeds (S), or from field censuses (F). Asterisk indicates data obtained from herbarium sheets. Endophytes include *Acremonium* taxa (Ac), *Atkinsonella hypoxylon* (Ah), *Atkinsonella texensis* (At), *Balansia cyperi* (Bc), *Balansia henningsiana* (Bh) and *Epichloë typhina* (Et). Estimates from different sites or different host populations are available from some host species.

Grass	Evidence	Frequency %	Fungus	Reference
Agrostis heimalis	H	> 75	Et	1
A. heimalis	H	86–97	Et	9
A. heimalis	H	45*	Ac	12
A. perennans	H	> 75	Et	1
A. perennans	H	60*	Ac	12
Ammophilia breviligulata	H	56	Ac	13
Brachyelytrum erectum	S	26	Et	1
Bromus anomalus	H	18*	Ac	12
Cinna arundinacea	H	71*	Ac	12
C. latifolia	H	46*	Ac	12
Cyperus virens	F	16–81	Bc	5
Danthonia spicata	F	15	Ah	4
Elymus canadensis	H	61*	Ac	12
E. hystrix	H	27	Et	3
E. villosus	H	22, 43	Et	2
E. virginicus	H	44, 45	Et	2
E. virginicus	H	47	Ac	12
Festuca arizonica	H	100*	Ac	10
F. eastwoodea	H	40*	Ac	10
F. ligulata	H	100*	Ac	10
F. obtusa	H	100	Ac	1
F. obtusa	H	100*	Ac	10
F. pacifica	H	67*	Ac	10
F. paradoxa	H	100*	Ac	10
F. scabrella	H	100*	Ac	10
F. subulata	H	67*	Ac	10
F. versuta	H	100*	Ac	10
Glyceria striata	F	24	Et	3
Panicum agrostoides	F	26–62	Bh	7
Poa alsodes	H	100	Ac	8
P. autumnalis	H	100	Et	3
P. autumnalis	H	100*	Ac	12
P. languida	H	100	Ac	8
P. palustris	H	53*	Ac	12
P. paucispicula	H	100*	Ac	12
P. sylvestris	H	96	Ac	3
P. wolfii	H	100	Ac	1
Sitanion longifolium	H	85*	Ac	12
Sphenopholis nitida	F	15	Et	3
S. nitida	H	> 10, < 50	Et	1

TABLE 9.2. *Continued.*

Grass	Evidence	Frequency %	Fungus	Reference
S. pallens	H	>10, <50	Et	1
Stipa eminens	H	100*	Ac	12
S. lobata	H	35*	Ac	12
S. leucotricha	F	5–66	At	6
S. robusta	H	82	Ac	8
S. robusta	H	100*	Ac	12
S. viridula	H	22*	Ac	12

References: 1, Clay & Leuchtmann 1989; 2, Leuchtmann & Clay 1993; 3, Bier & Clay, unpublished data; 4, Clay 1994; 5, Clay 1986; 6, Fowler & Clay 1995; 7, Clay *et al.* 1989; 8, Clay, unpublished data; 9, White & Chambless 1991; 10, White & Cole 1985; 11, Shelby & Dalrymple 1987; 12, White 1987; 13, White *et al.* 1992.

and Clements 1986; Latch *et al.* 1987; Francis & Baird 1989; Wilson *et al.* 1991). Infection frequencies of tall fescue (*Festuca arundinacea*) and perennial ryegrass (*Lolium perenne*), the two most important endophyte-infected pasture grasses, are especially well studied in the US and New Zealand; for example, recent estimates suggest that tall fescue covers 15 million ha in the eastern US and that 80% is infected (Coley *et al.* 1995). However, the grasses are introduced and infection frequencies in sown communities may reflect the initial composition of the seed.

Several points deserve comment. First, there are many grasses where infection occurs at 100% frequency, implying a selective advantage for infection. Second, in most other cases infection rates are substantial, ranging around 50%. Finally, infection rates are generally higher in grasses infected by asymptomatic, seed-borne endophytes. This is not surprising given that hosts infected by stromata-forming endophytes suffer from reduced fertility or complete sterility (Clay 1988b; White 1988). Strictly vertically transmitted symbionts are expected to be more mutualistic than related symbionts that spread contagiously (Bull, *et al.* 1991) and this will be reflected in higher infection rates within host populations.

EFFECTS ON HERBIVORES

Clavicipitaceous endophytes of grasses have gained their notoriety primarily from their toxic effects on herbivores. Laboratory and field studies have shown that herbivores are frequently deterred from feeding on infected plants or exhibit reduced survival and growth (Clay 1991; Latch 1993). Given that many endophytes are seed transmitted, seed predation may also be affected (Madej & Clay 1991; Knoch, *et al.* 1993). The number of herbivores so affected is large, as is the number of infected grasses, suggesting that protection from herbivory may be an important ecological force in natural, as well as agricultural, communities. Acquired chemical defences of grasses resulting from endophyte infection can be viewed as a type of symbiotic pesticide providing a selective advantage under herbivore pressure.

Fungal alkaloids present in host-plant tissues provide a basis for anti-herbivore effects (Clay 1988a). Like related ergot fungi (*Claviceps* spp.), systemic endophytes produce a range of ergot and other alkaloids (Dahlman *et al.* 1991; Porter 1994). Evidence that the alkaloids result from the fungi and not the plant include their production in pure culture, their absence in uninfected plants, and maternal inheritance of toxicity in grasses, consistent with seed transmission (Funk *et al.* 1983; Porter 1994). Feeding trials incorporating pure alkaloids into herbivore diets also mimic the effects of infected grasses (Rowan *et al.* 1986). Alkaloid concentrations can be very high. Toxic sleepy grass (*Stipa robusta*) from the southwestern US, which is infected by an *Acremonium* endophyte, contains up to 28 mg kg^{-1} of the ergot alkaloids lysergic and isolysergic acid amides (Petroski *et al.* 1992).

Two consequences of endophyte alkaloids relevant to the dynamics of endophyte-infected grasses in natural communities are differential damage to infected vs. uninfected plants and effects on herbivore population sizes. Both would provide an advantage to infected plants and to enhance their dominance of multispecies assemblages. Most research on host-plant selection comes from the agriculturally important tall fescue and perennial ryegrass. Latch (1993) and Rowan and Latch (1994) list 38 insect species deterred from feeding on these two species. A range of vertebrates, from cattle to sparrows, also discriminate against infected plants (or seeds) (Bacon & Siegel 1988; Clay 1991; Madej & Clay 1991; Van Santen 1992). It seems likely that herbivore deterrence is widespread given that alkaloid production by endophytes is widespread (Porter 1994). Data from other grasses support this view; for example, fall armyworm (*Spodoptera frugiperda*) discrimination between infected vs. uninfected red fescue (*Festuca rubra*) was stronger than for either tall fescue or perennial ryegrass (Clay, *et al.* 1993). An old literature exists on livestock avoidance of notoriously toxic wild grasses now known to be endophyte infected (Shaw 1873; Hance 1876; Bailey 1903; Bor 1960). Bailey (1903) stated 'that after one good dose of sleepy grass, horses will never touch it again'. Exceptions to the general pattern of herbivore avoidance occur (see Kirfman *et al.* 1986; Lewis & Clements 1986; Latch 1993) but the majority of published studies have found differences in host-plant selection.

Avoidance of, or reduced performance on, infected grasses should ultimately result in lower herbivore population sizes, and less damage to plants. A large body of data has shown significantly impaired performance of herbivores (e.g. reduced survival, growth rates and/or fecundity) in feeding studies (Bacon & Siegel 1988; Clay 1988a; Latch 1993). Again most research is from tall fescue and perennial ryegrass. Poisoning of domestic livestock on infected tall fescue and perennial ryegrass pastures is a major agricultural problem (see Prestidge, this volume).

Mammalian herbivores with reduced performance on infected grasses include cows, horses, oxen, sheep, deer, goats, rabbits, rats and mice (Clay 1991). Recent research has shown that pastures of endophyte-infected tall fescue contain significantly lower populations of non-domesticated small mammals than comparable endophyte-free pastures (Giuliana *et al.* 1994; Coley *et al.* 1995) and several granivorous birds exhibit increased weight losses when confined to diets containing infected (vs. uninfected) tall fescue seed (Madej & Clay 1991). A wide variety of insects and nematodes also exhibit reduced population sizes and/or reduced performance on infected grasses (Prestidge *et al.* 1982; Funk *et al.* 1983; Kirfman *et al.* 1986; Cheplick & Clay 1988; Kimmons *et al.* 1990; Clay 1991; Latch 1993). Over 20 wild grasses have also been shown to have detrimental effects on insects, with many exhibiting more dramatic effects than tall fescue or perennial ryegrass (Cheplick & Clay 1988; Siegel *et al.* 1990; Clay 1991). A number of wild grasses also result in mammalian toxicity (Bailey 1903; Nobindro 1934; Bor 1960; Bacon *et al.* 1986), suggesting that tall fescue and perennial ryegrass are not atypical endophyte hosts. More data, including negative data, are needed from a wider variety of host species.

EFFECTS OF COMPETITIVE INTERACTIONS AMONG PLANTS

A large body of data indicates that infected host plants are vegetatively vigorous, regardless of alterations in their reproductive biology. This vigour is evidenced by greater survival and growth rates in both natural populations and in controlled environments (Harberd 1961; Clay 1994; Clay *et al.* 1989; Rice *et al.* 1990; Hill *et al.* 1991). The best-studied endophyte host is tall fescue. This introduced grass is planted widely in pastures, roadsides and turf, but also is common in a wide range of unmanaged habitats in the US (Bacon & Siegel 1988). Table 9.3 summarizes numerous studies comparing the growth of infected vs. uninfected plants. This interaction may vary with life stage and resource levels; we have shown that infected tall fescue seedlings grow significantly slower than uninfected seedlings under low nutrient conditions (Cheplick *et al.* 1989). Experiments, field samples and anecdotal observations of many other grasses suggest that growth enhancement is a common characteristic of endophyte hosts (Bradshaw 1959; Clay 1986; Clay *et al.* 1989), although exceptions exist (see Clay 1990b, Lewis & Clements 1990). Endophyte infections of grasses therefore resemble other mutualistic plant–microbe symbioses, such as mycorrhizas and root nodules, that enhance plant growth. Like those symbioses, the benefit of endophyte infection may vary with environmental conditions.

Different types of fungal infections have been shown to both positively and negatively affect plant competitive ability in many species (Fitter 1977; Paul &

TABLE 9.3. Relative productivity of endophyte infected tall fescue (where uninfected = 1) in a variety of studies.

Reference	Relative productivity	Variable measured	Environment
Read & Camp 1986	1.55	Available forage	Field
Clay 1987	1.20	Tiller number	Greenhouse
Clay, 1987	1.24	Herbage	Greenhouse
Clay 1990a	1.45	Tiller number	Field
Arachevaleta et al. 1989	1.54	Herbage	Greenhouse
De Battista et al. 1990	1.18	Herbage	Greenhouse
Hill et al. 1990	1.05	Herbage	Greenhouse
Hill et al. 1990	0.91	Tiller number	Greenhouse
Rice et al. 1990	1.41	Seed number	Field
Hill et al. 1991	1.29	Herbage	Field
Hill et al. 1991	1.26	Seed weight	Field
Clay et al. 1993	1.22	Herbage	Greenhouse
Clay et al. 1993	1.42	Herbage	" w/herbivory
Bouton et al. 1993	1.26	Herbage	Field
Bouton et al. 1993	0.99	Herbage	Field
West et al. 1993	1.26	Herbage	Field
West et al. 1993	1.18	Tiller density	Field

Ayres 1986). This has important implications in that the structure and dynamics of plant communities may be conditional upon the presence or absence of fungal infection. We have conducted several competition experiments comparing the performance of infected vs. uninfected plants. In a field experiment with *Danthonia spicata* and *Anthoxanthum odoratum*, plants of the former species infected with *Atkinsonella hypoxylon* significantly outperformed uninfected plants when grown in competition with the latter species (Kelley & Clay 1987). In a greenhouse experiment, endophyte-infection enhanced competitive ability of tall fescue seedlings in nine of ten experimental combinations; however, infection suppressed competitive ability of perennial ryegrass seedlings (*Lolium perenne*) in seven of ten combinations (Marks *et al.* 1991). Another greenhouse experiment examined intra-specific competition in tall fescue under two density and two moisture conditions. There were highly significant effects of all three factors and infection resulted in greater productivity of infected plants in all eight experimental combinations but, unexpectedly, the competitive advantage of infected plants was greater under high moisture conditions (Clay 1993). Past research has suggested that the relative advantage of infected tall fescue is enhanced by drought stress (West 1994), as may also be the case with perennial ryegrass (Cunningham *et al.* 1993). In another greenhouse experiment with three endophyte-infected species, there was a general enhancement of performance with infection but it was herbivore dependent (see below; Clay *et al.* 1993). New Zealand studies of perennial ryegrass indicated that endophyte-infected plants competitively suppressed clover (*Trifolium repens*) to a significantly greater extend than uninfected plants (Sutherland & Hoglund 1990) although Prestidge *et al.* (1992) and Lewis and Clements (1990) did not find any effect of infection of ryegrass on clover growth. Competitive effects of endophyte infection may vary with environmental conditions or the genetic background of host plants.

INTERACTIONS BETWEEN ENDOPHYTE INFECTION, COMPETITION AND HERBIVORY

Competition among plants does not occur independently of herbivory in most plant populations yet the interaction of these two processes is less well studied than their direct effects. Differential herbivory may provide a competitive advantage to the least-damaged plants (or species) growing in competition with more damaged plants (Crawley 1983; Louda *et al.* 1990). A plant's competitive environment may also affect the consequences of herbivory (Bentley & Whittaker 1979). The importance of the interaction between competition and herbivory is illustrated by experiments where significant changes in plant composition occur where herbivores are excluded by the construction of physical exclosures (Watt 1960) or by applications of chemical pesticides (Brown & Gange 1989). The interaction between competition and herbivory may be especially pronounced where endophyte-infected and uninfected grasses grow in mixtures.

Several studies suggest an interaction among endophyte infection, competition and herbivory. Bradshaw (1959) found that frequency of infection of *Agrostis* species by *Epichloë typhina* was often very high in heavily grazed pastures in Britain, where infected plants form large, spreading clones. He suggested that morphological changes induced by infection were responsible for these patterns but the anti-herbivore effects of endophyte infection represent another possible explanation. Watson *et al.* (1993) also described an interaction among endophyte infection of perennial ryegrass, competition with clover and invertebrate herbivory. In uninfected pastures yields of clover were higher due to herbivore damage to ryegrass while, in contrast, in infected pastures ryegrass suffered less herbivore damage and yields of clover were depressed.

We have conducted several competition experiments in controlled environments where different combinations of infected and uninfected grasses have been grown in pots in DeWit-type designs with or without insect herbivory by the fall armyworm (*Spodoptera frugiperda*). In a greenhouse experiment where tall fescue seedlings were grown at two densities and three mixtures, infected fescue out-yielded un-infected fescue under all conditions (Fig. 9.1). At low density (16 plants per pot) herbivory resulted in a near defoliation of all plants, greatly reducing biomass compared to no herbivory treatments (Fig. 9.1a) but did not enhance the relative

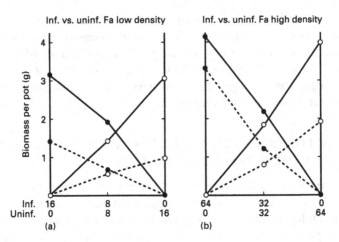

FIG. 9.1. Intra-specific competition between infected and uninfected tall fescue (*Festuca arundinacea*, Fa) with (dotted lines) or without (solid lines) herbivory at (a) low and (b) high densities. Seedlings were grown in 12.5-cm pots at a density of 16 or 64 per pot. There were five replicate pots of pure stands for each density/herbivory combination and 10 replicate pots for mixtures for each density/herbivory combination. Above-ground biomass was harvested after 12 weeks.

advantage of infected plants in intraspecific combination. At high density (64 plants per pot) there was a proportionally greater decline in yield of uninfected plants with herbivory but the difference was small (Fig. 9.1b). Overall, there were highly significant effects of herbivory, density, planting combination and herbivory by density interaction, but not of other interactions.

In another series of greenhouse experiments the effects of endophytes infection and herbivory on intra- and inter-specific competition were examined in three endophyte-infected grass species (Clay *et al.* 1993). Intra-specific competition experiments were analysed with two-way ANOVA where herbivory and planting combination were the two main effects. We expected that the relative advantage of infected plants in mixtures would be increased by herbivory. For perennial ryegrass, there were significant main effects but no interaction while in tall fescue there was only a significant effect of planting combination. However, in red fescue there were significant effects of herbivory and the interaction of herbivory with combination (Clay *et al.* 1993). Uninfected plants outyielded infected plants in pure stands in the absence of herbivory, indicating a cost of infection, but the relationship was reversed when herbivores were present. Three inter-specific competition experiments were also conducted with perennial ryegrass and *Poa pratensis*, tall fescue and *Dactylis glomerata*, and red fescue and perennial ryegrass. These species commonly co-occur in a wide range of agricultural and natural habitats. There were highly significant interactions between herbivory and planting combinations

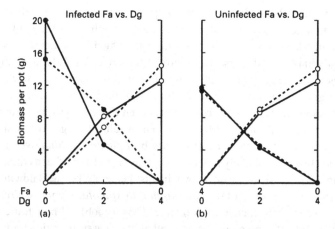

FIG. 9.2. Inter-specific competition between (a) infected or (b) uninfected tall fescue (*Festuca arundinacea*, Fa) with *Dactylis glomerata* (Dg), with (dotted lines) or without (solid lines) herbivory. Plants were grown in 15-cm pots at a density of four per pot. There were 40 replicate pots of the pure stands for each herbivory treatment and 80 replicate pots of mixtures for each treatment. Above-ground biomass was harvested after 20 weeks.

in all three pairs of grasses, indicating that herbivory effects varied with the endophyte status of perennial ryegrass and the two fescues (Clay *et al.* 1993). The results of one combination are illustrated in Fig. 9.2. With uninfected tall fescue, *D. glomerata* was the superior competitor and herbivory did not change that result (Fig. 9.2). With infected fescue *D. glomerata* was again the superior competitor in herbivore-free treatments but with herbivores tall fescue was the superior competitor in mixtures. These results suggest that herbivory can reverse the competitive hierarchy of species where some are endophyte infected.

There is a need for additional studies where both endophyte infection and levels of herbivory are controlled, especially in field environments. Endophyte infection can be controlled by sowing seeds of known infection frequency, and herbivory can be controlled by fencing and/or chemical applications. The increasing utilization of endophyte-infected grasses in managed habitats offers many opportunities, but herbivory is generally not manipulated experimentally. However, there are many situations where grazing is controlled by fences where the same set of grasses occur inside and outside of the fence. Comparisons of infection levels in historically grazed vs. ungrazed areas would prove informative. Insect herbivory would not necessarily differ, however, requiring more intensive approaches.

CHANGES IN INFECTION FREQUENCY OVER TIME

Increasing frequency of infection in host populations over time is consistent with the hypothesis that endophyte-infected grasses are at a selective advantage in mixed communities. For purely seed-transmitted endophytes, fitness differences of hosts vs. non-hosts are the only mechanism by which frequencies can change within populations. Seed dispersal from surrounding areas could also alter infection frequencies if seedling recruitment occurs. In contrast, increasing infection frequencies in grass populations where the fungi sporulate on host plants may also reflect contagious spread to uninfected plants even if the fungus has detrimental effects on host fitness (Lenski & May 1994). The results of Large (1952), showing a rapid increase in infection of *D. glomerata* by *Epichloë typhina* over a 7-year period in seed fields, most likely reflect contagious spread as neither asymptomatic infection nor seed transmission are known in this species (White & Baldwin 1992). Infection rates in marked cohorts of *Danthonia spicata*, *Cyperus virens* and *Sporobolus poiretii* also increased over time (Clay 1990b). Although the fungal species are capable of contagious spread in all of these cases, infection frequencies increased due to the higher mortality of uninfected plants.

Increases in infection rates of purely seed-transmitted endophytes have been quantified in several turf and pasture grass species by monitoring relative survival rates within cohorts, comparing infection rates in populations of different ages, and by sampling infection levels within a single population over time. In one study, a cohort of tall fescue initially having a 50% infection rate increased to 64% infection after 3 years (Clay 1990b). Read and Walker (1990) established experimental plots of infected and uninfected plants of three cultivars of tall fescue. From 1984 to 1989 one cultivar increased from 7% infection to 28% while another increased from 60% to 73%. In cultivar Ky-31, infected plots increased from 43% to 83% while low endophyte plots when from 5% infection frequency to 23% over the 6-year period. In an earlier study (Read & Camp 1986) infected plots in three pastures lost an average of 4% cover during a summer drought while uninfected plots lost 54% cover in the same pastures, leading to greatly increased infection frequencies over a single season. Bouton *et al.* (1993) also compared infection frequencies in infected and uninfected tall fescue plots at three sites over 3 years in Georgia, USA. Infected plots maintained high infection frequencies over the period in two sites while uninfected plots at one site climbed from < 1% to 11% infection but remained low at the other site. In the third site all uninfected plants died over the first growing season while infected plants maintained high survival and cover. In another study Thompson *et al.* (1989) compared infection frequencies in a series of pastures established with 0%, 15%, 30%, 45%, 60% and 75% infected seed over a 2-year period. All plots showed at least a slight increase in infection and the 15% and 30% pastures increased to 60% infection. All of these studies reported shorter-term fluctuations in infection frequency. It is not clear whether infection frequencies actually fluctuate this rapidly or whether the accuracy of sampling techniques varies with the stage of plant growth.

Considerable data are also available for perennial ryegrass. Lewis and Clements (1986) found infection occurred at 14 of 52 pastures surveyed in the UK and 12 of these were at least 15 years old. Ryegrass has been established in the UK for at least several hundred years. Guy (cited in Cunningham *et al.* 1993) surveyed 30 ryegrass pastures in Tasmania ranging in age from 1 to 25 years of age and found a positive linear correlation between pasture age and infection level with pastures 7 years old or older having infection frequencies ≥ 83%. In New Zealand, Francis and Baird (1989) documented rapid increases in the frequency of infection in recently sown ryegrass pastures while Cunningham *et al.* (1993) reported that in Australia two cultivars planted in 1984 with 78% and 80% infection frequencies had increased to 100% 4 years later. They also provided data on seedling establishment of three cultivars with both high-infection (66–80% infection frequency) and endophyte-free forms. After 6 weeks from sowing the infected forms had 1.51, 1.53 and 1.30 times higher seedling density compared to endophyte-free seedlings. Funk *et al.* (1983), in the US, compared loss of cover in turfgrass plots under extreme insect pressure over one summer season and found that plots maintaining 87–98% cover had infection levels ≥ 88% while plots with covers between 12% and 30% were essentially endophyte free.

Increases in infection frequency of other grasses infected by seed-transmitted endophytes have also been noted. Saha *et al.* (1987) found that hard fescue (*F. longifolia*) turves changed from 48% infection frequency to 84% over 7 years while red fescue turves increased from 94% to 97% over the same period. Several anecdotal reports of increasing frequency of infection in grasses over time, or greater prevalence in older sites, also exist (Diehl 1950; Bradshaw 1959; Harberd 1961; Bacon *et al.* 1986). The large number of native grasses with 100% infection frequency (Table 9.2) implies that at some point in the history of the species infection frequencies increase to fixation, but there have been no empirical studies censusing populations over time.

The data reviewed here suggest that infection frequencies tend to increase within host populations that have been examined. There are no published studies to my knowledge documenting significant long-term declines in infection frequencies. Increases in infection frequencies may occur very rapidly during seedling establishment or over a single season (Francis & Baird 1989; Bouton *et al.* 1993) or more gradually (Clay 1990a; Read & Walker 1990). Increases in infection frequency often are related to biotic or abiotic stresses, although certain stresses might decrease infection frequencies (Cheplick *et al.* 1989). In systems lacking contagious spread, the dynamics of endophyte infection mimic those of a selectively advantageous, maternally inherited gene (Clay 1993). Indeed, the presence of purely seed-transmitted endophytes is in itself evidence of their selective benefit since detrimental, seed-transmitted endophytes could not indefinitely persist in host populations (Bull *et al.* 1991). Environmental heterogeneity causing variable selection could cause changes in infection frequencies in both directions, but such studies have not yet been conducted.

MULTITROPHIC INTERACTIONS AMONG GRASSES, FUNGAL ENDOPHYTES AND HERBIVORES

The results reviewed here suggest that multitrophic interactions between grasses, endophytes and herbivores can result in an increase in the dominance of infected grasses within species and communities. Three specific processes are involved (Fig. 9.3a–c). First, the frequency of infection within host grass species increases over time and this increase will be accelerated under increasing herbivore pressure (Fig. 9.3a). Second, with increasing infection frequencies the importance of host species will increase in the community as a result of their increased resistance to herbivory and enhanced competitive ability (Fig. 9.3b). Third, the richness and importance of other species in the community will decline, resulting in lower diversity and dominance by one or a few endophyte infected species (Fig. 9.3c).

While individual components of this simple model are well documented in tall fescue, perennial ryegrass and other species, there have not been any comprehensive field experiments at the community level. However, several studies and anecdotal observations support the processes outlined in Fig. 9.3. For example, Shaw (1873) described how the introduction of sheep and cattle into South Africa resulted in

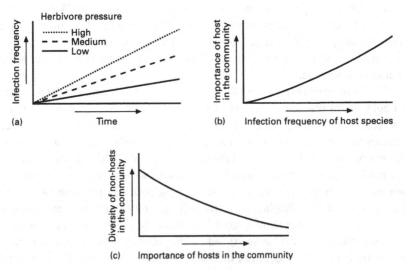

FIG. 9.3. Proposed relationships between infection frequency and (a) time (b) host dominance and (c) host dominance and non-host diversity.

large changes in vegetation and increasing dominance by the unpalatable drunk grass (*Melica decumbens*), which is endophyte infected (White 1987). Several *Stipa* species show similar patterns. Bor (1960) reported that *S. sibirica* forms almost pure stands in pine forests of Kashmir where cattle will not eat it. Hance (1876) described *S. inebrians* from the same area and stated that the grass was extremely common around Gulmuz, Kashmir where cattle also would not eat it. Naive horses, however, would eat the grass and die from the effects unless quickly treated. Bailey (1903) reported that another *Stipa*, sleepy grass (*S. robusta*), occurs in abundance in the Sacramento Mountains of New Mexico; despite many free-ranging cattle and horses which closely cropped other grasses, the sleepy grass was untouched. All of these species are infected by asymptomatic *Acremonium* endophytes at high, or 100%, frequencies (White 1987; Petroski *et al.* 1992; Bruehl *et al.* 1994). Bacon *et al.* (1986) also report that the unpalatable smut grass (*Sporobolus poiretii*) often dominates overgrazed pastures in the southeastern US where up to 60% of plants are infected by *Balansia epichloë*.

Several of the studies cited earlier also are consistent with the idea that endophyte-infected species exhibit a proportional increase in grassland communities. Harberd (1961) found that largest and most widespread red fescue clones in British pastures were endophyte infected. The greater cover of infected perennial ryegrass and tall fescue plots reported by Funk *et al.* (1983) and Read and Camp (1986) resulted in a reduced weed flora, while endophyte infection and herbivore pressure on ryegrass in New Zealand resulted in reduced clover cover (Watson *et al.* 1993).

Experiments where the entire plant community is quantified, and where herbivory and infection are manipulated, are needed. At a minimum, the establishment of replicated plots differing in the level of endophyte infection in one or more species is required. Herbivory can be manipulated by fencing or insecticide applications or some combination of both. The frequency of infection within host species and the relative importance of all plant species in the community should then be followed over several years. Ideally, experiments should be conducted with a range of host grasses at a range of sites. However, data from these types of experiments are not yet available.

REFERENCES

Arachevaleta, M., Bacon, C. W., Hoveland, C. S. & Radcliffe, D. E. (1989). Effect of tall fescue endophyte on plant response to environmental stress. *Agronomy Journal*, 81, 83–90.

Bacon, C. W. & Siegel, M. R. (1988). Endophyte parasitism of tall fescue. *Journal of Production Agriculture*, 1, 45–55.

Bacon, C. W., Porter, J. K. & Robbins, J. D. (1986). Ergot toxicity from endophyte infected weed grasses: a review. *Agronomy Journal*, 78, 106–16.

Bailey, V. (1903). Sleepy grass and its effect on horses. *Science*, 17, 392–393.

Bentley, S. & Whittaker, J. B. (1979). Effects of grazing by chrysomelid beetle, *Gastrophysa viridus*, on competition between *Rumex obtusifolius* and *Rumex crispus*. *Journal of Ecology*, **67**, 79–90.

Bor, N. (1960). *The Grasses of Burma, Ceylon, India, and Pakistan*. Pergamon Press, New York.

Bouton, J. H., Gates, R. N., Belesky, D. P. & Owsley, M. (1993). Yield and persistence of tall fescue in the southeastern coastal plain after removal of its endophyte. *Agronomy Journal*, **85**, 52–55.

Bradshaw, A. D. (1959). Population differentiation in *Agrostis tenuis* Sibth. II. The incidence and significance of infection by *Epichloë typhina*. *New Phytologist*, **58**, 310–315.

Brown, V. K. & Gange, A. C. (1989). Differential effects of above- and below-ground insect herbivory during early plant succession. *Oikos*, **54**, 67–76.

Bruehl, G. W., Kaiser, W. J. & Klein, R. E. (1994). An endophyte of *Achnatherum inebrians*, an intoxicating grass of northwest China. *Mycologia*, **86**, 773–776.

Bull, J. J., Molineux, I. J. & Rice, W. R. (1991). Selection of benevolence in host–parasite system. *Evolution*, **45**, 875–882.

Carter, W. (1973). *Insects in Relation to Plant Disease*. John Wiley & Sons, New York.

Cheplick, G. P. & Clay, K. (1988). Acquired chemical defences of grasses: the role of fungal endophytes. *Oikos*, **52**, 309–318.

Cheplick, G. P., Clay, K. & Marks, S. (1989). Interactions between infection by endophytic fungi and nutrient limitation in the grasses *Lolium perenne* and *Festuca arundinacea*. *New Phytologist*, **111**, 89–97.

Clay, K. (1986). Induced vivipary in the sedge *Cyperus virens* and the transmission of the fungus *Balansia cyperi* (Clavicipitaceae). *Canadian Journal of Botany*, **64**, 2984–2988.

Clay, K. (1987). Effects of fungal endophytes on the seed and seedling biology of *Lolium perenne* and *Festuca arundinacea*. *Oecologia*, **73**, 358–362.

Clay, K. (1988a). Fungal endophytes of grasses: a defensive mutualism between plants and fungi. *Ecology*, **69**, 10–16.

Clay, K. (1988b). Clavicipitaceous fungal endophytes of grasses: Coevolution and the change from parasitism to mutualism. *Co-evolution of Fungi with Plants and Animals* (Ed. by D. L. Hawksworth & K. Pirozynski), pp. 79–105. Academic Press, London.

Clay, K. (1990a). Comparative demography of three graminoids infected by systemic, clavicipitaceous fungi. *Ecology*, **71**, 558–570.

Clay, K. (1990b). Fungal endophytes of grasses. *Annual Review of Ecology and Systematics*, **21**, 275–297.

Clay, K. (1991). Ecological interactions among fungal endophytes, grasses, and herbivores. *Multitrophic Interactions among Microorganisms, Plants and Insects* (Ed. by P. Barbosa, V. Krischik & C. Jones), pp. 199–226. John Wiley and Sons, New York.

Clay, K. (1993). The ecology and evolution of endophytes. *Agriculture, Ecosystems and Environment*, **44**, 39–64.

Clay, K. (1994). Hereditary symbiosis in the grass genus *Danthonia*. *New Phytologist*, **126**, 223–231.

Clay, K. & Leuchtmann, A. (1989). Infection of woodland grasses by fungal endophytes. *Mycologia*, **81**, 805–811.

Clay, K., Cheplick, G. P. & Wray, S. M. (1989). Impact of the fungus *Balansia henningsiana* on the grass *Panicum agrostoides*: frequency of infection, plant growth and reproduction, and resistance to pests. *Oecologia*, **80**, 374–380.

Clay, K., Marks, S. & Cheplick, G. P. (1993). Effects of insect herbivory and fungal endophyte infection on competitive interactions among grasses. *Ecology*, **74**, 1767–1777.

Coley, A. B., Fribourg, H. A., Pelton, M. R. & Gwinn, K. D. (1995). Effects of tall fescue infestation on relative abundance of small mammals. *Journal of Environmental Quality*, **24**, 472–475.

Crawley, M. J. (1983). *Herbivory: The Dynamics of Animal–Plant Interactions*. University of California Press, Berkeley, CA.

Cunningham, P. J., Foot, J. Z. & Reed, K. F. M. (1993). Perennial ryegrass (*Lolium perenne*) endophyte (*Acremonium lolii*) relationships: The Australian experience. *Agriculture, Ecosystems and Environment*, **44**, 157–168.

Dahlman, D. L., Eichenseer, H. & Siegel, M. R. (1991). Chemical perspectives on endophyte–grass interactions and their implications to insect herbivory. *Microorganisms, Plants and Herbivores* (Ed. by C. Jones, V. Krischik & P. Barbosa), pp. 227–252. John Wiley and Sons, New York.

DeBattista, J. P., Bouton, J. H., Bacon, C. W. & Siegel, M. R. (1990). Rhizome and herbage production of endophyte-removed tall fescue clones and populations. *Agronomy Journal*, **82**, 651–654.

Diehl, W. W. (1950). *Balansia and the Balansiae in America.* USDA, Washington, DC.

Fitter, A. H. (1977). Influence of mycorrhizal infection on competition for phosphorus and potassium by two grasses. *New Phytologist*, **79**, 119–125.

Fowler, N. L. & Clay, K. (1995). Environmental heterogeneity, fungal parasitism and the demography of the grass *Stipa leuchotricha*. *Oecologia*, **103**, 55–62.

Francis, S. M. & Baird, D. B. (1989). Increase in the proportion of endophyte-infected perennial ryegrass plants in over-drilled pastures. *New Zealand Journal of Agricultural Research*, **32**, 437–440.

Funk, C. R., Halisky, P. M., Johnson, M. C., Siegel, M. R., Stewart, A. V., Ahmad, S., Hurley, R. H. & Harvey, I. C. (1983). An endophytic fungus and resistance to sod webworms: association in *Lolium perenne*. *Bio/Technology*, **1**, 189–191.

Giuliana, W. M., Elliott, C. L. & Sole, J. D. (1994). Significance of tall fescue in the diet of the eastern cottontail. *Prairie Naturalist*, **26**, 53–60.

Halisky, P. M., Cappellini, R. A. & McBeath, J. H. (1990). Fungal endophytes of Alaskan grasses. *International Symposium on* Acremonium/*Grass Interactions Proceedings*, **1**, 18.

Hance, H.F. (1876). On a Mongolian grass producing intoxication in cattle. *Journal of Botany*, **14**, 210–212.

Harberd, D. J. (1961). Note on choke disease of *Festuca rubra*. *Scottish Plant Breeding Station Report*, **1961**, 47–51.

Hill, N. S., Stringer, W. C., Rottinghaus, G. E., Belesky, D. P., Parrot, W. A. & Pope, D. D. (1990). Growth, morphological, and chemical component responses of tall fescue to *Acremonium coenophialum*. *Crop Science*, **30**, 156–161.

Hill, N. S., Belesky, D. P. & Stringer, W. C. (1991). Competitiveness of tall fescue and influenced by *Acremonium coenophialum*. *Crop Science*, **31**, 185–190.

Kelley, S. E. & Clay, K. (1987). Interspecific competitive interactions and the maintenance of genotypic variation within the populations of two perennial grasses. *Evolution*, **41**, 92–103.

Kimmons, C. A., Gwinn, K. D. & Bernard, E. C. (1990). Nematode reproduction on endophyte-infected and endophyte-free tall fescue. *Plant Disease*, **74**, 757–761.

Kirfman, G. W., Brandenburg, R. L. & Garner, G. B. (1986). Relationship between insect abundance and endophyte infestation level in tall fescue in Missouri. *Journal of the Kansas Entomological Society*, **59**, 552–554.

Knoch, T. R., Faeth, S. H. & Arnott, D. L. (1993). Endophytic fungi alter foraging and dispersal by desert seed-harvesting ants. *Oecologia*, **95**, 470–475.

Large, E. C. (1952). Surveys for choke (*Epichloë typhina*) in cocksfoot seed crops, 1951. *Plant Pathology*, **1**, 23–28.

Latch, G. C. M. (1993). Physiological interactions of endophytic fungi and their hosts. Biotic stress tolerance imparted to grasses by endophytes. *Agriculture, Ecosystems and Environment*, **44**, 143–156.

Latch, G. C. M., Potter, L. R. & Tyler, B. F. (1987). Incidence of endophytes in seeds from collections of *Lolium* and *Festuca* species. *Annals of Applied Biology*, **111**, 59–64.

Lenski, R. E. & May, R. M. (1994). The evolution of virulence in parasites and pathogens: reconciliation between two competing hypotheses. *Journal of Theoretical Biology*, **169**, 253–265.

Leuchtmann, A. (1992). Systematics, distribution and host specificity of grass endophytes. *Natural Toxins*, **1**, 150–162.

Leuchtmann, A. & Clay, K. (1993). Nonreciprocal compatibility interactions between *Epichloë typhina* and four host grasses. *Mycologia*, **85**, 157–163.

Lewis, G. C. & Clements, R. O. (1986). A survey of ryegrass endophyte (*Acremonium loliae*) in the U.K. and its apparent ineffectuality on a seedling pest. *Journal of Agricultural Science*, 107, 633–638.

Lewis, G. C. & Clements, R. O. (1990). Effect of *Acremonium lolii* on herbage yield of *Lolium perenne* at three sites in the United Kingdom. *Proceedings of the International Symposium on Acremonium/Grass Interactions* (Ed. by S. A. Quisenberry & R. E. Joost), pp. 160–162. Louisiana Agricultural Experiment Station, Baton Rouge, FL.

Louda, S. M., Keeler, K. H. & Holt, R. D. (1990). Herbivore influences on plant performance and competitive interactions. *Perspectives on Plant Competition* (Ed. by J. B. Grace & D. Tilman), pp. 413–444. Academic Press, New York.

Madej, C. W. & Clay, K. (1991). Avian seed preference and weight loss experiments: the effect of fungal endophyte-infected tall fescue seeds. *Oecologia*, 88, 296–302.

Marks, S., Clay, K. & Cheplick, G. P. (1991). Effects of fungal endophytes on interspecific and intraspecific competition in the grasses *Festuca arundinacea* and *Lolium perenne*. *Journal of Applied Ecology*, 28, 194–204.

Nobindro, U. (1934). Grass poisoning among cattle and goats in Assam. *Indian Veterinary Journal*, 10, 235–236.

Paul, N. D. & Ayres, P. G. (1986). Interference between healthy and rusted groundsel (*Senecio vulgaris* L.) within mixed populations of different densities and proportions. *New Phytologist*, 104, 257–269.

Petroski, R. J., Powell, R. G. & Clay, K. (1992). Alkaloids of *Stipa robusta* (Sleepygrass) infected with an *Acremonium* endophyte. *Natural Toxins*, 1, 84–88.

Porter, J. K. (1994). Chemical constituents of grass endophytes. *Biotechnology of Endophytic Fungi of Grasses* (Ed. by C. W. Bacon & J. F. White), pp. 103–123. CRC Press, Boca Raton, FL.

Power, A. (1991). Virus spread and vector dynamics in genetically diverse plant populations. *Ecology*, 72, 232–241.

Prestidge, R. A., Pottinger, R. P. & Barker, G. M. (1982). An association of *Lolium* endophyte with ryegrass resistance to Argentine stem weevil. *Proceedings of the New Zealand Weed Pest Control Conference*, 199–222.

Prestidge, R. A., Thom, E. R., Marshall, S. L., Taylor, M. J., Willoughby, B. & Wildermoth, D. D. (1992). Influence of *Acremonium lolii* infection in perennial ryegrass on germination, emergence, survival and growth of white clover. *New Zealand Journal of Agricultural Research*, 35, 225–234.

Ramsell, J. & Paul, N. D. (1990). Preferential grazing by molluscs of plants infected by rust fungi. *Oikos*, 58, 145–150.

Read, J. C. & Camp, B. J. (1986). The effect of fungal endophyte *Acremonium coenophialum* in tall fescue on animal performance, toxicity, and stand maintenance. *Agronomy Journal*, 78, 848–850.

Read, J. C. & Walker, D. W. (1990). The effect of the fungal endophyte *Acremonium coenophialum* on dry matter production and summer survival of tall fescue. *Proceedings of the International Symposium on Acremonium/Grass Interactions* (Ed. by S. S. Quisenberry & R. E. Joost), pp. 181–184. Louisiana Agricultural Experiment Station, Baton Rouge, FL.

Rice, J. S., Pinkerton, B. W., Stringer, W. C. & Undersander, D. J. (1990). Seed production in tall fescue as affected by fungal endophyte. *Crop Science*, 30, 1303–1305.

Rowan, D. D. & Latch, G. C. M. (1994). Utilization of endophyte-infected perennial ryegrasses for increased insect resistance. *Biotechnology of Endophytic Fungi of Grasses* (Ed. by C. W. Bacon & J. F. White), pp. 169–183. CRC Press, Boca Raton, FL.

Rowan, D. D., Hunt, M. B. & Gaynor, D. L. (1986). Peramine, a novel insect feeding deterrent from ryegrass infected with the endophyte *Acremonium loliae*. *Journal of the Chemical Society D Chemical Communications*, 142, 935–936.

Saha, D. C., Johnson-Cicalese, J. M., Halisky, P. M., Van Heemstra, M. I. & Funk, C. R. (1987). Occurrence and significance of endophytic fungi in the fine fescues. *Plant Disease*, 71, 1021–1024.

Sampson, K. (1933). The systematic infection of grasses by *Epichloë typhina* (Pers.) Tul. *Transactions of the British Mycological Society*, **18**, 30–47.

Schardl, C. L., Liu, J., White, J. F., Finkel, R. A., An, Z. & Siegel, M. R. (1991). Molecular phylogenetic relationships of nonpathogenic grass mycosymbionts and clavicipitaceous plant pathogens. *Plant Systematics and Evolution*, **178**, 27–41.

Shaw, J. (1873). On the changes going on in the vegetation of South Africa. *Botanical Journal of the Linnean Society*, **14**, 202–208.

Shelby, R. A. & Dalrymple, L. W. (1987). Incidence and distribution of the tall fescue endophyte in the United States. *Plant Disease*, **71**, 783–785.

Siegel, M. R., Latch, G. C. M., Bush, L. P., Fannin, N. F., Rowan, D. D., Tapper, B. A., Bacon, C. W. & Johnson, M. C. (1990). Fungal endophyte-infected grasses: alkaloid accumulation and aphid response. *Journal of Chemical Ecology*, **16**, 3301–3315.

Sutherland, B. L. & Hoglund, J. H. (1990). Effects of ryegrass containing the endophyte *Acremonium lolii* on associated white clover. *Proceedings of the International Symposium on Acremonium/ Grass Interactions* (Ed. by S. S. Quisenberry & R. E. Joost), pp. 67–71. Louisiana Agricultural Experiment Station, Baton Rouge, FL.

Thompson, R. W., Fribourg, H. A. & Reddick, B. B. (1989). Sample intensity and timing for detecting *Acremonium coenophialum* incidence in tall fescue pastures. *Agronomy Journal*, **81**, 966–971.

Van Santen, E. (1992). Animal preference of tall fescue during reproductive growth in the spring. *Agronomy Journal*, **84**, 979–982.

Watson, R. N., Prestidge, R. A. & Ball, O. J. P. (1993). Suppression of white clover by ryegrass infected with *Acremonium* endophyte. *Proceedings of the Second International Symposium on Acremonium/Grass Interactions* (Ed. by D. E. Hume, G. C. M. Latch & H. S. Easton), pp. 218–221. AgResearch Grasslands Research Centre, Palmerston North.

Watt, A. S. (1960). The effects of excluding rabbits from acidophilous grassland in Breckland. *Journal of Ecology*, **48**, 601–604.

West, C. P. (1994). Physiology and drought tolerance of endophyte-infected grasses. *Biotechnology of Endophytic Fungi of Grasses* (Ed. by C. W. Bacon & J. F. White), pp. 87–101. CRC Press, Boca Raton, FL.

West, C. P., Izekor, E., Turner, K. E. & Elmi, A. A. (1993). Endophyte effects on growth and persistence of tall fescue along a water-supply gradient. *Agronomy Journal*, **85**, 264–270.

White, J. F. (1987). Widespread distribution of endophytes in the Poaceae. *Plant Disease*, **71**, 340–342.

White, J. F. (1988). Endophyte–host associations in forage grasses. XI. A proposal concerning origin and evolution. *Mycologia*, **80**, 442–446.

White, J. F. & Baldwin, N. A. (1992). A preliminary enumeration of grass endophytes in west central England. *Sydowia*, **44**, 78–84.

White, J. F., Jr & Chambless, D. A. (1991). Endophyte–host associations in forage grasses. XV. Clustering of stromata-bearing individuals of *Agrostis hiemalis* infected by *Epichloë typhina*. *American Journal of Botany*, **78**, 527–533.

White, J. F. & Cole, G. T. (1985). Endophyte–host associations in forage grasses. I. Distribution of fungal endophytes in some species of *Lolium* and *Festuca*. *Mycologia*, **77**, 323–327.

White, J. F., Jr, Halisky, P. M., Sun, S., Morgan-Jones, G. & Funk, C. R., Jr (1992). Endophyte–host associations in grasses. XVI. Patterns of endophyte distribution in species of the tribe Agrostideae. *American Journal of Botany*, **79**, 472–477.

White, J. F., Morgan-Jones, G. & Morrow, A. C. (1993). Taxonomy, life cycle, reproduction and detection of *Acremonium* endophytes. *Agriculture, Ecosystems and Environment*, **44**, 13–37.

Wilson, A. D., Clement, S. L., Kaiser, W. J. & Lester, D. G. (1991). First report of clavicipitaceous anamorphic endophytes in *Hordeum* species. *Plant Disease*, **75**, 215.

10. A CATCH 22: THE UTILIZATION OF ENDOPHYTIC FUNGI FOR PEST MANAGEMENT

R. A. PRESTIDGE AND O. J-P. BALL

AgResearch, Ruakura Research Centre, Private Bag 3123, Hamilton, New Zealand

INTRODUCTION

Grasses have long been recognized for their strategic importance in food production, soil development, stabilization and conservation, and for enhancing recreational and environmental areas. Forage grasslands are the highest single land use in many countries and turf grasslands play a major role in providing recreation and enjoyment for people in many countries; for example, grassland used for forage production makes up more than 50% of land in New Zealand while in the USA the turf grass industry is worth US$25 billion annually. Large areas of natural grasslands include much of Australia, the pampas of South America, the prairies of North America, the steppes of Eurasia and the savannahs of Africa. Barnard (1966) estimated that approximately 20% of the land surface of the world was in permanent pastures or meadows.

Many species of grass contain symptomless non-sporulating endophytic fungi which rely on their host for nutrition, transmission to a new host and shelter. The fungal endophytes considered within this context exclude the mycorrhizas which can live both internally and externally. New endophyte/host associations are being reported every year and this trend will undoubtedly continue as interest in endophytes grows. Most published research has focused on the *Acremonium* spp. endophytes of tall fescue (*Festuca arundinacea*) and perennial ryegrass (*Lolium perenne*) in northern Europe, New Zealand, Australia and the southeastern USA. Several reviews have been published (Siegel *et al.* 1987; Carroll 1988; Clay 1988a, b, 1990; Latch 1993; Bacon & White 1994; Breen 1994). Relatively less is known about the endophytes of other pasture grass species, lawn grasses and cultivated grains.

In countries reliant on perennial ryegrass and tall fescue as the major forage species (such as New Zealand and the US, respectively) *Acremonium* endophytes have important implications for the agricultural industry. Through the production of a number of different alkaloids, *A. coenophialum* in tall fescue and *A. lolii* in perennial ryegrass have an adverse effect on animal production and health. However, these and other alkaloids confer on their host several advantages such as improved biomass and persistence through a reduction in invertebrate herbivory, particularly under harsh environmental or management conditions. Hence the presence of endophytes in agriculturally important grasses presents farmers with a Catch 22 situation. They either use endophyte-infected grasses and have an adverse effect on livestock production, or they use endophyte-free grasses and have reduced pasture production and persistence. In the turf industry where grasses are not usually grazed by animals, animal health is a minor issue and virtually all new turf grass cultivars are now infected with endophytic fungi (Funk *et al.* 1993).

This chapter discusses the utilization of endophytic fungi in perennial ryegrass and tall fescue forage and turf grasses. Because much of the recent information on *Acremonium* endophytes in perennial ryegrass and tall fescue has been published in the *Proceedings of the First and Second Symposiums on* Acremonium *Grass Interactions* (Hume *et al.* 1993; Joost Quisenberry 1993) respectively, reference is made to those papers which represent new information or those which require further emphasis.

ENDOPHYTE-PRODUCED ALKALOIDS

Grasses differ from dicotyledons in that they have simple architecture and generally lack a variety of secondary compounds that deter herbivory in dicotyledons (Tscharntke & Greiler 1995). To date, four main classes of secondary metabolites, the indole diterpenes, ergot alkaloids, pyrollopyrazines and loline alkaloids, have been identified in endophyte-infected perennial ryegrass and tall fescue. Detailed information is only available for a few key compounds and there is little published information on the occurrence and concentrations of most metabolites in various

TABLE 10.1. Alkaloids produced by *Acremonium*-infected tall fescue and perennial ryegrass. (From Gallagher and Hawkes 1986; Siegel *et al.* 1990; Powell *et al.* 1993; Miles *et al.* 1993; Penn *et al.* 1993; Rowan & Popay 1993; Porter 1994.)

Group	Alkaloid	Concentration (mg g^{-1})
Indole diterpenes	Paxilline	0.1–10
	Lolitrem A–D	0.1–25
	Lolitriol	0.1–10
Ergot alkaloids	Ergolines	0.1–2
	Lysergides	0.1–2
	Ergopeptides	0.1–2
	Ergopeptines	0.1–2
Pyrollopyrazine alkaloids	Peramine	10–50
Pyrrolizidine alkaloids	N-acetylloline	50–8000
	N-formylloline	50–8000

grass–endophyte associations or detailed studies of their effects on invertebrates and domestic animals. Table 10.1 summarizes the major alkaloids produced in endophyte-infected ryegrass and tall fescue and provides key references for interested readers. The effects of the alkaloids can be generalized as detrimental to animal health and production (indole diterpenes and ergot alkaloids) or beneficial to the production and longevity of the host (pyrollopyrazine and loline alkaloids). The known impact of specific alkaloids on invertebrates and grazing animals is discussed in later sections.

TRANSMISSION OF ENDOPHYTES TO NEW HOSTS

Acremonium spp. endophytes are not horizontally transmitted to new, previously uninfected hosts. The only known form of transmission from one host to another is vertically via the seed of an infected plant.

It is possible to remove an endophyte from its natural host plant and artificially inoculate another host with the same strain, even across different plant genera, for example *A. coenophialum* to perennial ryegrass (Latch & Christensen 1985). Not all inoculations result in successful novel grass/endophyte associations. Some transfers fail completely, while in others, the endophyte does not thrive (Christensen 1995). Our knowledge of why some associations are compatible while others are not is slim.

Not all *Acremonium* species or isolates produce lolitrems, ergovaline or peramine in their natural hosts and the levels of alkaloid produced may also vary greatly in grass species and cultivars infected with different strains of endophyte (Siegel *et al.* 1990). Plant genotype plays a major role on the outcome of any plant/endophyte association. Thus the transfer of one endophyte isolate to a new host may result in the production of the same spectrum of alkaloids but in different concentrations (Latch 1994). Seasonal factors and the growth status of the plant are also major factors influencing the growth of the endophyte mycelium and the production of alkaloids (Ball *et al.* 1995).

The ability to artificially transfer an endophyte from its natural host plant to another host has been invaluable for testing the effects of different alkaloids on animals, as it is possible to have the same grass host cultivar infected with a range of endophyte isolates and endophyte species (Christensen 1995). Thus, different arrays of alkaloids produced in genetically similar plants can be evaluated against pests and grazing animals. Although it has been possible to transfer endophytes into uninfected seedlings, it is not yet routinely possible to inoculate different strains of endophyte into the same plant genotype as the inoculation procedure has been most successful with seedling hosts. However, limited success has been achieved through the inoculation of callus cultures (Kearney *et al.* 1991) and plantlets derived from single meristems (O'Sullivan & Latch 1993).

EFFECTS OF ENDOPHYTES ON VERTEBRATES

Domestic livestock

Although fungal endophytes of grasses have been known for a long time (Vogl 1898), their economic importance was not recognized until the late 1970s when an association was made between *A. coenophialum* and a toxicity syndrome in cattle consuming tall fescue. Since then, a range of behavioural, physiological and reproductive effects have been recorded in animals fed on endophyte-infected perennial ryegrass and tall fescue herbage or seed (Table 10.2).

In the southern and eastern states of the USA, tall fescue is grown on more than 14 million hectares. Tall fescue is primarily used for cattle grazing and it has been estimated that the negative impact of endophyte infection on beef production exceeds US$750 million annually (Hoveland 1993). In New Zealand, perennial ryegrass is the preferred grass species and is sown on more than 7 million hectares of land. The negative impact of the endophyte of perennial ryegrass on animal production, mainly sheep, probably exceeds NZ$120 million annually. Although these *Acremonium*–plant relationships are well defined for tall fescue and perennial ryegrass in terms of their effect on livestock, it is highly probable, based on the widespread occurrence of fungal endophytes in grasses (Clay, this volume), that there are many other undefined endophyte–plant relationships.

At least three categories of animal problems are now recognized as being associated with the consumption of *A. coenophialum*-infected tall fescue (Schmidt & Osborne 1993). These include fescue foot, fat necrosis and fescue toxicosis. The mechanisms and relationships among the disorders have not been fully established. Fescue foot is a gangrenous condition of the feet and tips of the tail and ears which most commonly occurs in cattle grazing tall fescue when ambient temperatures are low. This condition occurs primarily because ingestion of ergot alkaloids causes constriction of veins and capillaries which reduces blood flow to the body extremities (Solomons *et al.* 1989). This condition may be debilitating to animals although its effects on the beef industry are probably slight.

Fat necrosis is characterized by the presence of hard or necrotic fat deposited in the adipose tissue of the abdominal cavity. The presence of necrotic fat deposits is probably indicative of metabolic disorders such as that associated with pancreatitis in humans (Banks 1982). Like fescue foot, the impact of fat necrosis to the beef industry has not yet been quantified.

Fescue toxicosis or 'summer slump' or 'summer syndrome' are terms used to describe the unthrifty animal appearance during warm summer periods when ambient daytime temperatures often exceed 30 °C. Symptoms of fescue toxicosis are reduced animal intake and animal performance, lethargy, reduced tolerance to high environmental temperatures, excessive salivation and rough hair coats (Stuedemann

TABLE 10.2. Vertebrates reported to be adversely affected by *Acremonium*-infected perennial ryegrass and tall fescue and their metabolites. (From Prestidge 1993; Schmidt & Osborne 1993; Stuedemann & Thompson 1993).

Animal	Endophyte/grass*	Clinical and subclinical effects
Sheep	*A. lolii/L.p.*	Ryegrass staggers, reduced growth rate, reduced hormonal levels, toxicity, reduced reproduction
Cattle	*A. lolii/L.p.*	Reduced milk production, reduced hormonal levels, ryegrass staggers
	A. coenophialum/F.a.	Fescue foot, bovine fat necrosis, fescue toxicosis, behavioural changes, reduced intake, hypothermia, elevated respiration, reduced liveweight gains, reduced milk production, reduced reproduction
Wapiti deer (*Cervus elaphus mannitobensis* L.)	*A. lolii/L.p.*	Ryegrass staggers, toxicity
Red deer (*Cervus elaphus* L.)	*A. lolii/L.p.*	Ryegrass staggers
	A. coenophialum/F.a.	Reduced reproduction, toxicity, agalactica
Goats	*A. lolii/L.p.*	Ryegrass staggers
Mice	*A. lolii/L.p.*	Excitation staggers, lowered reproduction, aberrant behaviour, cannibalistic behaviour, reduced growth rates
Rats	*A. coenophialum/F.a.*	Excitation, lowered reproduction, aberrant behaviour, reduced milk production, reduced growth rates
Rabbits	*A. coenophialum/F.a.*	Lowered intake, reduced growth rates, lowered reproduction
Granivorous birds	*A. coenophialum/F.a.*	Lowered intake, reduced weight gain

* *L.p., Lolium perenne; F.a., Festuca arundinacea.*

& Hoveland 1988). Behavioural changes are also apparent, with affected animals standing in ponds and troughs and mainly eating in the evening when ambient temperatures are lower. These effects are primarily caused by ingestion of ergot alkaloids (Stuedemann & Thompson 1993).

In New Zealand, sheep grazing perennial ryegrass dominant pasture have been known to exhibit ryegrass staggers particularly in summer and autumn, since about the turn of the century. It was only in 1981 that the association between an endophytic fungus and ryegrass staggers was demonstrated (Fletcher & Harvey 1981). Isolation and purification of the tremor-inducing fractions of *A. lolii*-infected perennial ryegrass during the 1980s demonstrated the importance of lolitrem B in causing ryegrass staggers. A number of other lolitrems and structurally related compounds may also be involved as many compounds are tremorgenic to test animals.

Ryegrass staggers is a neuromuscular disease which has been recorded in New Zealand, Australia, Europe and the USA. The earliest symptoms are fine head tremors and trembling of the muscles in the neck and limbs. As severity increases, mild postural uncoordination, head nodding and swaying while standing, develops. Exercise exacerbates the symptoms and severely affected animals may walk with a stiff-legged gait and prance on fore, hind or all four limbs. This usually ends in collapse on the ground where the animal suffers severe muscular spasms. These usually last about 5–10 s, after which the animal quickly regains a standing position and walks away apparently unaffected. Sheep, horses and deer are more susceptible than cattle to ryegrass staggers because they graze closer to the crown of the plant thereby ingesting more basal sheath material. Mortality attributed to ryegrass staggers is generally low and most animals recover rapidly when non-toxic feed is offered. Since the original link between *A. lolii* and ryegrass staggers was proven, the range of documented effects of endophyte in perennial ryegrass on grazing sheep has grown to include reduced growth, faecal scouring and faecal soiling of wool, increased incidence of flies living in the wool and increased rectal temperatures and respiration rates. It appears likely that some of these effects are due to the presence of ergot peptide alkaloids which at 1–5 μg g^{-1} are widespread in *A. lolii*-infected perennial ryegrass and ergovaline is usually predominant (Powell *et al.* 1993). Animal experiments evaluating the effects of ergot peptide alkaloids are currently under way and further insights into the cause of the debilitating effects will be gained. The true costs of animals grazing *A. lolii*-infected perennial ryegrasses have not yet been fully quantified as more research on subclinical effects is required.

Non-domestic animals

Small animals, particularly mice and rats, have often been used to model the effects of *Acremonium* sp.-infected seed and herbage, and tremorgenic mycotoxins in grazing animals. Feeding experiments have provided insight into the effects of *Acremonium* sp. on the behavioural, lactational and reproductive physiology of animals; for example, mice fed a diet of 50% *A. lolli*-infected perennial ryegrass seed were more excitable and had fewer surviving pups because of cannibalistic and other aberrant behaviour of the parents compared with mice fed a diet of *A. lolii*-free seed. Mice also have reduced growth and reproductive performance when fed *A. lolii*-infected diets. Intra-peritoneal injection of mice with tremorgenic mycotoxins induced symptoms of ryegrass staggers similar to that in grazing animals.

Changes in the population dynamics and behaviour of wild, free-ranging mammalian populations are poorly documented, but reports that feral rabbit populations were lower on tall fescue compared to other pasture species are common. Chemical extracts from *A. coenophialum*-infected tall fescue have been shown to affect rabbits (Daniels *et al.* 1984). Farmed deer in New Zealand suffer severely from ryegrass staggers and most farmed deer in New Zealand have originated from captured feral populations. Laboratory studies have also revealed adverse effects of infected tall fescue on rabbits and rats, suggesting that feral populations could also be adversely affected in habitats dominated by infected tall fescue. Graminivorous bird species have exhibited reduced weight gains with increasing proportions of infected tall fescue seed in the diet, suggesting that some bird populations may also be reduced as a function of increasing endophyte infection in grasslands (Madej & Clay 1991).

EFFECTS OF ENDOPHYTES ON INVERTEBRATES

Impetus for much of the recent interest in grass endophytes has originated from the recognition that perennial ryegrass infected with *A. lolii* was resistant to attack by Argentine stem weevil (*Listronotus bonariensis* (Coleoptera: Curculionidae) (Prestidge *et al.* 1982). In the past 14 years, research has produced a list of more than 40 species of insects that are adversely affected by endophytes in grasses and sedges (Table 10.3). Several different insect orders are involved as are various modes of feeding, including stem borers, foliage feeders, phloem suckers, seed feeders and root feeders. Not all insects tested are adversely affected and often closely related species such as *Rhopalosiphum padi* and *R. maidis* have different responses. Some insect species for which grass is not their natural food are adversely affected by endophytes (Popay *et al.* 1993) and there are even a few reports of insects preferring to feed on endophyte-infected grasses; for example, Argentine stem weevil fed significantly more on perennial ryegrass infected with a *Gliocladium*-like endophyte than on endophyte-free grass (Gaynor *et al.* 1983).

The mechanisms of resistance arise from deterrence, which causes the insect to avoid infected host grasses, or from antibiosis, where insects feeding on the grasses suffer from growth, developmental or reproductive disorders. Usually for those species for which grass is a natural host, such as Argentine stem weevil, fall armyworm, *R. padi*, etc., antibiosis is accompanied by deterrency to endophyte infection. Thus, acute antibiosis is rarely observed for these species as a reduction in feeding generally occurs first. However, some insect species are killed when

TABLE 10.3. Insects adversely affected by *Acremonium*-infected tall fescue and perennial ryegrass and alkaloids associated with infected grasses. (From Clement *et al.* 1993.)

Insect	Grass*	Effects on insect
Coleoptera		
Listronotus bonariensis (Argentine stem weevil)	*L.p.*, *F.a.*	Feeding and oviposition deterrence, reduced survival/ovarian development, lower densities (*L.p.*)
Heteronychus arator (black beetle)	*L.p.*	Feeding deterrence, lower densities
Sphenophorus parvulus (bluegrass billbug)	*L.p.*, *F.a.*	Lower densities (*L.p.*), reduced survival
S. venatus	*L.p.*, *F.a.*	Reduced survival
S. inaequalis	*L.p.*, *F.a.*	Reduced survival
S. minimus	*L.p.*, *F.a.*	Reduced survival
Chaetocnema pulicaria (corn flea beetle)	*F.a.*	Lower densities
Tribolium sp. (flour beetle)	*F.a.*	*Lower densities*
T. castaneum (red flour beetle)	*L.p.*, *F.a.*	Reduced survival, reduced population growth rates
Popillia japonica (Japanese beetle*)*	*F.a.*	Feeding deterrence, lower densities
Cyclocephala lurida (southern masked chafer)	*F.a.*	Lower weight gain
Costelytra zealandica (grass grub)	*F.a.*	Reduced survival
Diptera		
Drosophila melanogaster (fruit fly)	*F.a.*	Reduced survival
Musca autumnalis (face fly)†	*F.a.*	Reduced oviposition, reduced pupation
Hemiptera		
Blissus leucopterus hirtus (hairy cinch bug)	*L.p.*	Feeding deterrence, reduced survival
Draculacephala spp. (sharpshooter leafhopper)	*F.a.*	Lower densities
Endria inimica (painted leafhopper)	*F.a.*	Lower densities
Agallia constricta (leafhopper)	*F.a.*	Lower densities
Exitianus exitiosus (graylawn leafhopper)	*F.a.*	Lower densities
Graminella nigrifrons (back-faced leafhopper)	*F.a.*	Lower densities
Prosapia bicincta (two-lined spittlebug)	*F.a.*	Lower densities
Oncopeltus fasciatus (large milkweed bug)	*F.a.*	Feeding deterrence, high mortality
Homoptera		
Balanococcus poae (pasture mealy bug)	*L.p.*	Lower densities
Rhopalosiphum padi (bird cherry-oat aphid)	*F.a.*	Feeding deterrence, reduced survival, reduced reproduction
R. maidis (corn leaf aphid)	*L.p.*	Feeding deterrence
Schizaphis graminum (greenbug aphid)	*L.p.*, *F.a.*	Feeding deterrence, reduced suvival
Diuraphis noxia (Russian wheat aphid)	*L.p.*, *F.a.*	Reduced survival

TABLE 10.3. *Continued.*

Insect	Grass*	Effects on insect
Sipha flava (yellow sugarcane aphid)	L.p.	Feeding deterrence
Aploneura lentisci (root aphid)	F.p.	No development
Rhopalomyzus poae (leaf aphid)	F.p.	Reduced development
Lepidoptera		
Crambus spp. (sod webworms)	L.p., F.a.	Lack of feeding, lower densities (L.p.), reduced survival (F.a.)
Parapediasia teterrella (bluegrass webworm)	L.p.	Feeding deterrence, reduced survival, lower weight gain
Ostrinia nubilalis (European corn borer)	F.a.	Altered feeding behaviour, lower weight gain
Spodoptera frugiperda (fall armyworm)	L.p., F.a.	Feeding deterrence, reduced survival, lower weight gain
S. eridania (southern armyworm)	L.p.	Reduced survival
Mythimna convecta (common armyworm)	L.p.	Feeding deterrence, delayed development
Agrotis infusa (cutworm)	L.p.	Fedding deterrence
Graphania mutans (cutworm)	L.p.	Feeding deterrence, reduced survival

* *L.p., Lolium perenne* (perennial ryegrass); *F.a., Festuca arundinacea* (tall fescue).
† Dougherty and Knapp (1994).

feeding on endophyte-infected grass apparently without being deterred, for example the Russian wheat aphid, *Diuraphis noxia* (Clement *et al.* 1992). Although we do not understand all the complexities between endophyte–grass associations in terms of the different alkaloids produced or of their effects on invertebrates, the effects of *A. lolii* infection in perennial ryegrass on Argentine stem weevil has been most studied and will be used as a model for further discussion.

Argentine stem weevil is a major pasture pest throughout New Zealand (Prestidge *et al.* 1991). If feeds predominantly on Gramineae and maintains a close association with its host plant throughout its life-cycle. The most critical factor determining host-plant resistance to Argentine stem weevil is the feeding and oviposition habit of the highly mobile adult stage. Adults have a large number of olfactory and gustatory sensors (Pilkington 1987) and they rely on these to detect the presence of *A. lolii*. Adults feed on the foliage of grasses and oviposit in the leaf sheath, by biting a hole and inserting eggs into the epidermal tissue. This behaviour is critical in the selection of nutritionally suitable oviposition sites. Larvae are stem borers and although they transfer from tiller to tiller as the need arises, they are not highly mobile between plants. In the field, the population dynamics of Argentine stem weevil are closely linked with the availability of *A. lolii*-free ryegrass tillers (Fig. 10.1). In pastures where all perennial ryegrass plants are infected by *A. lolii*,

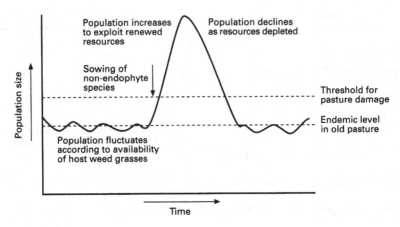

FIG. 10.1. Changes in population size of Argentine stem weevil in relation to changes in the abundance of *A. lolii*-free tillers. (From Barker *et al.* 1989.)

TABLE 10.4. Argentine stem weevil adult feeding and oviposition response on *A. lolii*-infected (+) and *A. lolii*-free (−) ryegrasses in choice and no-choice tests. (From Barker *et al.* 1984a, b.)

	Feedings scars per weevil at 96 hrs				Eggs oviposited per female at 96 hrs			
	Choice		No choice		Choice		No choice	
Ryegrass cultivar	+	−	+	−	+	−	+	−
Ellett	1.3	8.2	0.9	4.2	0.7	2.9	0.7	2.1
Nui	0.9	7.6	4.7	7.9	0.9	4.6	0.4	2.4
LSD5%*	3.8		3.0		1.8		1.8	

*LSD5%, least significant difference at $P < 0.05$.

as in old pastures, population size is theoretically nil. In reality, this is unlikely because of the presence of volunteer weed grasses. In pasture resown with endophyte-free ryegrasses, the weevils respond with increased reproductive effort, resulting in high larval numbers and high tiller mortality. This effect is primarily mediated through the adult stage which is deterred from feeding and ovipositing on *A. lolii*-infected ryegrass tillers and therefore selectively feeds on *A. lolii*-free tillers (Table 10.4). This is mainly due to the presence of peramine which has been found to deter adult weevils from feeding on artificial diet at a concentration of 0.1 μg g^{-1}. The ergopeptine alkaloids may also contribute to adult deterrency, with ergovaline and ergotamine reducing feeding in choice tests at 0.1 and 1.0 μg g^{-1}, respectively. Paxilline may also deter adult feeding on diet at concentrations of 1.0 μg g^{-1} (Rowan

& Popay 1993). Other important endophyte-related factors influencing Argentine stem weevil populations include reduced larval survival (Barker *et al.* 1984b), reduced larval development rates (Prestidge & Gallagher 1988), and increased flight muscle development facilitating dispersal (Barker *et al.* 1989) when larvae and adults are exposed to *A. lolii* over a period of time. Reduced larval survival and development rates are probably caused by lolitrem B (toxic at 5 μg g^{-1}) and paxilline (at 10 μg g^{-1}). There is little known about the biological effects of other metabolites in the *A. lolii*/ryegrass association; however, recent studies have shown that some of these may have considerable activity against larvae (Prestidge & Ball 1993).

Three other insect pasture pest species in New Zealand, black beetle (*Heteronychus arator*), cutworm (*Graphania mutans*) and pasture mealy bug (*Balanococcus poae*), now have confirmed sensitivity to *A. lolii*-infected perennial ryegrass (Prestidge *et al.* 1994). The effects of *A. lolii* and associated metabolites on these insect pests are less well known. Studies incorporating the major alkaloids into artificial diets have shown that adult black beetle were unaffected by peramine (up to 60 μg g^{-1}), lolitrem B (up to 20 μg g^{-1}), and paxilline (up to 20 μg g^{-1}), but were very sensitive to ergotamine at 5 μg g^{-1} (Prestidge & Ball 1993). It now appears likely that ergovaline is responsible for the observed black beetle feeding deterrency. In addition to the above insect species, sod webworm (*Crambus* spp.) (Funk *et al.* 1983) and black field cricket (*Teleogryllus commodus*) (Van Heeswijck & McDonald 1992) may also be adversely affected by *A. lolii*, although this has yet to be confirmed for the New Zealand populations. In all of these cases of *A. lolii*-conferred resistance, endophyte infection adversely affects the insect stages that feed on the above-ground parts of plants, particularly those stages that feed near the base of the plant where the concentrations of endophyte mycelium and most of the known alkaloids are greatest. There have been no reports of any major adverse effect on the plant root feeding stages of susceptible insects such as *H. arator* or on other root feeding invertebrates with the possible exception of plant-feeding nematode species, such as *Pratylenchus* sp. (Prestidge *et al.* 1993). This is in spite of the detection of mycelium and associated alkaloids in the roots of endophyte-infected plants. However, in artificial *Acremonium*/grass associations, root feeding larvae of *Costelytra zealandica* White (Coleoptera: Scarabaeidae) were adversely affected by endophyte (Popay *et al.* 1993).

There is no single alkaloid responsible for the observed effects of endophyte infection on insects; for example, from the New Zealand studies different metabolites are responsible for the recorded effects on major pastoral pests (Table 10.5). If there is a major difference between the types of insects affected by *A. lolii*-infected perennial ryegrass and *A. coenophialum*-infected tall fescue, it is that root-feeding insects may be adversely affected to a greater extent when feeding on tall fescue. This is probably due to the levels of loline alkaloids found in the roots of

TABLE 10.5. The effect of the major alkaloids produced by *Acremonium* spp.-infected perennial ryegrass and tall fescue on four pastoral pest species and three domestic livestock species in New Zealand (+, adversely affected; – no affect; nt, not tested; (?), unknown effects. (From Prestidge *et al.* 1994; Hume *et al.* 1993.)

Insect pest/domestic animal	Stage	Peramine	Lolitrem B	Paxilline	Ergopeptine	Loline	Effects
Argentine stem weevil (*Listronotus bonariensis*)	Adult	+	–	+	+	nt	Feeding deterrent
	Larva	+	+	+	–	+	Increased mortality, reduced growth rates, feeding deterrent
Black beetle (*Heteronychus arator*)	Adult	–	–	–	+	nt	Feeding deterrent
Cutworm (*Graphania mutans*)	Larva	+	nt	nt	+	nt	Reduced growth rate, increased mortality
Grass grub (*Costelytra zealandica*)	Larva	–	–	–	nt	+	Reduced growth rate, increased mortality
Sheep		–	+	+	+	(?)	Ryegrass staggers, low production, increased mortality
Cattle		–	+	+	+	(?)	Ryegrass staggers, low production, increased mortality
Horses		–	+	(?)	+	(?)	Ryegrass staggers, agalatica, increased mortality

A. coenophialum-infected grasses; hence, root feeding stages of Japanese beetle *Popillia japonica* and *Cyclocephala* spp. are adversely influenced by *A. coenophialum* in tall fescue (Oliver *et al.* 1990; Potter *et al.* 1992).

In addition to having a direct effect on primary consumers, endophyte infection status may have multitrophic level effects; for example, parasitoid wasp (*Euplectus comstockii*) larvae reared on fall armyworm larvae that had been fed endophyte-free tall fescue initiated pupation sooner than those reared on fall armyworm larvae fed *A. coenophialum*-infected tall fescue (Bultman *et al.* 1993). In some of our more recent studies predatory invertebrates, particularly spiders, were more abundant on *A. lolii*-infected ryegrass plots particularly during summer and autumn (Prestidge & Ball, unpublished data). Endophyte-infection status may also have important effects on detritivores and other invertebrates involved with litter burial and removal; for example earthworm (*Lumbricus rubellus*) populations were significantly greater on *A. lolii*-free perennial ryegrass plots compared to *A. lolii*-infected plots during autumn (Prestidge & Ball, unpublished data). Face flies (*Musca autumnalis*) oviposited more eggs in dung collected from cattle fed endophyte-free tall fescue than on endophyte-infected tall fescue. Pupation of face flies was also reduced on dung of endophyte-infected tall fescue (Dougherty & Knapp 1994). Thus, endophyte-infection may have an adverse effect on litter removal and soil fertility status in the longer term.

APPROACHES TO USING ENDOPHYTES FOR PEST MANAGEMENT

It is obvious that in some areas of the world, endophytes play a greater role in assisting the persistence of pasture grasses than in others. In New Zealand, endophyte-free perennial ryegrass will not persist in many regions primarily due to predation by Argentine stem weevil. In Europe, the commonly used cultivars of perennial ryegrass have little endophyte infection; however, common pest species such as frit fly (*Oscinella frit*) are tolerant of *A. lolii* infection (Lewis & Clements 1986) and may have coevolved with endophyte. Insects appear to be less important in pastures in Europe and none has been shown to be affected greatly by endophyte. However, *A. lolii* presumably confers some advantage to the host because the percentage of endophyte infection is higher in older pastures than in younger pastures (Clements & Lewis 1993).

The use of endophytic fungi in perennial ryegrass and tall fescue for pest management and improved persistence under adverse environmental conditions is already being realized as many commercially available cultivars with high endophyte levels are available. Farmers and agriculturalists are using endophyte-infected forage grasses and have developed management strategies to minimize the adverse effects on domestic livestock. Maintenance of an intermediate level of endophyte infection

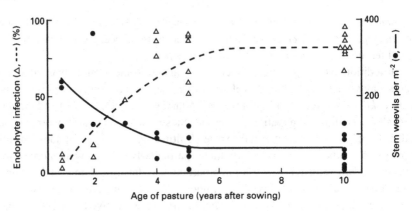

FIG. 10.2. Percentage composition of *A. lolii*-infected perennial ryegrass and density of Argentine stem weevil in pastures of different age in March/April 1984, excluding pastures sown with high-endophyte (> 80%) seedlines of ryegrass.

is not a viable farming option because selective herbivory on endophyte-free plants means that pastures quickly revert to a high endophyte status (Fig. 10.2). The research fraternity, however, has been developing an 'ideal' endophyte by a number of different approaches. The ideal endophyte is defined as an endophyte which protects the plant from attack by herbivorous invertebrates but does not cause adverse animal health effects. Although it may appear a simple goal, the development of an ideal endophyte in perennial ryegrass in New Zealand has proven difficult to achieve primarily because of the number and diversity of alkaloids involved in animal toxicity, the lack of knowledge of the effects of the alkaloids on animals, and the dual beneficial and detrimental effects of some alkaloids on animals and invertebrates (Table 10.4).

Reducing the diversity of alkaloids

After a programme of searching and chemically screening a wide range of naturally occurring grass/endophyte associations from throughout the world, a strain of endophyte, which did not produce lolitrem B but which protected its host from Argentine stem weevil attack, was discovered. This isolate was artificially inoculated into a range of ryegrass genotypes and evaluated at a range of sites throughout New Zealand (Fletcher *et al.* 1991). The results (Fig. 10.3) were an outstanding success and indicated that lambs did not develop ryegrass staggers and that the grass was resistant to Argentine stem weevil in the field. The results, however, gave new insights into the interaction between ryegrass genotype and *A. lolii* isolates in terms of the quantities of alkaloids produced. Although animals did not develop ryegrass staggers on the trials, animal weight gains on the selected

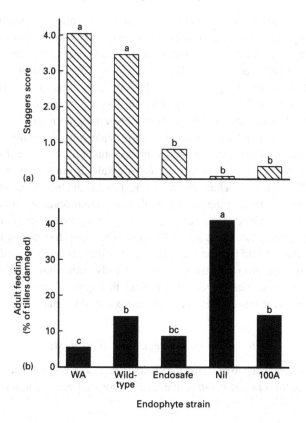

FIG. 10.3. Ryegrass staggers scores for (a) lambs and (b) percentage tillers damaged by Argentine stem weevil adults during February/March on perennial ryegrass with different endophyte treatments. WA, selected ecotype which produces very high levels of lolitrem B and peramine; wild-type, unselected strain with 'typical' levels of alkaloids; endosafe, selected strain which produces very low levels of lolitrem B and 'typical' levels of other alkaloids; nil, endophyte-free; 100A, selected ecotype which produces very low levels of lolitrem B and low levels of peramine. Treatments with the same letters are not significantly different at $P = 0.05$. (From Fletcher *et al.* 1991.)

endophyte/plant genotype were similar to weight gains on wild type endophyte. Apparently, the clinical effects of ryegrass staggers caused by ingestion of lolitrem B were masking animal production effects on livestock grazing *A. lolii*-infected perennial ryegrass. The lack of animal weight gain has subsequently been shown to be associated with a further alkaloid (ergovaline) and it now appears that in this particular endophyte/host association where lolitrem B was not produced, ergovaline levels were elevated to the point where animal production was adversely affected.

Current research is attempting to develop an *A. lolii*-infected perennial ryegrass that does not produce either lolitrem B and ergovaline. However, from an ecological viewpoint, there are several disadvantages with the approach of reducing the number of alkaloids. First, as observed in New Zealand, when some alkaloids are absent other metabolites may be present in higher concentrations, or, as yet unknown alkaloids may be produced. Second, different insect pests are adversely affected by different alkaloids; for example in New Zealand representatives from the four main classes of alkaloid will be required to provide control against a number of different pests (Table 10.5). It is apparent that much of the current endophyte-mediated plant resistance is due to different alkaloids either acting together or affecting different insects. Third, by narrowing the suite of alkaloids available within perennial ryegrass, there is the possibility that as yet unrecognized insects, that are already being maintained at low densities by the existing commercial combinations, may develop pest status. If the suite of alkaloids becomes too narrow and there is reliance on one alkaloid for pest management then there is the increased possibility that pests may develop resistance, as has already been shown for *Drosophila melanogaster* (Pless *et al.* 1993). Nevertheless, this approach may provide a short-term solution to the insect resistance/animal toxicity dilemma.

Reducing the concentration of alkaloids

An alternative approach to reducing the diversity of alkaloids is to reduce the concentration of alkaloids to a low and stable level. A cocktail of alkaloids at low, stable

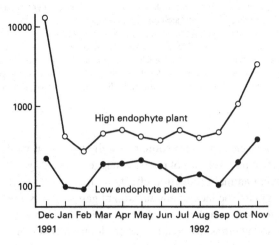

FIG. 10.4. Monthly *A. lolii* concentrations in the basal component of representative perennial ryegrass plants with high and low endophyte concentrations. (From Ball *et al.* 1995).

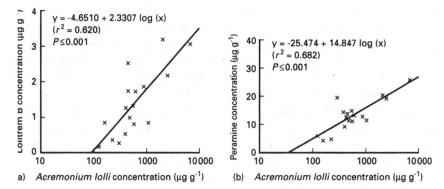

FIG. 10.5. Relationship between the yearly mean concentrations of *A. lolii* mycelium, and (a) lolitrem B and (b) peramine in individual wild type *A. lolii*-infected perennial ryegrass plants during 1991–1992. (From Ball *et al.* 1995).

concentrations may alleviate potential problems of animal ill health and insects developing resistance, and decrease the possibility of unknown insects developing pest status. Evidence to support this approach is based on the great variation in the concentration of *A. lolii* mycelium and associated alkaloids within different wild-type perennial ryegrass plants. Although, there are large seasonal fluctuations, the relative concentration of mycelium and alkaloids in plants with high and low levels is maintained over time (Fig. 10.4). There is also evidence indicating that relative alkaloid concentrations are inherited with the endophyte from one plant generation to the next (Ball, unpublished data). In addition, in natural associations, as the mycelium concentration within a plant increases, the alkaloid concentrations within the same plant also usually increases (Fig. 10.5). Obvious exceptions to this relationship do occur (indeed this is the basis for the alkaloid reductionist approach above), and there are likely to be plant and endophyte genotype interactive effects. If this approach is to be successful the real challenge facing agriculturalists is to determine the lower thresholds at which these alkaloids are required for pest control, and the upper thresholds when animal health and productivity begin to be adversely affected.

Potential of biotechnology

Endophytes can be genetically manipulated or used as carriers for genetically engineered bioinsecticides such as the *Bacillus thuringensis* (Bt) toxin gene. An important first step in the development of gene transfer systems in endophytes has been the recent demonstration of transformation in both *A. lolii* (Murray *et al.* 1992) and *A. coenophialum* (Tsai *et al.* 1992). In both cases, the marker gene for hygromycin resistance was introduced under the control of fungal promoters and terminators. Once the transformation system has been developed, gene targeting (gene replacement) methods will be required so that genes in specific biochemical pathways can be altered or inactivated. This approach has been proposed for perennial ryegrass by Gurney & Mantle (1993) who have used chemically induced mutations to block the secondary metabolite pathway in the production of neurotoxic indole-diterpenoids such as lolitrem B and paxilline.

Murray *et al.* (1992) demonstrated that endophytes could also serve as surrogate hosts to introduce foreign genes into grasses. Using this transformation process, Van Heeswijck *et al.* (1993) have attempted to introduce a Bt toxin gene and a proteinase inhibitor gene from *Nicotiana alata* into perennial ryegrass.

Successful transformation of *A. lolii* and *A. coenophialum* and inoculation into superior genotypes of perennial ryegrass and tall fescue will allow the production of specific grasses for specific purposes. For turf grasses, for example, the plant breeder desires a genotype with a dense sward, of even dark green colour which is resistant to environmental stresses, pests and pathogens. In forage grasses the primary goal may be increased digestibility, high animal production, and enhanced resistance to environmental stresses, pests and pathogens. In crop plants, where *Acremonium* endophytes have been discovered in close relatives of major grain-producing species, such as *Hordeum* spp. (Wilson *et al.* 1991), inoculation of crop plants with selected endophytes appears to be possible. Clay (1993) has highlighted the potential problems that may arise from genetically manipulated endophyte/ grass associations. The most obvious problem may be that the modified grasses will escape from their intended habitat (lawns, pastures) and will invade 'natural' habitats, with negative consequences for the resident fauna and flora. Grasses artificially inoculated with selected endophytes effectively become different plants in the same way that genetically engineered plants become different and the same public concerns of the release of genetically engineered plants may apply. The inability to predict the impact of new genetically engineered organisms in a new environment can sometimes result in major catastrophes.

Unanswered questions

Many questions remain to be answered about the role of *A. lolii* in perennial ryegrass and whether utilization of *Acremonium* is desirable; for example, what is the potential of Argentine stem weevil, black beetle and other invertebrates to evolve resistance or tolerance to fungal alkaloids in grasses? We have only just begun to examine the effects of *A. lolii* on parasitoids and insect predators. What are the effects of *A. lolii* and related alkaloids on invertebrates and micro-organisms in-volved with ruminant nutrition, litter and dung decomposition, and nutrient turnover? Considering the importance of white clover (*Trifolium repens*) in biologically fixing nitrogen, what are the effects of *A. lolii* on companion legume species?

Although a number of key alkaloids involved with animal health and insect pest resistance have been identified, we are aware of unknown (unidentified) alkaloids with significant biological activity. These compounds often occur in low concentrations and chemical instrumentation in often measuring alkaloid levels at the lower level of detection. Many of these compounds could be highly toxic but at this stage are relatively ignored because they occur at such low levels.

ACKNOWLEDGEMENTS

Our research has benefited greatly from interaction with the multidisciplinary team researching endophyte/grass interactions. The research was supported by the New Zealand Lotteries Grants Board for purchase of the ELISA plate reader, and the Foundation for Research, Science and Technology.

REFERENCES

Bacon, C. W. & White, J. F. (1994). *Biotechnology of Endophytic Fungi in Grasses.* CRC Press, Ann Arbor.

Ball, O. J-P., Prestidge, R. A. & Sprosen, J. M. (1995). Interrelationships between *Acremonium lolii*, peramine, and lolitrem B in perennial ryegrass. *Applied Environmental Microbiology*, **61**, 1527–1533.

Banks, P. A. (1982). Metabolic complications of pancreatitis. *Complications of Pancreatitis, Medical and Surgical Management* (Ed. by G. L. Bradley & W. G. Saunders), pp. 176–202. Saunders, Philadelphia.

Barker, G. M., Pottinger, R. P. & Addison, P. J. (1984a). Effect of *Lolium* endophyte fungus infections on survival of larval Argentine stem weevil. *New Zealand Journal of Agricultural Research*, **27**, 279–281.

Barker, G. M., Pottinger, R. P., Addison, P. J. & Prestidge, R. A. (1984b). Effect of *Lolium* endophyte fungus infections on behaviour of adult Argentine stem weevil. *New Zealand Journal of Agricultural Research*, **27**, 271–278.

Barker, G. M., Pottinger, R. P. & Addison, P. J. (1989). Population dynamics of the Argentine stem weevil (*Listronotus bonariensis*) in pastures of Waikato, New Zealand. *Agriculture, Ecosystems and Environment*, **26**, 79–115.

Barnard, C. (1966). *Grasses and Grasslands.* St Martins Press, New York.

Breen, J. P. (1994). *Acremonium* endophyte interactions with enhanced plant resistance to insects. *Annual Review of Entomology*, **39**, 401–423.

Bultman, T. L., Coudron, T. A., Crowder, R. J. & Davies, A. P. (1993). Effect of *Acremonium coenophialum* in the diet of fall armyworm on the development of the parasitoid *Euplectrus comstockii. Proceedings of the 2nd International Symposium on* Acremonium/*Grass Interactions* (Ed. by D. E. Hume, G. C. M. Latch & H. S. Easton), pp. 151–154. AgResearch, Palmerston North.

Carroll, G. C. (1988). Fungal endophytes in stems and leaves: from latent pathogen to mutualistic symbiont. *Ecology*, **69**, 2–9.

Christensen, M. J. (1995). Variation in the ability of *Acremonium* endophytes of *Lolium perenne, Festuca arundinacea* and *F. pratensis* to form compatible associations in the three grasses. *Mycological Research*, **99**, 466–470.

Clay, K. (1988a). Clavicipitaceous endophytes of grasses: coevolution and the change from parasitism to mutualism. *Coevolution of Fungi with Plants and Animals* (Ed. by D. L. Hawksworth & K. Pirozynski), pp. 79–105. Academic Press, London.

Clay, K. (1988b). Fungal endophytes of grasses: a defensive mutualism between plants and fungi. *Ecology*, 69, 10–16.

Clay, K. (1990). Fungal endophytes of grasses. *Annual Review of Ecology and Systematic*, 21, 275–297.

Clay, K. (1993). The potential role of endophytes in ecosystems. *Biotechnology of Endophytic Fungi of Grasses* (Ed. by C. W. Bacon & J. F. White Jr), pp. 73–86. CRC Press, Ann Arbor.

Clement, S. L., Lester, D. G., Wilson, A. D. & Pike, K. S. (1992). Behaviour and performance of *Diuraphis noxia* (Homoptera: Aphididae) on fungal endophyte-infected and uninfected perennial ryegrass. *Journal of Economic Entomology*, 85, 583–588.

Clement, S. L., Kaiser, W. J. & Eichenseer, H. (1993). *Acremonium* endophytes in germplasms of major grasses and their utilisation for insect resistance. *Biotechnology of Endophytic Fungi of Grasses* (Ed. by C. W. Bacon & J. F. White Jr), pp. 185–199. CRC Press, Ann Arbor.

Clements, R. O. & Lewis, G. C. (1993). Ryegrass endophyte research in the UK. *Proceedings of the Sixth Australian Conference on Grassland Invertebrate Ecology* (Ed. by R. A. Prestidge), pp. 169–173. AgResearch, Hamilton.

Daniels, L. B., Ahmed, A., Nelson, T. S., Piper, E. L. & Beasley, J. N. (1984). Physiological responses in pregnant white rabbits given a chemical extract of toxic tall fescue. *Nutrition Reports International*, 29, 505–510.

Dougherty, C. I. & Knapp, F. W. (1994). Oviposition and development of face flies in dung from cattle on herbage and supplemented herbage diets. *Veterinary Parasitology*, 55, 1–2.

Fletcher, L. R. & Harrey, I. C. (1981). An association of a *Lolium* endophyte with ryegrass staggers. *New Zealand Veterinary Journal*, 29, 185.

Fletcher, L. R., Popay, A. J. & Tapper, B. A. (1991). Evaluation of several lolitrem-free endophyte/ perennial ryegrass combinations. *Proceedings of the New Zealand Grassland Association*, 53, 215–220.

Funk, C. R., Halisky, P. M., Johnson, M. C., Siegel, M. R., Stewart, A. V., Ahmad, S., Hurley, R. H. & Harvey, I. C. (1983). An endophytic fungus and resistance to sod webworms: an association in *Lolium perenne* L. *Bio/Technology*, 1, 189–191.

Funk, C. R., White, R. H. & Breen, J. B. (1993). Importance of *Acremonium* endophytes in turf-grass breeding and management. *Agriculture, Ecosystems and Environment*, 44, 196–215.

Gallagher, R. I. & Hawkes, A. D. (1986). The potent tremorgenic neurotoxins lolitrem B and aflatrem: a comparison of the tremor response in mice. *Experientia*, 42, 823–825.

Gaynor, D. L., Rowan, D. D., Latch, G. C. M. & Pilkington, S. (1983). Preliminary results on the biochemical relationship between adult Argentine stem weevil and two endophytes in ryegrass. *Proceedings of the New Zealand Weed and Pest Control Conference*, 36, 220–224.

Gurney, K. A. & Mantle, P. G. (1993). Mutagenesis of ryegrass endophyte (*Acremonium* sp.) via protoplasts. *Proceedings of the 2nd International Symposium on* Acremonium/Grass Interactions (Ed. by D. E. Hume, G. C. M. Latch & H. S. Easton), pp. 45–50. AgResearch, Palmerston North.

Hoveland, C. S. (1993). Importance and economic significance of the *Acremonium* endophytes to performance of animals and grass plants. *Agriculture, Ecosystems and Environment*, 44, 3–12.

Hume, D. E., Latch, G. C. M., Easton, H. S. (1993). *Proceedings of the 2nd International Symposium on* Acremonium/Grass Interactions. AgResearch, Palmerston North.

Joost, R. & Quisenberry, S. (1993). *Acremonium*/grass interactions. *Agriculture, Ecosystems and Environment*, 44, 1–324.

Kearney, J. F., Parrott, W. A. & Hill, N. S. (1991). Infection of somatic embryos of tall fescue with *Acremonium coenophialum. Crop Science*, 31, 979–984.

Latch, G. C. M. (1993). Physiological interactions of endophytic fungi and their hosts. Biotic stress tolerance imparted to grasses by endophytes. *Agriculture, Ecosystems and Environment*, 44, 143–156.

Latch, G. C. M. (1994). Influence of *Acremonium* endophytes on perennial grass improvement. *New Zealand Journal of Agricultural Research*, 37, 311–318.

Latch, G. C. M. & Christensen, M. J. (1985). Artificial infection of grasses with endophytes. *Annals of Applied Biology*, 107, 17–24.

Lewis, G. C. & Clements, R. O. (1986). A survey of ryegrass endophyte (*Acremonium lolii*) in the UK and its apparent ineffectuality on a seedling pest. *Journal of Agricultural Science*, Cambridge, 107, 633–638.

Madej, C. W. & Clay, K. (1991). Avian seed preference and weight loss experiments: the role of fungal endophyte-infected all fescue seeds. *Oecologia*, 88, 296–302.

Miles, C. O., Munday, S. C., Wilkens, A. L., Ede, R. M., Hawkes, A. D., Embling, P. P. & Towers, N. R. (1993). Large scale isolation of lolitrem B, structure determination of some minor lolitrems, and tremorgenic activities of lolitrem B and paxilline in sheep. *Proceedings of the 2nd International Symposium on Acremonium/Grass Interactions* (Ed. by D. E. Hume, G. C. M. Latch & H. S. Easton), pp. 85–88. AgResearch, Palmerston Worth.

Murray, F. R., Latch, G. C. M. & Scott, D. B. (1992). Surrogate transformation of perennial ryegrass, *Lolium perenne*, using genetically modified *Acremonium* endophyte. *Molecular Genetics*, 233, 1–9.

O'Sullivan, B. D. & Latch, G. C. M. (1993). Infestation of plantlets, derived from ryegrass and tall fescue meristems, with *Acremonium* endophytes. *Proceedings of the 2nd International Symposium in Acremonium/Grass Interactions* (Ed. by D. E. Hume, G. C. M. Latch & H. S. Easton), pp. 16–17. AgResearch, Palmerston North.

Oliver, J. B., Pless, C. D. & Gwinn, K. D. (1990). Effect of endophyte, *Acremonium coenophialum*, in 'Kentucky 31' tall fescue, *Festuca arundinacea*, on survival of *Popillia japonica*. *Proceedings of the International Symposium on Acremonium/Grass Interactions* (Ed. by R. Joost & S. Quisenberry), p. 73. Baton Rouge, LA.

Penn, J., Garthwaite, I., Christensen, M. J., Johnson, C. M. & Towers, N. R. (1993). The importance of paxilline in screening for potentially tremorgenic *Acremonium* isolates. *Proceedings of the 2nd International Symposium on* Acremonium/*Grass Interactions* (Ed. by D. E. Hume, G. C. M. Latch & H. S. Easton), pp. 88–92. AgResearch, Palmerston North.

Pilkington, S. (1987). *The Behavioural Biology of Argentine Stem Weevil in Relation to Host-plant.* MSc Thesis, Massey University, New Zealand.

Pless, C. D., Gwinn, K. D., Cole, A. M., Chalkey, D. B. & Gibson, V. C. (1993). Development of resistance by *Drosophila melanogaster* (Diptera: Drosophilidae) to toxic factors in powdered *Acremonium*-infected tall fescue seed. *Proceedings of the 2nd International Symposium on* Acremonium/*Grass Interactions* (Ed. by D. E. Hume, G. C. M. Latch & H. S. Easton), pp. 170–173. AgResearch, Palmerston North.

Popay, A. J., Mainland, R. A. & Saunders, C. J. (1993). The effect of endophytes in fescue grass on growth and survival of third instar grass grub larvae. *Proceedings of the 2nd International Symposium on* Acremonium/*Grass Interactions* (Ed. by D. E. Hume, G. C. M. Latch & H. S. Easton), pp. 174–176. AgResearch, Palmerston North.

Porter, J. K. (1994). Chemical constituents of grass endophytes. *Biotechnology of Endophytic Fungi of Grasses* (Ed. by C. W. Bacon & J. F. White Jr), pp. 103–124. CRC Press, Ann Arbor.

Potter, D. A., Patterson, C. G. & Redmond, C. T. (1992). Influence of turfgrass species and tall fescue endophyte on feeding ecology of Japanese beetle and southern masked chafer grubs (Coleoptera: Scarabaeidae). *Journal of Economic Entomology*, 85, 900–909.

Powell, R. G., Tepaske, M. R., Plattner, R. D. & Petroski, R. J. (1993). Recent progress in the chemistry of grass/fungal interactions. *Proceedings of the 2nd International Symposium on* Acremonium/*Grass Interactions* (Ed. by D. E. Hume, G. C. M. Latch & H. S. Easton), pp. 85–88. AgResearch, Palmerston North.

Prestidge, R. A. (1993). Causes and control of perennial ryegrass staggers in New Zealand. *Agriculture, Ecosystems and Environment*, 44, 283–300.

Prestidge, R. A. & Ball, O. J-P. (1993). The role of endophytes in alleviating plant biotic stress in New Zealand. *Proceedings of the 2nd International Symposium on* Acremonium/Grass *Interactions* (Ed. by D. E. Hume, G. C. M. Latch & H. S. Easton), pp. 141–151. AgResearch, Palmerston North.

Prestidge, R. A. & Gallagher, R. T. (1988). Endophyte fungus confers resistance to ryegrass: Argentine stem weevil larval studies. *Ecological Entomology*, **13**, 429–435.

Prestidge, R. A., Pottinger, R. P. & Barker, G. M. (1982). An association of *Lolium* endophyte with ryegrass resistance to Argentine stem weevil. *Proceedings of the New Zealand Weed and Pest Control Conference*, **35**, 119–122.

Prestidge, R. A., Barker, G. M. & Pottinger, R. P. (1991). The economic cost of Argentine stem weevil in pastures in New Zealand. *Proceedings of the New Zealand Weed and Pest Control Conference*, **44**, 165–170.

Prestidge, R. A., Watson, R. N. & Thom, E. R. (1993). Invertebrate pests of endophyte-infected and endophyte-free perennial ryegrass swards in northern New Zealand. *Proceedings of the International Grassland Congress*, **17**, 934–935.

Prestidge, R. A., Popay, A. J. & Ball, O. J-P. (1994). Biological control of pastoral pests using *Acremonium* spp. endophytes. *Proceedings of the New Zealand Grassland Association*, **56**, 33–38.

Rowan, D. D. & Popay, A. J. (1993). Endophytic fungi as mediators of plant-insect interactions. *Insect–Plant Interactions* (Ed. by E. A. Bernays), pp. 83–103 CRC Press, Ann Arbor.

Schmidt, S. P. & Osborne, T. G. (1993). Effects of endophyte-infected tall fescue on animal performance. *Agriculture, Ecosystems and Environment*, **44**, 233–262.

Siegel, M. R., Latch, G. C. M. & Johnson, M. C. (1987). Fungal endophytes of grasses. *Annual Review of Phytopathology*, **25**, 293–315.

Siegel, M. R., Latch, G. C. M., Bush, L. P., Fannin, N. F., Rowan, D. D., Tapper. B. A., Bacon, C. W. & Johnson, M. C. (1990). Fungal endophyte-infected grasses: alkaloid accumulation and aphid response. *Journal of Chemical Ecology*, **16**, 3301–3315.

Solomons, R. N., Oliver, J. W. & Linnabary, R. D. (1989). Reactivity of dorsal pedal vein of cattle to selected alkaloids associated with *Acremonium coenophialum*-infected fescue grass. *American Journal of Veterinary Research*, **2**, 235–238.

Stuedemann, J. A. & Hoveland, C. S. (1988). Fescue endophyte history and impact on animal agriculture. *Journal of Production Agriculture*, **1**, 39–44.

Stuedemann, J. A. & Thompson, F. N. (1993). Management strategies and potential opportunities to reduce the effects of endophyte-infested tall fescue on animal performance. *Proceedings of the 2nd International Symposium on* Acremonium/Grass *Interactions* (Ed. by D. E. Hume, G. C. M. Latch & H. S. Easton), pp. 103–114. AgResearch, Palmerston North.

Tsai, H. F., Siegel, N.R. & Schardl, C. L. (1992). Transformation of *Acremonium coenophialum*, a protective fungal symbiont of the grass *Festuca arundinacea. Current Genetics*, **22**, 399–406.

Tscharntke, T. & Greiler, H-J. (1995). Insect communities, grasses and grasslands. *Annual Review of Entomology*, **40**, 535–558.

Van Heeswijck, R. & McDonald, G. (1992). *Acremonium* endophytes in perennial ryegrass and other pasture grasses in Australia and New Zealand. *Australian Journal of Agricultural Research*, **43**, 1608–1709.

Van Heeswijck, R., McDonald, G., Woodward, J. & Huxley, H. (1993). Strategies for pasture pest control with a genetically modified endophyte. *Proceedings of the 2nd International Symposium on* Acremonium/Grass *Interactions* (Ed. by D. E. Hume, G. C. M. Latch & H. S. Easton), pp. 55–56. AgResearch, Palmerston North.

Vogl, A. (1898). Mehl und die anderan Mehlprodukkt der Cerealien und Leguminosen. *Zeitschrift Nahrungsmittel-Untersuchung, Hygiene und Warenkunde*, **12**, 25–29.

Wilson, A. D., Clement, S. L., Kaiser. W. J. & Lester, D. G. (1991). First report of clavicipitaceous anamorphic endophytes in *Hordeum* species. *Plant Diseases*, **75**, 215–217.

CONCLUDING REMARKS

J. B. WHITTAKER

The core of the studies reviewed in this chapter is summarized in Fig. 1 (modified from Allen 1991). Hatcher and Ayres concentrated on insect herbivore–plant pathogen–plant interactions whilst Clay and Prestidge and Ball dealt with endophytes and their impact on plants and herbivores in grasslands and agricultural communities respectively (the 'above-ground' of Fig. 1). Gange and Bower discussed the mycorrhizal–insect interaction and Roncadori the nematode–mycorrhizal component (the 'below-ground' of Fig. 1). Of course, one limitation of such a diagram is that the components are represented as if they were coexistent in time or space whereas Hatcher and Ayres emphasized the importance of indirect interactions. These may not necessarily be coincident in time or space but through modification of the substrate or environment, affect events elsewhere. In the case of mycorrhizas and nematodes too (Roncadori), the effects of each group on the other may be mediated through host changes rather than directly. The significance of spatial coincidence of mycorrhizal fungi and insects is a theme of Gange and Bower's presentation.

These considerations make the design of experiments involving multitrophic interactions problematical. In the case of insects and pathogens interacting on a single host, for example, it may well be that the sequence of inoculation is crucial. This can be a special problem when the lead time for establishment is different for different organisms. Colonization by mycorrhizal fungi, for example, may take 2–4 weeks whereas parasitic nematodes may establish in hours or days. Similarly, *Gastrophysa viridula*, the chrysomelid beetle studied by Hatcher and Ayres is present on its food plants (*Rumex* spp.) from April to October whereas field infection by the interacting rust *Uromyces rumicis* usually occurs from August to October.

A further consideration is that some of the effects of the various organisms may be short term and transient. Nevertheless, there is increasing evidence that the tolerance or resistance of a host to a particular organism may be considerably modified by the presence (not necessarily simultaneously) of other heterotrophs. It is not self evident that such interactions will be additive. Much depends on the extent to which host-plant chemistry is affected and whether the plant response involves the same broad-spectrum host defences or different signals.

Thus fungal endophytes, which are often toxic to vertebrate herbivores, may confer improved productivity and persistence of hosts by also reducing invertebrate, especially insect herbivory (Prestidge and Ball). However, they may also increase the vegetative vigour of infected plants, independent of any herbivore effect (Clay).

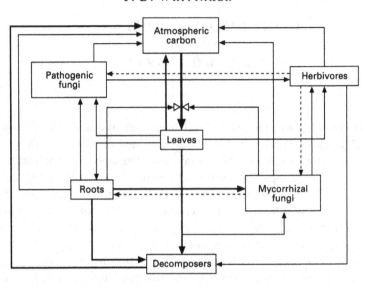

FIG. 1. The interrelationship between plants, pathogens, herbivores and mycorrhizas. (Adapted from Allen 1991.)

It is known (Whittaker 1979; Cottam *et al*. 1986) that insect herbivory can interact with plant competition to change its outcome and in this symposium Clay adds another dimension by showing that the outcome of competition may be further modified by the presence of endophyte infection.

The experimental work summarized here falls into two categories: experiments in which multitrophic interactions have been constructed as a feature of the experimental design (see Hatcher & Ayres and Clay) and the converse, the elimination or at least manipulation, of interactions within natural communities (Gange & Bower) to test the effects of exclusion of one or more participants. Both types of experiments are important – the first to allow study in detail of the mechanisms of multitrophic interactions and the second to enable us to determine the relevance of the findings from the more strictly controlled experimental designs to the field. It is becoming increasingly clear through the studies reviewed here that each component considered is likely to have more profound effects as part of a complex multitrophic system than in the simple conditions of single-factor experiments. Because a potential interaction can be conceived and even shown to occur under experimental conditions is no guarantee that it has important ecological relevance or (more difficult to establish) ecological generality. It is clear that in our attempts to understand the mechanisms at work in communities we cannot hope to investigate all possible interactions. The challenge that remains is to identify those components that are having important effects on the dynamics of each other and to broaden the scope of the pioneering studies described here to demonstrate their wider ecological and perhaps evolutionary significance and application in biological control.

REFERENCES AND FURTHER READING

Allen, M. F. (1991). *The Ecology of Mycorrhizae*. Cambridge University Press, New York.

Cottam, D. A., Whittaker, J. B. & Malloch, A. J. C. (1986). The effects of chrysomelid beetle grazing and plant competition on the growth of *Rumex obtusifolius*. *Oecologia*, **70**, 452–456.

Crawley, M. J. (1983). *Herbivory. The Dynamics of Animal–Plant Interactions*. Blackwell Scientific Publications, Oxford.

Summerhayes, V. S. & Elton, C. S. (1923). Contributions to the ecology of Spitzbergen and Bear Island. *Journal of Ecology*, **11**, 214–286.

Varley, G. C. (1957). Ecology as an experimental science. *Journal of Animal Ecology*, **26**, 251–261.

Warnock, A. J., Fitter, A. H. & Usher, M. B. (1982). The influence of a springtail *Folsomia candida* (Insecta, Collembola) on the mycorrhizal association of leek (*Allium porrum*) and the vesicular-arbuscular mycorrhizal endophyte *Glomus fasciculatus*. *New Phytologist*, **90**, 285–292.

Whittaker, J. B. (1979). Invertebrate grazing, competition and plant dynamics. *Population Dynamics* (Ed. by R. M. Anderson, B. D. Turner & L. R. Taylor), pp. 207–222. Symposia of the British Ecological Society, **20**. Blackwell Scientific Publications, Oxford.

PART 3
PLANT–ANIMAL
INTERACTIONS

INTRODUCTORY REMARKS

R. KARBAN

Department of Entomology, University of California, Davis, CA 95616, USA

Recently, there has been a renewed interest in understanding the factors that shape ecological communities, including those community modules centred on plants, their herbivores, and the predators and parasites of herbivores. The current debate compares the importance of top-down effects (control exerted by predators on lower trophic levels) and bottom-up effects (control exerted by resources available to each trophic level). This is not a new debate, but reflects a long development of ideas about multitrophic interactions. The notion that populations of herbivores are controlled by predators and parasites was earlier championed by Hairston *et al.* (1960) among others, whilst the idea that limited resources control herbivore populations was argued by Lindeman (1942), White (1978) and many others.

The current debate is more sophisticated than previous discussions. Almost all parties seem to agree that both predators and food can be important. To take an extreme point of view, I believe that a careful and skilful experimenter can demonstrate that almost any single process is statistically significant for a particular system. Such a demonstration is not very interesting. A far more useful exercise involves assessing the relative importance of several forces in determining community structure (Quinn & Dunham 1983; Welden & Slauson 1986; Karban 1989; Underwood & Petrotis 1993; Polis 1994). None the less, most of the participants in this symposium and in the larger debate first acknowledge this fact and then go on to emphasize either top-down or bottom-up forces to the exclusion of the other. Instead, I urge future workers to try not to advocate one position or the other, but to consider pluralistic and conditional models of the multiple processes that organize communities. By pluralistic, I mean that hypotheses should consider several processes acting simultaneously; predation, disease, competition for resources, competition for enemy free space, limitations imposed by abiotic conditions, etc. can all play roles, and these roles can be sorted out with experiments using appropriate statistical designs. By conditional, I mean that under certain conditions (ecosystems, environments, particular histories) some processes are likely to be more important and under other conditions other processes are likely to be more important. This lack of universality does not mean that generality is impossible and that we should all go about describing our particular systems without regard to the bigger picture. It does imply that very simple, universal models are not likely to be valid and that certain conditions will need to be specified before generalizations are meaningful. Pluralistic and conditional hypotheses are more difficult to formulate and less satisfying than simple, single-factor models and they have rarely been attempted in the past (see Schoener 1986, for an early attempt to specify conditions).

11. INDUCED RESPONSES IN TREES: MEDIATORS OF INTERACTIONS AMONG MACRO- AND MICRO-HERBIVORES?

S. H. FAETH AND D. WILSON

Department of Zoology, Box 871501, Arizona State University, Tempe, AZ 85287-1501, USA

INTRODUCTION

It is now well established that most, if not all, plants respond to mechanical and herbivore damage by changes in allelochemistry, nutrition, cell structure and growth, physiology, morphology and phenology (Faeth 1987, 1988, 1991a; Karban & Myers 1989; Tallamy & Raupp 1991). Changes in allelochemistry have been of particular interest to ecologists, because such responses may be another line of plant defence against herbivores, beyond the repertoire of constitutive chemical defences. The notion of induced defences is particularly appealing, because costly chemical defences are mobilized by the plant when attacked by herbivores, rather than constantly maintained, as in the cases of constitutive defences. The idea that induced responses can be effective defences against herbivores is generally accepted both on empirical (see references in reviews by Faeth 1988; Karban & Myers 1989) and theoretical (Edelstein-Keshet & Rauscher 1989; Riessen 1992; Adler & Karban 1994) grounds, despite some methodological problems (Fowler & Lawton 1985), complications caused by environmental variation (Faeth 1994; Bruin *et al.* 1995, Hunter & Schultz 1995; Karban & Niiho 1995) and alternative hypotheses for plant changes that occur after herbivory (e.g. carbon/nutrient balance, Tuomi *et al.* 1990; Bryant *et al.* 1991). Further, others have hypothesized that induced responses may indirectly increase plant fitness by attracting natural enemies of herbivores (Price 1986; Whitman 1988; Dicke *et al.* 1990; Turlings *et al.* 1990, 1991; Dicke & Sabelis 1992; Vet & Dicke 1992) or by serving as warning signals to nearby, non-damaged plants of imminent attack by herbivores (Baldwin & Schultz 1983; Bruin *et al.* 1995). Empirical tests for the latter two hypothesis, however, are currently scarce (Faeth 1994; Bruin *et al.* 1995).

Although the question of whether induced responses directly or indirectly increase plant fitness remains largely unresolved, it is clear that allelochemical, physiological, morphological and phenological changes can have important consequences for herbivores that feed upon host plants. Generally, changes in host plants during or following herbivory by invertebrates and vertebrates can affect behaviour, population dynamics and community structure of concurrent and subsequent herbivores (Faeth 1986, 1991a; Karban & Myers 1989; Tallamy & Raupp 1991), although the magnitude and direction of effects are not readily predictable (Faeth 1988, 1994). Janzen (1973) first explicitly recognized the potential for competition between temporally and spatially separated species feeding on the same host plant. Induced responses in host plants set the stage for the spectrum of positive and negative ecological interactions between herbivores, such as competition, amensalisms and mutualisms. This suite of interactions mediated by induced responses has been documented for many invertebrate and vertebrate herbivore–host plant systems (Faeth 1988, 1990, 1991a; Karban & Myers 1989; Tallamy & Raupp 1991).

While evidence is accumulating that invertebrate and vertebrate macroherbivores interact via induced changes in shared host plants, there are relatively few studies that examine potential interactions between micro-herbivores (such as bacteria, fungi and viruses) and macro-herbivores. This paucity is surprising considering that plants generally harbour enormous numbers and great species diversity of micro-organisms on plant surfaces (Fokkema & Van Den Heuvel 1986; Juniper & Southwood 1986; Andrews & Hirano 1991) and in plant tissues (Petrini 1986), either as mutualists, pathogens, saprophytes or commensalists. While some groups of micro-organisms such as mutualistic mycorrhizas and disease-causing microbes, are well studied because of their economic importance, few ecologists have considered the potential interactions between micro- and macro-herbivores via the induced changes they can cause in their mutual host plants. These interactions are likely to be very common since macro-herbivores, such as folivorous insects, contact, consume or introduce micro-organisms with literally every bite of a leaf. Indeed, Fowler and Lawton (1985) criticized induced defence studies because insect herbivory introduces micro-organisms into host plants from the plant surface and in saliva, which in turn may be the causative agent of induced responses, rather than the macro-herbivore damage *per se*.

Like macro-herbivores, micro-organisms can greatly alter chemistry, nutrition, physiology and morphology of host plants; for example, the elegant experiments of Karban and Carey (1984) and Karban *et al.* (1987) showed that a plant fungal pathogen, verticillum wilt, can cause induced changes that negatively affect herbivorous mites on cotton. Similarly, one mite species can alter the plant for later feeding mite species (Karban & Carey 1984; Karban *et al.* 1987; Karban 1991). Other studies have demonstrated that diseased plants present altered resources to insect herbivores (see references in Horsfall & Cowling 1980; Andrews & Hirano 1991) presumably because of induced changes occurring in the host plant. In some cases, micro-herbivores enhance plant quality for macro-herbivores (e.g. Berenbaum 1988; Johnson & Whitney 1994).

One group of micro-organisms that has recently attracted the attention of ecologists is the endophytic fungi – fungi that live internally and asymptomatically within plant tissues. In some grasses, fungal endophytes are specialized and thought to be plant mutualists because they produce mycotoxins that either kill or deter herbivores (Clay 1988, 1990; Carroll 1988, 1991; Breen 1994). The presence of endophytes has been viewed as a supplementary defence against herbivores, similar to induced defences. Endophytes in grasses have been termed 'acquired defences' (Cheplick & Clay 1988) and those in woody plants as evolutionary ways that long-lived plants can keep pace with short-lived herbivores (Carroll 1988). Fungal endophytes also are known to increase drought and flooding resistance and enhance germination success and competitive abilities of grasses (Clay 1992; Clay *et al.* 1993). While endophytic fungi are diverse and abundant in many woody plants (Petrini 1986), much less is known about their effects on these plants or on macro-herbivores sharing these host plants (Saikkonen & Neuvonen 1993; Saikkonen *et al.* 1996). In Douglas fir, the endophyte *Rhabdocline parkeri* (Sherwood-Pike *et al.* 1986) (Discomycete: Rhytismales) appears to have negative effects on several species of gall midges in the genus *Contrarinia* (Carroll 1988, 1991). The endophyte *Discula quercina* (Westd.) (Coelomycete) increases mortality of the gall former *Besbicus mirabilis* on Garrey oak (Wilson & Carroll 1994; Wilson 1995a). However, the endophytic fungus *Rhytisma acerinum* (Pers.) in sycamore positively affects populations of two aphid species (Gange 1996). There are relatively few studies of interaction between endophytes of woody plants and macro-herbivores relative to studies of endophytes in grasses (Carroll 1986, 1991).

Here, we report results of our long-term studies of interactions among fungal endophyte species inhabiting foliage of Emory oak *Quercus emoryi* (Fagaceae), and a dominant macro-herbivore, the leafminer *Cameraria* sp. nov. (Davis) (Lepidoptera: Gracilariidae).

STUDY ORGANISMS

Full descriptions of Emory oak, fungal endophyte species and the leaf miner are provided elsewhere (Faeth 1990, 1991b; Hammon & Faeth 1992, Faeth & Hammon 1996, 1997a, b). For sake of brevity, only salient features of each are included here.

Emory oak is a common evergreen oak at mid to high elevations in the southwestern USA. It flushes new leaves in April–May, just as 1-year-old leaves are abscised. *Cameraria* sp. nov., a univoltine leafminer, is the dominant herbivore on Emory oak at our study site in central Arizona, sometimes reaching densities of 10–20 mines per leaf (Faeth 1990, 1991b). After mating, females lay eggs on new leaf surface, eggs hatch in a few days, and larvae mine leaves for the next 11 months. Pupation occurs in the mine just before abscission and adults emerge in April–May.

More than 12 different species of fungal endophytes inhabit leaf tissues of Emory oak at the study site. However, three species comprise more than 90%

of all infections. These species are an ascomycete, *Ophiognomia cryptica*, Wilson and Barr, and two coelomycetes, *Plectophomella* sp. and *Asteromella* sp. For convenience, we call these three endophytes, QE2, QE7 and QE1, respectively. All three species are likely to colonize leaves via spores transmitted from rainsplash from other infected leaves (Wilson 1996). Infection levels of the three species vary seasonally (Fig. 11.1). New leaves are uninfected until summer rains when QE1 appears, while QE7 appears in the fall months, followed by peaks in QE2 during late winter rains. We have termed these species 'endophytes' because surface sterilization and subsequent growth on media indicate their presence inside leaves and they are asymptomatic for all, or at least a part, of their life-cycle. Others have used the term endophyte more restrictively to apply for fungi that interact mutualistically with the host plant (Wennström 1994). However, since very little is known of the ecological role of the vast majority of such fungi, it seems premature to use 'endophyte' to imply a specific ecological interaction (Wilson 1993, 1995b).

In this system, we address three basic questions relating to the potential for interactions within and among micro- (endophytic fungi) and macro-herbivores (leafminers) on a shared host plant (Emory oak):

1 Do fungal endophytes directly or indirectly alter population dynamics of leafminers?

2 Do leafminers alter abundances or diversity of fungal endophytes?

3 Do fungal endophyte species interact on shared host leaves?

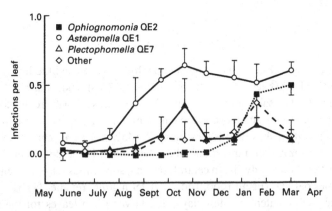

FIG. 11.1. Mean (± SE) seasonal changes in infection frequency of the three most common endophytic fungal species in the 1992–1993 growing season. Each point represents mean per cent infection of leaves from five Emory oak trees. The 'other' category is mean infection per leaf of all other endophytic fungal species. At least 50 leaves from each tree at each date were surface sterilized and cultured on nutrient agar.

FUNGAL ENDOPHYTES – EFFECTS ON LEAFMINERS

We have monitored infection levels of endophytes within leaves of Emory oak trees since 1989 (Hammon & Faeth 1992; Faeth & Hammon 1996, 1997a, b). Infection levels of individual endophytic species and all endophytes combined are significantly and positively associated with mines of *Cameraria* sp. nov. in two of the four growing seasons. Further, infection levels of leaves with dead leafminers are higher than those with live leafminers for one of these seasons, 1991–92 (Faeth & Hammon 1997a). Thus, we hypothesized that the presence of endophytes negatively affect leafminer survival, particularly in early instars (instars 1–4) where a large amount of unexplained mortality occurs (Faeth 1990, 1991b).

However, subsequent, and more detailed, correlative analyses of infection levels with leafminer survival and mortality in later growing seasons showed no relationship between infection levels and leafminer mortality (Faeth & Hammon 1997a). Presence of at least one endophyte species, *Asteromella* sp. (QE1), was still correlated with leafminers and finer-scale sampling showed endophyte infections were highest in the mined parts of leaves. However, neither unexplained mortality in early instars nor any other mortality factor in later instars (predation or parasitism) was associated with level of infection (Faeth & Hammon 1997a).

Similarly, experimental manipulations of endophyte infections did not provide strong evidence that endophytes affect leafminer mortality (Faeth & Hammon 1997b). In one experiment, we injected known quantities of spores of QE1, QE7 and another endophytic micro-organism, a filamentous yeast, directly into early instar mines. Control mines were injected with distilled water. There were no differences in survival of injected and control leafminers (Fig. 11.2). However, two endophyte species, QE7 and the filamentous yeast, were associated with delayed development (Fig. 11.3), since more leafminers in this treatment were still in prepupal stages at time of leaf abscission. Because larvae generally must complete final instars before leaf abscission in April and May, in order to survive, delayed development might result in increased mortality indirectly through leaf abscission. In another recent experiment where endophyte infections were manipulated on whole branches by protecting branches from rainsplash and spraying spore suspensions on the surface of leaves, we found increased abscission of heavily mined leaves (Wilson & Faeth, unpublished data). We are now testing the hypothesis that endophyte infections delay development and promote leaf abscission (fungal endophytes may accelerate leaf abscission through nutritional and chemical changes induced in the leaf). However, based upon results thus far, we conclude that endophytes are associated with mining activity, but they have only indirect effects on leafminer mortality. Preszler and Boecklen (1994) and Preszler *et al.* (1996) found similar results for another oak–endophyte–leafminer system. Early instar mortality was not strongly correlated with endophyte frequency. They did find,

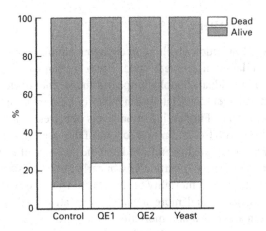

FIG. 11.2. Relative proportion of living and dead *Cameraria* larvae, prepupae or pupae in leaves with treatments of injections of sterilized, distilled water (control) or spore suspensions of QE1, QE7 or an endophytic yeast at the end of the 1992–1993 growing season. None of the treatments differed significantly from the control ($P > 0.05$).

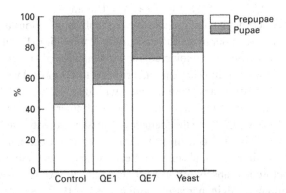

FIG. 11.3. Relative proportion of prepupae and pupae of living *Cameraria* leafminers in spore injection treatments of QE1, QE7, an endophytic yeast or sterilized, distilled water (controls) at the end of the 1992–1993 growing season. There were significantly more prepupae in the QE7 and yeast treatments relative to the controls ($P < 0.05$), indicating longer development time in these treatments.

however, a positive correlation between endophyte frequency at high elevations and increased parasitism of leafminers.

LEAFMINERS – EFFECTS ON ENDOPHYTES

It is clear that leafminers, or more likely leafmining activity, is associated with increased infection levels of all endophytic species. This association holds for spatial scales both within (mined areas) and between (mined vs unmined) leaves. We proposed that this association was mainly caused by physical and chemical changes induced by leafmining activity that facilitated infection of the leaf (Faeth & Hammon 1997a, b). *Cameraria* larvae mine upper leaf epidermal layers initially and then deeper into palisade layers during later instars. Surfaces of mines are brown in colour and the waxy cuticle that likely prevents most fungal penetration is diminished in mined areas.

In experiments where we artificially damaged upper leaf surfaces either by a pinprick or scraping a 1-mm^2 area of cuticle, infection levels of endophytes were higher than in paired control leaves (Wilson, Faeth & Anderson, unpublished data). These artificially damaged leaves did not differ from mined leaves in infection frequencies. Thus, it appears that leafmining activity and other herbivory or physical damage enhances localized colonization and presence of endophytic fungi in leaves of Emory oak. The interaction between endophytes and leafminers is thus most accurately classified as amensalistic, where the presence of the macro-herbivore positively influences populations of endophytes but the endophytes have little or no effect on the macro-herbivore population.

ENDOPHYTES AND LEAFMINERS – EFFECTS ON OTHER ENDOPHYTES

We addressed the question of how endophytic species sharing a leaf interact with other endophytic species through correlational and experimental studies. In tests of pairwise associations between endophytic species (QE1, QE2, QE7 and all other species combined) on five Emory oak trees in 1992–93 we found that the presence of one endophytic species is always negatively associated with another, and usually significantly so (Table 11.1), at least at the whole leaf spatial scale. The preponderance of negative associations suggests that infection by one endophyte species inhibits that of other species within a leaf.

However, inferring biological interactions from associational data can be misleading and sometimes erroneous (Sokal & Rohlf 1981; Ludwig & Reynolds 1988). Further, from previous studies of Emory oak, we know that leaves damaged either by macro-herbivores or artificially show significant and sustained increases in tannin content, particularly condensed tannins (Faeth 1986). Condensed tannins are known to be particularly effective antifungal agents (Zucker 1983). Thus, if the presence of one endophytic species inhibits colonization or growth of others in the same leaf, then this inhibition could be caused by one or more mechanisms. These include:

1 An endophyte may directly interfere in colonization of other fungi, either by direct competition for leaf space or by producing mycotoxins that inhibit germination

TABLE 11.1. Association of endophytic fungal species in leaves of five Emory oak trees in the 1992–1993 growing season based on presence/absence data from cultures of surface sterilized leaves. Results of *Chi square* tests for each sample data ($n > 50$ leaves/tree/date) are shown. All significant chi square values indicate that endophyte species are more negatively associated with each other than expected by chance. 'Other' represents collectively all fungal species other than *Asteromella* sp. (QE1), *Ophiognomia* sp. (QE2) and *Plectophomella* sp. (QE7).

| | Species pairs | | | | | |
Date	QE1/QE7	QE1/other	QE7/other	QE2/other	QE1/QE2	QE2/QE7
Aug 18	2.80 NS	a	a	a	a	a
Sept 19	4.80*	40.95***	a	a	a	a
Oct 18	64.80***	18.04***	17.29***	a	a	a
Nov 14	4.23*	23.08***	a	a	a	a
Dec 18	10.84**	36.38***	0.28 NS	0.01 NS	a	a
Jan 13	56.02***	83.46***	33.15***	18.59***	118.95***	54.13***
Feb 23	0.05 NS	11.70**	a	3.45 NS	35.89***	17.81***

a, too few observations (five or less in any cell) for a valid Chi square test; NS, not significant; $P > 0.05$; df = 1; *$P < 0.05$, df = 1; **$P < 0.01$, df = 1; ***$P < 0.001$, df = 1.

or growth of other fungi. Direct competition for leaf space is unlikely in our system because infection foci are very localized and growth is minimal in living leaf tissue.

2   An endophyte may indirectly inhibit other fungi by inducing the plant to increase tannins or other inducible allelochemicals in other parts of the leaf. Infections in leaves of other plants are known to cause increases in tannins and phytoalexins (Bell & Mace 1980; Fokkema & Van Den Heuvel 1986; Pirozynski & Hawksworth 1988).

3   For endophytes introduced by physical or macro-herbivore damage, the damage itself may result in increases in defensive compounds such as condensed tannins, in other parts of the leaf and thus inhibit colonization. This hypothesis suggests that macro-herbivores such as leafminers increase fungal colonization locally in the area of damage, but interfere with colonization in the leaf at other spatial scales.

To test these hypotheses, we artificially damaged one leaf of 190 Emory oak leaf pairs (6-mm diameter hole punches). All leaves were first surface-sterilized with 30% H_2O_2 to prevent inoculation of phyllosphere micro-organisms. Each leaf pair was on a terminal twig and was similarly sized and otherwise undamaged by macro-herbivores. At the end of the growing season (March 1995), we collected all remaining leaf pairs ($n = 112$, the remainder were missing one or both leaves or tags were missing), and surface-sterilized the leaves, divided each leaf into six equal segments, measured area of the segments, and plated the segments on agar medium. Endophytes were scored and identified for each leaf segment every 2 weeks for a 10-week period.

We used ANOVA, paired *t*-tests and non-parametric equivalents when data were non-normal to compare frequency of total endophyte infections per cm^2 of leaf. Damaged segments have higher mean levels of infection than undamaged segments from the same leaf (Table 11.2A), as we had predicted based upon results presented above. Physical damage, such as leafminer or other macro-herbivore damage, breaches the leaf surface and facilitates colonization of oak leaves by spores of endophytic fungi. On the basis of the entire leaf, damaged leaves have higher total infections per cm^2 than do the undamaged, paired leaves (Table 11.2B). Thus, relative to hypothesis 3, physical or macro-herbivore damage positively affects the presence of endophytes both within (damaged areas) but also between leaves.

We then compared levels of infection in undamaged segments of hole punched leaves and the undamaged pair (Table 11.2C). Despite higher infection levels overall, hole punched leaves have lower infection levels in undamaged segments than do undamaged leaves. This result suggests that damage increases infection probability locally (at the site of damage) and for the leaf as a whole, but may reduce probability of infection in undamaged leaf parts. Mechanical or macro-herbivore damage *per se* may either induce plant defences such as tannins, or the infections in the damage segment either directly or indirectly inhibit colonization and growth of other endophytic species in undamaged segments.

We had hoped to distinguish these alternatives in the final comparison (Table 11.2D). We compared infection frequencies of undamaged segments of hole-punched leaves for those leaves with and without infections in the damaged segment. If the damage *per se* inhibits infection then we expect no difference in infection levels from leaves with and without infections in the hole-punched segment. However, if the presence of the endophyte in a damaged segment contributes to inhibition in intact segments, then those leaves with infected, damaged segments should have lower infections in intact segments. This first prediction is supported. Infection levels do not differ significantly (Table 11.2D). However, there were only six leaves without infections in the damaged segment and these leaves did have higher mean

TABLE 11.2. Statistical tests of infection levels differences for (A) damaged segments vs. undamaged segments (same leaf); (B) damaged leaves vs. undamaged (control) leaves; (C) undamaged segments of damaged leaves vs. corresponding segments of undamaged (control) leaves; and (D) undamaged segments of damaged leaves when damaged segment is infected vs. uninfected.

Comparison	Mean infection (cm^{-2})	Mean infection (cm^{-2})	N	*t*	*P*	Test
	(Damaged)	(Undamaged)				
A	2.32 ± 1.27	0.71 ± 0.92	112	10.83	< .001	ANOVA
B	1.07 ± 0.54	0.92 ± 0.45	112	2.26	0.02	ANOVA
C	0.85 ± 0.50	0.97 ± 0.50	112	1.70	0.01	Signed rank
	(Infected)	(Uninfected)				
D	0.84 ± 0.56	0.97 ± 0.39	112, 6	0.57	0.57	ANOVA

infection level in the undamaged segments, although not significantly so. Thus, our results suggest damage *per se* inhibits infection by endophytes in undamaged parts of Emory oak leaves, but conclusive tests of direct endophyte inhibition or simply induction caused by damage, await other experiments.

DISCUSSION

It is clear from our studies of endophytes and leafminers on Emory oak that complexity is the norm, and categorization of interactions among and within micro-herbivore and macro-herbivore species is tenuous at best. At one spatial scale, the entire leaf, leafminer or other simulated macro-herbivore damage facilitates colonization of endophytes, probably by breaching otherwise inhospitable leaf surfaces. However, at finer spatial scales, damage may inhibit colonization or growth in undamaged segments, either by induction of plant chemical defences such as tannins, or by endophyte colonization in damaged segments that directly or indirectly inhibit colonization of other endophytes in other parts of the leaf.

Unlike earlier conventional ideas that many fungal endophytes are plant mutualists (Carroll 1988; Clay 1988, 1990; Strong 1988), we find no strong evidence that fungal endophytes in Emory oak directly defend plants against macroherbivores, such as the leaf miner, *Cameraria*. While we have not tested other macro-herbivores thoroughly, we *a priori* selected *Cameraria* as a test organism because we predicted effects of endophytes should be more detectable on this leafminer than on other macro-herbivores. *Cameraria* has a long larval development time of 11 months within leaves and thus long exposure to fungal endophytes. Larvae are unable to behaviourally avoid leaves (eggs and larvae are confined to a single leaf, selected by the ovipositing female) that may be lower quality due to the presence of fungal endophytes or their by-products. There is evidence that endophytes may slow development of *Cameraria* and thus indirectly increase mortality from premature leaf abscission, an induced phenological response (Faeth 1987, 1988, 1991a). Similarly, Preszler *et al.* (1996) found no correlation of endophyte frequency and direct mortality of another leafminer in Gambel's oak, but did find some evidence that endophytes are correlated with increased hymenopteran parasitism of the leafminer at high elevations. However, endophytes from other woody plants apparently can increase mortality of some insects, such as gall midges in Douglas fir (Carroll 1988, 1991) and a gall former in Garrey oak (Wilson & Carroll 1994; Wilson 1995a), while others increase growth and fecundity (Gange 1996).

We do not know if the presence of endophytes in Emory oak alters some other aspect of plant fitness, such as increased resistance to other macro-herbivores, or resistance to drought and water stress, as has been found for some endophytes of grasses. One fundamental difference between endophytes of some grasses and endophytes of woody plants is that endophytic fungi, such as *Acremonium* in fescue grasses, generally do not sporulate in nature but colonize offspring plants by growing into seeds on infected, maternal plants. Plant reproduction and fungal reproduction are thus closely linked, and in such cases of vertical transmission, mutualistic interactions are evolutionarily more probable (Ewald 1983, 1987, 1994; Williams & Neese 1991; Hammon & Faeth 1992). For endophytes of oaks, transmission is horizontal via spores carried by rainsplash. In cases of horizontal transmission, mutualistic interactions between endophyte and plant may be less likely (Hammon & Faeth 1992; Saikkonen *et al.* 1996).

Clearly, endophytes benefit from the presence of macro-herbivores at least as evidenced by higher infection levels in damaged segments and in damaged leaves overall. It seems this benefit is most likely due to a physical breach of the leaf by macro-herbivore feeding. However, there could be direct transmission of spores in saliva, since mobile macro-herbivores are likely to encounter countless spores on the surface of leaves. Macro-herbivores could thus not only provide routes for spore penetration but also serve as vectors and inoculators of spores from one leaf, or one plant, to another. Fowler and Lawton's (1985) suggestion, that induced chemical changes after macro-herbivore damage may be caused indirectly by the micro-organisms they introduced, certainly merits further examination.

From our study as well as others (e.g Saikkonen & Neuvonen 1993; Gange 1996; Saikkonen *et al.* 1996), it is apparent that endophytic fungi and other endophytic micro-organisms, such as bacteria and viruses, can interact positively, negatively or not at all with other microbial species and with macro-herbivores. Although in some cases these interactions may be direct, we suspect that many will prove to be indirect by inducing allelochemical, nutritional, phenological, physiological or morphological changes in the shared host plant. 'Indirect' does not equate to 'unimportant'. To the contrary, indirect interactions are likely to be much more frequent than direct ones in most communities. Further, these indirect interactions between and among micro- and macro-herbivores are often ignored by ecologists, have important implications for ecological and evolutionary theories of host range, temporal and spatial feeding patterns, and abundance and diversity of insect–plant interactions (Hammon & Faeth 1992); for example, premature leaf abscission, an important source of mortality for sedentary folivorous insects, may be mediated by the introduction of micro-organisms rather than by insect damage *per se*. Insects may preferentially oviposit and feed on rapidly growing vigorous plants (the 'plant vigour' hypothesis, (Price 1991)) because of the low endophyte infection status of vigorous tissues. Within leaves, insect damage may be over-dispersed not only because of 'halos' of induced plant defences (Edwards & Wratten 1983, 1985), but because insect feeding either directly or indirectly enhances localized infections of endophytes, which in turn alter plant tissues in the region of damage. Similarly, temporal patterns of folivorous insect feeding, such as concentration of feeding early in the season on young leaves, was hypothesized to be caused by changes in phytochemistry (Feeny 1970). However, the virtual absence of endophyte infections in new leaves provides an alternative explanation (Hammon & Faeth 1992). At leat one recent study (Gange 1996) provides empirical support for this alternative 'endophyte' explanation.

The leaf surface and interior provide a wide arena for interactions among micro-herbivores, such as the endophytic fungi. These interactions are likely to vary widely in time and space and with the great diversity of microscopic organisms. The negative association between endophytic fungi on Emory oak suggests competitive interactions among some species, but our preliminary experiments are not conclusive. We should not expect easy or simple answers since the endo- and epiphytic community on the surface and interior of a single leaf may be more complex than the macro-species community in an entire intertidal zone or a deciduous forest. Similarly, the classic categorization of micro-organism species associated with plants as mutualistic, symbiotic, pathogenic or saprophytic is probably unrealistic since these interactions are not fixed in time and space.

ACKNOWLEDGEMENTS

We thank R. Anderson, J. Burton, S. Coleman, K. Craven, C. Hudson and H. Triplett for assistance in this research. R. A. Prestidge and an anonymous reviewer provided valuable comments on the manuscript. This research was supported by NSF grants BSR 91-07296 and DEB 94-06934 to SHF.

REFERENCES

Adler, F. R. & Karban, R. (1994). Defended fortresses or moving targets? Another model of inducible defenses inspired by military metaphors. *American Naturalist*, **144**, 813–832.

Andrews, J. H. & Hirano, S. S. (1991). *Microbial Ecology of Leaves*. Springer, New York.

Baldwin, I. T. & Schultz, J. C. (1983). Rapid changes in tree leaf chemistry induced by damage: evidence for communication between plants. *Science*, **221**, 277–279.

Bell, A. A. & Mace, M. E. (1980). Biochemistry and physiology of resistance *Fungal Wilt Diseases of Plants* (Ed. by E. Mace, A. A. Bell & C. H. Beckman), pp. 431–486. Academic Press, New York.

Berenbaum, M. R. (1988). Allelochemicals in insect–microbe–plant interactions; agents provocateurs in the coevolutionary arms race. *Novel Aspects of Insect–Plant Interactions* (Ed. by P. Barbosa & D. K. Letourneau), pp. 97–123. John Wiley & Sons, New York.

Breen, J. P. (1994). *Acremonium* endophyte interactions with enhanced plant resistance to insects. *Annual Review of Entomology*, **39**, 401–423.

Bruin, J., Sabelis, M. W. & Dicke, M. (1995). Do plants tap SOS signals from their infested neighbors? *Trends in Ecology and Evolution*, **10**, 167–170.

Bryant, J. P., Danell, K., Provenza, F., Reichardt, P. B., Clausen, T. A. & Werner, R. A. (1991). Effects of mammal browsing on the chemistry of deciduous woody plants. *Phytochemical Induction by Herbivores* (Ed. by D. W. Tallamy & M. J. Raupp), pp. 135–154. John Wiley & Sons, New York.

Carroll, G. C. (1986). The biology of endophytism with particular reference to woody perennials.

Microbiology of the Phyllosphere (Ed. by J. Fokkema & J. Van Den Heuvel), pp. 205–222. Cambridge University Press, Cambridge.

Carroll, G. C. (1988). Fungal endophytes in stems and leaves: from latent pathogen to mutualistic symbiont. *Ecology*, 69, 2–9.

Carroll, G. C. (1991). Fungal associates of woody plants as insect antagonists in leaves and stems. *Microbial Mediation of Plant–Herbivore Interactions* (Ed. by P. Barbosa, V. A. Krischik & C. G. Jones), pp. 253–271. John Wiley & Sons, New York.

Cheplick, G. P. & Clay, K. (1988). Acquired chemical defenses of grasses: the role of fungal endophytes. *Oikos*, 52, 309–318.

Clay K. (1988). Fungal endophytes of grasses: a defensive mutualism between plants and fungi. *Ecology*, 69, 10–16.

Clay, K. (1990). Fungal endophytes of grasses. *Annual Review of Ecology and Systematics*, 21, 275–297.

Clay, K. (1992). Fungal endophytes of plants: biological and chemical diversity. *Natural Toxins*, 1, 147–149.

Clay, K., Marks, S. & Cheplick, G. P. (1993). Effects of insect herbivory and fungal endophyte infection on competitive interactions among grasses. *Ecology*, 74, 1767–1777.

Dicke, M. & Sabelis, M. W. (1992). How plants obtain predatory mites as bodyguards. *Netherlands Journal of Zoology*, 38, 148–165.

Dicke, M., Sabelis, M. W., Takabayashi, J., Bruin, J. & Posthumus, M. A. (1990). Plant strategies of manipulating predator–prey interactions through allelochemicals: prospects for application in pest control. *Journal of Chemical Ecology*, 16, 3091–3118.

Edelstein-Keshet, L. & Rauscher, M. (1989). The effects of inducible plant defenses on herbivore populations. I. Mobile herbivores in continuous time. *American Naturalist*, 133, 787–810.

Edwards, P. J. & Wratten, S. D. (1983). Wound-induced defenses in plants and their consequences for patterns of insect grazing. *Oecologia*, 59, 88–93.

Edwards, P. J. & Wratten, S. D. (1985). Induced plant defences against insect grazing: fact or artefact? *American Zoologist*, 44, 70–74.

Ewald, P. W. (1987). Host–parasite relations, vectors, and the evolution of disease severity. *Annual Review of Ecology and Systematics*, 14, 465–485.

Ewald, P. W. (1983). Transmission modes and the evolution of the parasitism–mutualism continuum. *Annals of the New York Academy of Science*, 503, 295–306.

Ewald, P. W. (1994). *Evolution of Infectious Disease*. Oxford University Press, Oxford.

Faeth, S. H. (1986). Indirect interactions between temporally separated herbivores mediated by the host plant. *Ecology*, 67, 479–494.

Faeth, S. H. (1987). Community structure and folivorous insect outbreaks: the roles of vertical and horizontal interactions. *Insect Outbreaks* (Ed. by P. Barbosa & J. C. Schultz), pp. 135–171. Academic Press, New York.

Faeth, S. H. (1988). Plant-mediated interactions between seasonal herbivores: enough for evolution or coevolution? *Chemical Mediation of Coevolution* (Ed. by K. C. Spencer), pp. 391–414. Academic Press, New York.

Faeth, S. H. (1990). Aggregation of a leafminer, *Cameraria* sp. nov. (Davis): consequences and causes. *Journal of Animal Ecology*, 59, 569–586.

Faeth, S. H. (1991a). Variable induced responses: direct and indirect effects on oak folivores. *Phytochemical Induction by Herbivores* (Ed. by M. J. Raupp & D. W. Tallamy), pp. 135–171. John Wiley & Sons, New York.

Faeth, S. H. (1991b). The effect of oak leaf size on abundances, dispersion, and survival of the leafminer, *Cameraria* sp. nov. (Lepidoptera: Gracilariidae). *Environmental Entomology*, 20, 196–204.

Faeth, S. H. (1994). Induced plant responses: effects on parasitoids and other natural enemies of phytophagous insects. *Parasitoid Community Ecology* (Ed. by B. A. Hawkins & W. Sheehan), pp. 245–260. Oxford Science Publications, Oxford.

Faeth, S. H. & Hammon, K. E. (1996). Fungal endophytes and phytochemistry of oak foliage: determinants of oviposition preference of leafminers? *Oecologia* (in press).

Faeth, S. H. & Hammon, K. E. (1997a). Fungal endophytes in oak trees. I. Long-term patterns of abundances and associations with leafminers. *Ecology* (in press).

Faeth, S. H. & Hammon, K. E. (1997b). Fungal endophytes in oak trees. II. Experimental analyses of interactions with leafminers. *Ecology* (in press).

Feeny P. (1970). Seasonal changes in oak leaf tannins and nutrients as a cause of spring feeding by winter moth caterpillars. *Ecology*, **51**, 565–581.

Fokkema, N. J. & Van Den Heuvel, J. (Eds) (1986). *Microbiology of the Phyllosphere*. Cambridge University Press, Cambridge.

Fowler, S. V. & Lawton, J. H. (1985). Rapidly induced defences and talking trees: the devil's advocate position. *American Naturalist*, **126**, 181–195.

Gange, A. C. (1996). Positive effects of endophyte infections on sycamore aphids. *Oikos*, **75**, 500–510.

Hammon, K. E. & Faeth, S. H. (1992). Ecology of plant-herbivore communities: a fungal component? *Natural Toxins*, **1**, 197–208.

Horsfall, J. G. & Cowling, E. B. (Eds) (1980). *Plant Disease – An Advanced Treatise*, Vol. 5. Academic Press, New York.

Hunter, M. D. & Schultz, J. C. (1995). Fertilization mitigates chemical induction and herbivore responses within damaged oak trees. *Ecology*, **76**, 1226–1232.

Janzen, D. H. (1973). Host plants as islands. II. Competition in evolutionary and ecological time. *American Naturalist*, **107**, 786–790.

Johnson, J. A. & Whitney, N. J. (1994). Cytotoxicity and insecticidal activity of endophytic fungi from black spruce (*Picea mariana*) needles. *Canadian Journal of Microbiology*, **40**, 24–27.

Juniper, B. & Southwood, T. R. E. (1986). *Insects and the Plant Surface*. Edward Arnold, London.

Karban, R. (1991). Inducible resistance in agricultural systems. *Phytochemical Induction by Herbivores* (Ed. by D. W. Tallamy & M. J. Raupp), pp. 402–419. John Wiley & Sons, New York.

Karban, R. & Carey, J. R. (1984). Induced resistance of cotton seedlings to mites. *Science*, **225**, 53–54.

Karban, R. & Myers, J. H. (1989). Induced plant responses to herbivory. *Annual Review of Ecology and Systematics*, **20**, 331–348.

Karban, R. & Niiho, C. (1995). Induced resistance and susceptibility to herbivory: plant memory and altered plant development. *Ecology*, **76**, 1220–1225.

Karban, R., Adamchak, R. & Schanthorst, W. C. (1987). Induced resistance and interspecific competition between spider mites and a vascular wilt fungus. *Science*, **235**, 678–680.

Ludwig, J. A. & Reynolds, J. F. (1988). *Statistical Ecology*. John Wiley & Sons, New York.

Petrini, O. (1986). Taxonomy of endophytic fungi of aerial plant tissues. *Microbiology of the Phyllosphere* (Ed. by N. J. Fokkema & J. Van Den Heuvel), pp. 175–187. Cambridge University Press, Cambridge.

Pirozynski, K. A. & Hawksworth, D. L. (1988). *Coevolution of Fungi with Plants and Animals*. Academic Press, New York.

Preszler, R. W. & Boecklen, W. J. (1994). A three-trophic level analysis of the effects of plant hybridization on a leaf-mining moth. *Oecologia*, **100**, 66–73.

Preszler, R. W., Gaylord, E. S. & Boecklen, W. J. (in press). Reduced parasitism of a leaf-mining moth on trees with high infection frequencies of an endophytic fungus. *Oecologia*.

Price, P. W. (1986). Ecological aspects of host plant resistance and biological control: interactions among three trophic levels. *Interactions of Plant Resistance and Parasitoid Predators of Insects* (Ed by D. J. Boethel & R. D. Eikenbarry), pp. 11–30. Ellis Horwood, Chichester.

Price, P. W. (1991). The plant vigor hypothesis and herbivore attack. *Oikos*, **62**, 244–251.

Riessen, H. P. (1992). Cost–benefit model for the induction of an antipredator defense. *American Naturalist*, **140**, 349–362.

Saikkonen, K. & Neuvonen, S. (1993). European sawfly and microbial interactions mediated by the host plant. *Sawfly Life History Adaptation to Woody Plants* (Ed. by M. Wagner & K. Raffa), pp. 431–450. Academic Press, San Diego.

Saikkonen, K., Helander, M., Ranta, H., Neuvonen, S., Virtanen, T., Suomela, J. & Vuorinen, P. (1996). Endophyte-mediated interactions between woody plants and insect herbivores? *Entomologia Experimentalis et Applicata*, 80, 269–271.

Sherwood-Pike, M., Stone, J. & Carroll, G. C. (1986). *Rhabdocline parkerii*, a ubiquitous foliar endophyte of Douglas fir. *Canadian Journal of Botany*, 64, 1849–1855.

Sokal, R. R. & Rohlf, F. J. (1981). *Biometry*, 2nd edn. W. H. Freeman, San Francisco.

Strong, D. R., Jr (1988). Endophytic mutualism and plant protection from herbivores. *Ecology*, 69, 1.

Tallamy, D. W. & Raupp, M. J. (Eds) (1991). *Phytochemical Induction by Herbivores*. John Wiley & Sons, New York.

Tuomi, J., Niemelä, P. & Siren, S. (1990). The Panglossian paradigm and delayed inducible accumulation of foliar phenolics in mountain birch. *Oikos*, 59, 399–410.

Turlings, T. C. J., Tumlinson, J. H. & Lewis, W. J. (1990). Exploitation of herbivore-induced plant odors by host-seeking parasitic wasps. *Science*, 250, 1251–1253.

Turlings, T. C. J., Tumlinson, J. H., Eller, F. J. & Lewis, W. J. (1991). Larval-damaged plants: source of volatile synomones that guide the parasitoid *Cotesia marginiventris* to the micro-habitat of its hosts. *Entomologia Experimentalis et Applicata*, 58, 75–82.

Vet, L. E. M. & Dicke, M. (1992). Ecology of infochemical use by natural enemies in a tritrophic context. *Annual Review of Entomology*, 37, 141–172.

Wennström, A. (1994). Endophyte – the misuse of an old term. *Oikos*, 71, 535–536.

Whitman, D. W. (1988). Allelochemical interactions among plants, herbivores, and their predators. *Novel Aspects of Insect-Plant Interactions* (Ed. by P. Barbosa & D. Letourneau), pp. 11–64. John Wiley & Sons, New York.

Williams, G. C. & Neese, R. M. (1991). The dawn of Darwinian medicine. *The Quarterly Review of Biology*, 66, 1–22.

Wilson, D. (1993). Fungal endophytes: out of sight but should not be out of mind. *Oikos*, 68, 379–384.

Wilson, D. (1995a). Fungal endophytes which invade insect galls: insect pathogens, benign saprophytes, or fungal inquilines? *Oecologia*, 103, 255–260.

Wilson, D. (1995b). Endophyte – the evolution of a term, and clarification of its use and definition. *Oikos*, 73, 274–276.

Wilson, D. (1996). Manipulation of infection levels of horizontally transmitted fungal endophytes in the field. *Mycological Research*, 100, 827–830.

Wilson, D. & Carroll, G. C. (1994). Infection studies of *Discula quercina*, an endophyte of *Quercus garryana*. *Mycologia*, 86, 635–647.

Zucker, W. V. (1983). Tannins: does structure determine function? *American Naturalist*, 121, 335–365.

12. HOST-PLANT MEDIATED INTERACTIONS BETWEEN SPATIALLY SEPARATED HERBIVORES: EFFECTS ON COMMUNITY STRUCTURE

G. J. MASTERS AND V. K. BROWN

International Institute of Entomology (An Institute of CAB International), 56 Queen's Gate, London SW7 5JR, UK

INTRODUCTION

Central questions in community ecology include how species interact and what are the consequences of such interactions for community structure and function? Attempts to answer such questions are now woven into standard ecological texts (e.g. Begon *et al.* 1990; Putman 1994). Inter- and intra-specific interactions are varied and complex, resulting in many attempts to define and categorize them (e.g. Abrams 1987). Simply, interactions can be classified as direct, involving pairwise interactions between species (e.g. predation/parasitism), or indirect, involving mediation by a third party. Direct interactions are well documented, with examples being found in this symposium (e.g. Jones *et al.*; Tscharntke, this volume). By their nature, these interactions are simply between or within trophic levels, and through their simplicity are commonly naive in terms of ecological communities. However, these direct interactions often serve as a cornerstone to more complex interactions, which do influence community structure and function. The occurrence of indirect interactions between species is equally well cited (e.g. Bach 1991), though their significance provides a challenge which is yet to be fully grasped (Strauss 1991). The importance of indirect interactions has been highlighted by the apparent competition models of Holt (1977), Holt and Lawton (1993) and a growing number of empirical studies, notably Faeth (1986), Karban (1986), Schmitt (1987), Strauss (1987) and Faeth and Wilson (this volume). Essentially, as the number of species increases within a community, the occurrence and significance of indirect interactions will also increase, thereby exacerbating the contribution of such interactions in the dynamics of communities. Even within the simplest of communities, there is scope for numerous indirect interactions. These, in turn, may be modified by abiotic factors.

In this review, we focus on indirect interactions and use as our model the co-existence of species exploiting a single host plant (Fig. 12.1). Direct interactions, between herbivore and plant (between trophic levels, Fig. 12.1a(i)) and between plants (within a trophic level, Fig. 12.1a(ii)) are the building blocks for more complex systems. The impact of a specialist herbivore (restricted to a single plant species) can have indirect effects on a competing plant species (Fig. 12.1b(i)). Likewise, a generalist herbivore, feeding on two or more plant species, can influence both direct and indirect interactions between plants (Fig. 12.1b(ii)). Hence, herbivory will clearly impact on the competitive ability of plant species within a population or community, as reviewed by Louda *et al.* (1990). We woefully omit further consideration of this topic since it has received attention elsewhere and is beyond the scope of this contribution. Rarely in nature is a plant only attacked by a single herbivore, be it a species or individual. Hence, the effects of herbivory by a species have the potential to affect other herbivores exploiting the same host plant. It is on this indirect interaction which we now focus.

Our critique is restricted to herbivorous insects, since they exhibit diverse feeding relations with host plants, many of which are well understood in terms of their effects on plant fitness. Phytophagous insects also show a range of interactions

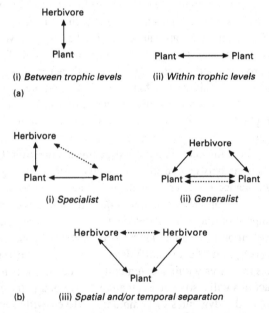

FIG. 12.1. Direct (solid lines) and indirect (dotted lines) interactions within simple communities. (a) Pairwise interactions; (i) between trophic levels (e.g. herbivory) and (ii) within trophic levels (e.g. plant competition). (b) Direct and indirect interactions between two trophic levels but three species; (i) specialist herbivore (feeding only on one plant species) and its indirect effect on a second non-host plant, (ii) generalist herbivore mediation of an indirect, or direct, interaction between two host plants, and (iii) host-plant mediation of an indirect interaction between two spatially and/or temporally separated herbivores.

with their host plants, and their relatively short life-cycles mean that they can be temporally separated, while their diversity of feeding strategies on different parts of the plant enables them to be spatially separated. In each case, the common denominator is the host plant, providing the potential for the study of a common, but often ignored, type of multitrophic interaction, namely herbivore–plant–herbivore (Fig. 12.1b(iii)). Because of the abundance of insects, such interactions are widespread in nature. Our emphasis is to assess the role of this type of interaction in the structure and functioning of communities. Initially, we consider the diversity of insect–related interactions on plants and their impact *per se*, before exploring a model system in depth. Finally, we detail evidence for such interactions at the individual, population and community level before speculating on their significance and implications in community structure and function.

INSECTS: DIVERSITY AND IMPACT ON PLANTS

Estimates of the total number of extant species differ by at least an order of magnitude (May 1988; Reid 1992), even so, few question that a large proportion of the world's fauna comprises phytophagous insects (Southwood 1978). Indeed, Hendrix (1980) estimated that 50% of insect species are phytophagous, equating to approximately 25% of all the species on earth. Insect herbivores exploit plants in a variety of ways, attacking all plant tissues, for example feeding on leaves, stems, roots, flowers, seeds and fruits, whilst living externally or internally, and feeding on the mesophyll or liquid phloem or xylem. Through their habit of exploiting a large number of plant resources, several insect species (comprising one, few or many individuals) can share the same plant at any one time or at different times during the plant's life-cycle. Thus, the potential for a host plant to provide feeding niches increases with the longevity of the plant and its structural complexity (Lawton & MacGarvin 1986).

The majority of work on insect herbivory concentrates on above-ground feeding. Herbivory has been found to reduce leaf area and number, leaf turnover, plant stature and fruit production and quality, all measures of plant performance (e.g. Brown *et al.* 1987; Gange 1990; Louda & Potvin 1995; Vranjic & Gullan 1990). Herbivory has deleterious effects on plant reproduction, mainly through reducing the size and number of seeds (e.g. Bentley *et al.* 1980; Louda 1984). Additionally, direct seed predation can reduce several fold the number of viable seeds per plant (e.g. Vaughton 1990) and limit recruitment in a seed-limited population (Louda & Potvin 1995). Thus, plant fitness can be modified by insect herbivory. In the extreme case, herbivory can induce plant mortality, although usually the effect is less dramatic.

Foliar feeding is known to induce physiological responses within the host plant (Harborne 1988) – be they changes in nutrient balance or water relations, or where the presence of the herbivore forms an additional sink for plant nutrients (i.e. diverting nutrients away from plant growth), or changes in secondary compounds. Defoliation (namely a reduction in leaf area) equates to a reduction in photosynthesis and may stimulate the host plant to compensate for this loss in an attempt to maintain plant growth and metabolism. Such compensatory responses generally include an increase in the photosynthetic rate in the remaining foliage to initiate regrowth. Alternatively (and/or additionally), plants which invest in storage can translocate carbon to regrowth areas (Caldwell *et al.* 1981). Another direct physiological response to herbivory is a change in plant nutrient levels. For example, feeding by sap-feeding, galling or mining insects can lead to the formation of an additional sink for plant nitrogen and carbon (Llewellyn 1982). This can often lead to a significant reduction of nitrogen and/or carbon within the host plant, and consequent alteration of the C:N ratio. Gange and Brown (1989) and Masters and Brown (1992) found that sap feeding or leaf mining resulted in a significant reduction in total foliar nitrogen. Herbivory can also lead to an increase in the concentration of secondary compounds within the host plant. Such chemicals are thought to have a defensive role against subsequent herbivore attack (Karban & Myers 1989). Hartley and Lawton (1987) investigated the effects of insect grazing, mining and artificial hole punching on birch leaves and found that the concentration of phenolics increased by 9–25% (depending upon the type of damage). Similar responses have been found in herbaceous species. Defensive responses are not restricted to phenolics; alkaloids, hydroxamic acids, proteinase inhibitors, tannins and terpenes can be produced in response to insect attack (Croxford 1990).

Until relatively recently, below-ground insect herbivory has been largely ignored by ecologists, although it has a long history in the agricultural and biocontrol literature. However, below-ground herbivory has now been reviewed (e.g. Andersen 1987; Brown & Gange 1990) and has become the focus of experimental work (e.g. Goldson *et al.* 1987; Brown & Gange 1989). The majority of below-ground structures are roots and rhizomes, but tubers, corms, bulbs, nodules and stems are all liable to attack by insects, albeit by a more restricted range of taxa. Like foliar herbivores, root feeding has detrimental effects on plant performance, reducing biomass, leaf area, stature and growth (e.g. Gange & Brown 1989; Wedderburn *et al.* 1990; Spike & Tollefson 1991; Masters & Brown 1992) and plant fecundity (Powell & Myers 1988; Müller-Scharer 1991).

Below-ground insect herbivory has dramatic effects on the physiological status of the plant. Insect root feeders generally prune, or probe (in the case of root-feeding Homoptera), the fine roots of plants which are essential for the uptake of water and nutrients. In addition, more specialist internal feeders may exploit the entire root system (Müller *et al.* 1989). Other below-ground insects can feed on mycorrhizas which are very important in nutrient uptake by the plant. Through decreasing the plant's ability to capture soil water and nutrients, stress may be induced within the host plant (Brown & Gange 1990; Masters *et al.* 1993), which is commonly manifest by a decrease in foliar water content (Ridsdill Smith 1977). Where stress occurs, plant growth is maintained by a variety of responses (Hsiao 1973). If the stress is not transient, then the majority of plants respond by mobilizing/ translocating amino acids (soluble nitrogen) and carbohydrates (sugars) from the roots to the foliage (Hsiao 1973; Mattson & Haack 1987). Root herbivory has been found to decrease foliar water content and increase the soluble nitrogen and carbohydrate levels found in the leaves (Gange & Brown 1989; Masters 1995). This can happen even before wilting occurs.

INTERACTIONS BETWEEN INSECT HERBIVORES: CLASSIFICATION AND DIVERSITY

A spatial interaction is defined here as one where two species share the same host plant at the same time, but utilize different feeding niches, such as that between insects feeding on the upper and lower surfaces of leaves, outside or inside of plant structures or above- and below-ground. We would include here insects exploiting the plant in a different way (namely the different feeding guilds of Root (1973), such as leaf chewer and leaf miner; stem galler and root sucker). Clearly, it is easy

	SEPARATION	EXAMPLE
SPATIAL	Niche	Leaf vs. stem
	Guild	Chewer vs. sucker
	Niche and guild	Root chewer vs. stem sucker
TEMPORAL	Feeding time	Early season leaf chewer ⟶ Late season leaf chewer
SPATIO-TEMPORAL	Niche	Early season leaf feeder ⟶ Late season flower feeder
	Guild	Early season chewer ⟶ Late season miner
	Niche and guild	Early season root sucker ⟶ Late season leaf chewer

FIG. 12.2. Classification of host-plant mediated interactions between insect herbivores. Insects can be spatially separated in terms of niches (structure fed on), guilds (resource fed on) or both. Interactions can be through time, denoted by arrows.

to envisage a temporal component in the interaction, such as a species feeding on a structure early in the season, whereas others feed later. We therefore recognize three categories of indirect interactions involving two (or more) insect herbivores and their common host plant:

1   *spatial interactions*: where the insect herbivores are of the same or different guilds, but exploit different niches on a common host plant at the same time;

2   *temporal interactions*: where the insect herbivores exploit the same feeding niche, but at different times;

3   *spatio-temporal*: where temporally separated insect herbivores are of different guilds and/or exploit different niches.

This classification is outlined in Fig. 12.2, which details the different host-plant mediated interactions between insect herbivores. All of these interactions can be beneficial, detrimental or neutral in terms of growth and performance of the insect herbivore.

Spatial interactions

Evidence is provided by an early study by Kidd *et al.* (1985) which illustrates the fine-scale utilization of niches that can occur between 'interacting' insects. Their study system was two aphid species (*Schizolachnus pineti* and *Eulachnus agilis*) feeding solely on pine needles, especially of the shared host, *Pinus sylvestris*. However, *S. pineti* occupied only the outer surface of a needle, whilst the inner surface was occupied by *E. agilis*. They found that *E. agilis* was twice as likely to be found on needles occupied by *S. pineti* as on unoccupied needles. This association was found to benefit *E. agilis* through greater survival, faster growth rates, larger size and potentially increased fecundity (Leather & Dixon 1984). This indirect and asymmetric beneficial interaction was suggested to be in response to differences in needle nutritional quality (Kidd *et al.* 1985). Chongrattanameteekul *et al.* (1991) found a similar interaction between the cereal aphids *Rhopalosiphum padi* and *Sitobion avenae* feeding on wheat. At low densities, *R. padi* increased the fecundity and longevity of *S. avenae*. *S. avenae* had no significant effect (though there was a trend towards a beneficial effect) on the performance of *R. padi*. This is suggestive of an indirect interaction beneficial to both species. However, competition was evident at high densities (four times the lowest density), with both species having reduced longevity and fecundity as a result of a combination of intra- and inter-specific effects.

Temporal interactions

The majority of studies of interactions between two or more herbivore species have generally involved herbivory at one point in time having some effect on a herbivore feeding later (e.g. Haukioja 1980). When such insects share a feeding site, their separation in time may involve physical or chemical changes in the structure they occupy. Most work to date has involved chewing insects feeding on trees; for example, defoliation of oak and birch early in the growing season has been shown to have adverse effects on the performance of insects feeding later (e.g. Hunter 1987). However, not all interactions are negative and thereby suggestive of induced defences. Williams and Myers (1984) investigated the interaction between western tent caterpillars (*Malacasoma californicum pluviale*) and fall webworms (*Hyphantria cunea*) mediated by the common host plant, red alder (*Alnus rubra*). Webworm larvae, feeding on leaves which had been subjected to moderate densities of previous herbivory by western tent caterpillars, grew faster and had greater pupal weights than those fed on undamaged leaves.

Surprisingly, there have been few experimental tests of the effects of the same species feeding at different times on the same host plant, which would indicate the potential for complex temporal interactions. Seedling herbivory by the multivoltine bug *Lygus rugulipennis* resulted in a similar reduction in reproductive biomass of *Tripleurospermum inodorum*, as when feeding throughout the life of the plant. There was, however, a more dramatic effect on vegetative biomass of the adult plant. Herbivory during the seedling stage resulted in a 71% reduction in adult plant biomass compared with only a 48% reduction when bug feeding occurred throughout the life of the plant (Cervantes Peredo, unpublished data). Clearly, these effects on seedling plants have potential effects for insects (of the same or different species) feeding later. This is an avenue of research which, to our knowledge, is largely unexplored.

Spatio-temporal interactions

Interactions combining both a spatial and temporal component are probably the most common in nature. Few natural systems will not involve some overlap, due to lack of synchrony of the life-cycles of the herbivores which exploit different niches or are of different guilds; for example, host-plant mediated inter-specific competition has been found in laboratory studies between the Colorado potato beetle (*Leptinotarsa decemlineata*), a leaf chewer, and the potato leafhopper (*Empoasca fabae*), a leaf sucker (Tomlin & Sears 1992). Leafhoppers were found to produce fewer eggs when feeding on plants that had been attacked by the leaf beetle, whilst beetle fitness (survival of second generation larvae to adult and larval weight gain) was significantly decreased when feeding on leaves damaged by the leafhoppers. In another example, defoliation of *Erigeron glaucus* by a plume moth (*Platyptilia williamsii*) had a detrimental effect on spittlebugs (*Philaenus spumaris*) feeding later in the season (Karban 1986).

Several studies have invoked an induced defence argument to explain inter-actions. For example, winter moth (*Opherophtera brumata*) feeding on oak (*Quercus robur*) reduced the fecundity, longevity and density of the aphid *Tuberculoides annulatus* sharing the same host plant (Silva-Bohorquez 1987). West (1985) determined that caterpillar feeding on oak in spring had a negative effect on the oviposition of the leaf miner *Phyllonorycter harrisella*. He inferred that the late appearance of this species could be due to asymmetric interactions (wound-induced changes and/or competition) with the caterpillars, thus forcing temporal separation. This study suggests a major change in community dynamics (leaf miners feeding later rather than earlier when the quality of the leaves is better) as a result of an indirect interaction beween two insect herbivores. Spatio-temporal interactions between different insect feeding guilds were further investigated by Leather (1993), who examined the effect of early season caterpillar feeding on aphids sharing bird cherry (*Prunus padus*). Although this experiment employed artificial defoliation, it demonstrated that early 'defoliation' reduced natural grazing by herbivores and the subsequent colonization of the tree by the aphid *Rhopalosiphum padi*. It is easy to explain such results by induced defences, but there is a growing body of evidence questioning their role and occurrence (Williams & Myers 1984); for example, a recent study questioned the hypothesis that host-plant mediated interactions be-tween herbivores are always negative (Martin *et al.* 1994). Specifically, these authors investigated the interaction between the caterpillars *Datana ministra* and *Malacosoma disstria* and the aphid *Symydobius americanus*, mediated via the host plant, paper birch. Caterpillar feeding had no measurable effect on the performance of the aphids. However, as the authors acknowledge, the phloem-feeding aphid might avoid induced defences caused by chewing insects.

Clearly, we need experiments in which we can manipulate interactions between species; these are particularly needed in the case of spatial separation. Here, we take a model system, of above- and below-ground insect herbivores, and by lab-oratory experiments aim to determine a mechanism for the interaction. We start with an appraisal of the limited evidence of this type of interaction and then define a hypothesis (by means of a conceptual model), before taking the system to the field where there is clearly more variability.

INTERACTIONS ABOVE- AND BELOW-GROUND

There is limited evidence of spatial interactions involving above- and below-ground herbivores. From what is known of the impact of these feeders on plant growth, performance and physiology, there is clearly potential for interesting indirect interactions. It is for this reason that we use this interaction as our model system and have designed a series of experiments to explore the interaction at the individual, population and community level. To our knowledge, there are no insect parasitoids or predators which will prey on both above- and below-ground insects. Hence, the interaction between these organisms can only be via the host plant. We consider this an exciting and novel area, highlighting an ecologically little-known group of insects (the root feeders) and their structural and physiological effect on the host plant, as well as their indirect effects on other herbivores.

Only two previous studies have investigated such interactions, those by Gange and Brown (1989) and Moran and Whitham (1990). No one has yet investigated temporal interactions between the two feeding groups (e.g. the effect of early root herbivory on later foliar feeders, or vice versa), though potentially interesting interactions could occur; for example, *Sitona* spp. can co-occur on the same host as larvae and adults, where the larvae are root feeders (Goldson *et al.* 1984) and the adults foliar feeders. This system requires further investigation, as the consequences of juvenile root feeding could affect the same individual when it becomes an adult feeding on the foliage; for example, if root feeding generates a change in the nutrient balance of the foliage, then adult fitness of the same individual might be altered, thus generating interesting coevolutionary scenarios. Gange and Brown (1989) conducted a pot trial investigating the effects of root herbivory by larvae of the scarab *Phyllopertha horticola* on the performance of black bean aphid (*Aphis fabae*) mediated through the host plant *Capsella bursa-pastoris*. Root feeding caused water stress within the plant which led to increased soluble nitrogen (amino acids, especially proline) in the foliage. This in turn resulted in an increase in the weight, growth rate, fecundity and adult longevity of the aphid. Subsequently, Moran and Whitham (1990) investigated the interaction between root-feeding and leaf-galling aphids mediated through the common host plant *Chenopodium album* in a series of garden and growth chamber experiments. Root feeding, by *Pemphigus betae*, had no measurable effect on its leaf-feeding counterpart *Hayhurstia atriplicis*. However, leaf galling was found to decrease root aphid populations by an average of 91% on susceptible plant genotypes. They attributed this asymmetric interaction to competition for phloem sap. Leaf-galling aphids have been shown to be a very strong sink for plant nutrients (e.g. Llewellyn 1982), such that the galler could have diverted the flow of nutrients from the leaves to the roots, hence restricting the food supply to the subterranean feeder and so reducing the size of their population.

These two studies have found that root herbivory, by chafer larvae, can increase the performance of a foliar-feeding aphid, whereas leaf galling by aphids can lead to a large reduction in the populations of root-feeding aphids. Masters *et al.* (1993) suggest that below-ground insect herbivory will result in a host-plant mediated beneficial effect on foliar-feeding insects, whilst above-ground insect herbivory may have an indirect negative effect on root feeders, as shown in Fig. 12.3. Root feeding limits the plant's ability to take up water and nutrients, thus leading to a stress response within the plant. Such responses generally lead to a mobilization

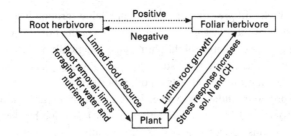

FIG. 12.3. Host plant-mediated interactions between above- and below-ground insect herbivores. (From Masters *et al.* 1993.)

or translocation of soluble nitrogen and carbohydrate to the foliage. These nutrients are essential for insect development and reproduction and may lead to an increase in the performance of foliar-feeding herbivores. Insect herbivory above ground generally results in a reduction in root biomass, thus limiting the amount of available resources for the below-ground insect herbivores. Hence, above- and below-ground insect herbivores indirectly interact, with root herbivory having a beneficial effect on leaf feeders and foliar herbivory having a detrimental effect on root feeders. In the following sections, we consider the evidence for such an interaction, addressing questions such as: does it occur, does it affect the population dynamics of either insect and what are the consequences for insect communities?

EVIDENCE FROM INDIVIDUALS, POPULATIONS AND COMMUNITIES

Individuals

The spatial interaction between a root-chewing chafer larva *Phyllopertha horticola* (Coleoptera: Scarabaeidae), and a leaf miner *Chromatamyia syngensiae* (Diptera: Agromyzidae), mediated by the annual forb *Sonchus oleraceus* (Compositae) was studied in a factorial design in the laboratory by Masters and Brown (1992). It was found that leaf miner pupal weight (an indicator of fecundity (Quiring & McNeil 1984)) was increased when the host plant was subjected to root herbivory. However, the mean relative growth rate of the chafer larvae was significantly reduced when the leaves of plants were also being mined. Hence, in this host-plant mediated interaction, root feeding had a beneficial effect on the leaf miners, whilst foliar feeding had a negative effect on the chafer larvae; an indirect plus–minus interaction. Subsequently, we conducted a series of laboratory experiments where the experimental root feeder was consistently larvae of *P. horticola*, but different foliar-feeding guilds were represented: leaf chewer *Mamestra brassicae* (Lepidoptera: Noctuidae), phloem feeder *Myzus persicae* (Homoptera: Aphididae) and leaf miner

C. syngensiae. In all cases, the growth rate of the root feeder was significantly decreased when the foliar feeder was present. Root herbivory significantly increased the performance of the phloem-feeding aphid and the leaf miner. However, there was no significant effect of root herbivory on any performance-related parameter of the external leaf chewer. Here, root herbivory led to a reduction in the rate of consumption of leaf material, hence caterpillars were able to maintain their growth and develop normally, but ate less in doing so. The relationship, between the leaf miner and chafer larvae, was found to be robust even when the density of both the above- and below-ground insect herbivores was increased (Masters 1995a). Root herbivory, at high and low density, led to a 41% increase in soluble nitrogen in the foliage of the host plant *S. oleraceus*. This in turn resulted in a significant increase in the pupal weight of leaf miners sharing the same host plant (Fig. 12.4a). Plants subjected to two generations of leaf mining had 54% less root biomass and chafers sharing these plants had greatly reduced growth rates (Fig. 12.4b).

In these examples, the relative water concentration of leaves was reduced, whilst the concentration of soluble nitrogen and carbohydrate was increased by root herbivory. Hence, root feeding does appear to produce a stress response within a plant (Fig. 12.3), which leads to an increase in the performance of foliar-feeding

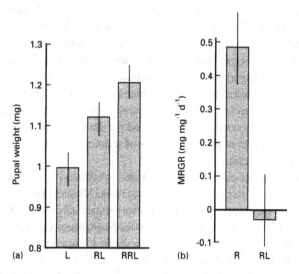

FIG. 12.4. Host plant-mediated interactions between above- and below-ground insect herbivores; (a) Effect of root herbivory by chafer larvae, at two densities, on leaf miner pupal weight in the absence of root feeders (L), with one chafer per pot (density of 70 m^{-2} (RL)) and with three chafers per pot (density of 210 m^{-2} (RRL)). (b) Effect of leaf mining for two generations on chafer larval growth rate (MRGR) in the absence (R) and presence (RL) of leaf miners. There were 10 replicates of each treatment. Bars indicate one standard error. (From Masters 1995.)

insects sharing the same host plant. Additionally, in the majority of the above examples, foliar herbivory reduced root biomass and resulted in reduced growth of the chafer larvae. The exception was when the foliar herbivore was the aphid *Myzus persicae*. However, in this case the herbivore pressure (one aphid per host plant) was minimal and resulted in no reduction of root biomass. However, a reduction in the growth of the root-feeding insect was still found. We therefore suggest that the mechanism of decreased root biomass correlating with a reduction in the available food for the root feeder may be applicable for a range of foliar-feeding insects. However, the occurrence of induced chemical responses cannot be ruled out and warrants further investigation (Seastedt *et al.* 1988). Interestingly, Holland and Detling (1990) have shown that above-ground grazing by prairie dogs not only leads to a decrease in root biomass but also affects carbon allocation and its role in nitrogen cycling responses to feeding. These authors found that above-ground grazing led to a greater allocation of carbon to the root system, thus altering foliar and root C:N ratios. An increase in carbon inputs to the roots in response to grazing would lead to a changed C:N ratio, and thus potentially increase below-ground insect performance (Sutherland 1971).

Populations

Results from controlled laboratory studies are important in elucidating mechanisms and hypotheses for testing in natural conditions. However, such results are only meaningful if they are also evident in field populations, where more natural variability can be expected. Moran and Whitham (1990) found that the presence of colonies of leaf-galling aphids had a substantial negative impact on populations of root-feeding aphids, occasionally driving them to extinction. They also found that large colonies of foliar-feeding aphids reduced root biomass. This could be very important, as roots do not only provide food for the root-feeding aphids but also feeding niches, and one of the implications of reduced root biomass is a reduction in root structure or architecture.

Contrary to the conclusions of Moran and Whitham (1990) and those of the laboratory studies of Masters and Brown (1992) and Masters (1995), some workers have speculated and found that foliar herbivory, especially within grassland communities, leads to enhanced densities of root-feeding insects (e.g. Roberts & Morton 1985; Seastedt *et al.* 1988). In both of these studies, above-ground herbivory was by vertebrate herbivores and root-feeder densities exhibited a curvilinear response, with maximum densities occurring at intermediate levels of grazing. These studies do not dispute that above-ground herbivory leads to a reduction in root biomass, since the positive response of the below-ground herbivores occurs in spite of a reduction of food quantity. Seastedt *et al.* (1988) invoke a resource quality argument. They present laboratory and field evidence combined with conceptual models which argue that moderate levels of vertebrate grazing leads to an increase in the quality (especially nitrogen concentration) of the roots, thus leading to greater performance of root feeders. However, these authors worked in tall grass prairie where interactions may be more complex and involve other organisms, such as fungi.

In another garden experiment, we monitored the impact of root herbivory on the colonization and population growth of foliar-feeding aphids throughout a single growing season. The shared host plant, *S. oleraceus*, was grown in small weeded plots within a ruderal site in the second year of succession from bare ground. Three treatments were imposed, a control where a granular soil insecticide was applied, natural densities of below-ground insect herbivores (approximating to 70 m^{-2}) and enhanced natural densities (×3) with introductions of a range of root-feeding insects naturally occurring in the field site. The dynamics of the foliar-feeding aphids are shown in Fig. 12.5. Throughout the season, the plants subjected to either level of root herbivory had significantly greater populations of aphids feeding on their foliage, stems and flower heads ($F_{2,209} = 5.54$, $P < 0.05$) (Masters 1995b). There appears to be a beneficial effect of root herbivory, similar to that recorded from the laboratory experiments detailed above. We can only assume that the mechanism is the same, since we were not able to sample the plants destructively. Root herbivory has also been found to enhance the success of a seed-feeding tephritid fly in field conditions (Rogers, Masters & Jones, unpublished data). This seed predator was found to visit marsh thistles (*Cirsium palustre*) which were subjected to below-ground insect herbivory in preference to those treated with soil insecticide. This led to a

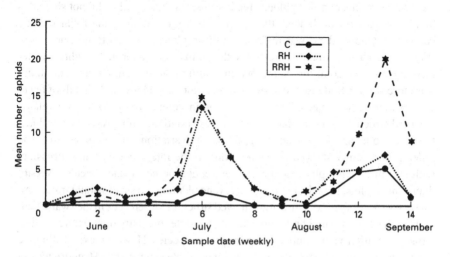

FIG. 12.5. Influence of root herbivory on the density of foliar-feeding aphids throughout the summer. Soil insecticide-treated plants (C), natural levels of root herbivory (RH) and enhanced (×3) levels of root herbivory (RRH) (from Masters 1995b).

difference in oviposition, as untreated plants had a higher proportion of flower heads attacked by the tephritid. Parasitism of the seed predator was also monitored, and 33% more parasitoids were found in flower heads of plants subjected to root herbivory, although this increase was not statistically significant. The impact of this type of indirect interaction on higher trophic levels clearly warrants investigation.

Communities

If effects, such as those described above, can be seen in populations in the field, then the way is clear to look for effects between spatially separated herbivores at the community level. We investigated the effect of below-ground insect herbivory on plant and foliar insect communities in an acidic grassland succession at Silwood Park, Ascot, Berks, UK (Brown & Gange 1989). Four sites of different successional ages were monitored, ruderal (2 years old), early successional (5 years old), mid-successional (11 years old) and a late successional pasture (> 18 years old). Within each site, there were replicated plots of two treatments, soil insecticide-treated and untreated controls (natural densities of below-ground insect herbivores). The plant assemblages were recorded using point quadrats (Brown & Gange 1989), whilst the insects were sampled by D-vac suction. Species richness, vegetative cover and the size of the individual plants were greater in soil insecticide-treated plots. Intuitively, one would predict more above-ground insects in these plots. However, the reverse was apparent, with plots containing below-ground insect herbivores generally supporting larger populations of above-ground insect herbivores. Although the effect was clear cut for phloem-feeding insects, other guilds did not show this trend (Masters 1992). Indeed, the community composition of the foliar-feeding Hemiptera (sap-feeders) was different in those plots subjected to root herbivory (Fig. 12.6). The communities treated with soil insecticide formed a tighter cluster than those subjected to root herbivory, indicating that the hemipteran community was more similar. Soil insecticide-treated plots contained fewer individuals of fewer species than those subjected to below-ground insect herbivory. Axis 1 was related to sward structure and explained 59.8% of the variation, whilst Axis 2 was related to successional age and explained 44.6% of the variation within the data set. The ruderal community showed differences due to the rate of vegetation succession, with soil insecticide plots having greater vegetation cover and species diversity than control plots, namely, being more advanced along the successional trajectory. The early successional communities showed structural differences in the sward, with plots with below-ground herbivores having a greater diversity of structures and so attracting a different assemblage of hemipteran species. However, even within the sap-feeding guild, there were anomalies, for example vetch aphid (*Megoura viciae*) population density was lower in plots subjected to below-ground insect herbivory. It is possible that such species may be responding more strongly to other factors,

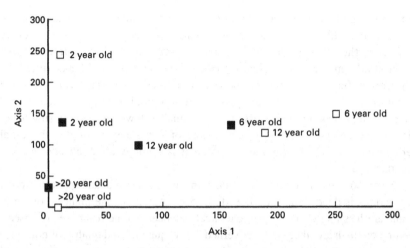

FIG. 12.6. Principle components analysis of the Hemiptera from four successional communities manipulated with soil insecticide. ■, soil insecticide-treated plots; □, plots subjected to root herbivory.

for example, to greater competition (for space, food, etc.) from other insects or from changes in vegetation structure.

By comparison, Evans (1991) conducted a factorial field experiment in Kansas tallgrass prairie. Soil insecticide application significantly increased the density of above-ground leafhoppers and planthoppers, but had no effect on the densities of the aphids. Clearly, these results are the opposite to those described above, but are likely to be related to the stated ineffectiveness of the soil insecticides (Evans 1991). Additionally, insect herbivores have variable responses to plants that are stressed, such that if root herbivory does cause plant stress, the foliar feeder may not necessarily increase in abundance (McQuate & Connor 1990). Unfortunately, Evans did not monitor the response of the below-ground insect herbivores.

MECHANISMS, SIGNIFICANCE AND IMPLICATIONS OF HERBIVORE–PLANT–HERBIVORE INTERACTIONS

As shown in Figs 12.1 and 12.2, interactions between species can take many forms. We have described one particular system, namely herbivore–plant–herbivore interactions, in terms of spatial separation. Herbivores can indirectly interact spatially (niches and resources) and temporally and any general mechanism may vary depending on the nature of the interaction. When two herbivores share a host plant at exactly the same time (Fig. 12.2 spatial), it is difficult to see how plant morphological changes could mediate the interaction. We propose that physiological changes, induced by herbivory, mediate spatial interactions. Herbivores of the same guild, but separated by different niches, would interact via the shared resource (e.g. phloem sap). Hence, any changes induced in this shared resource will affect the other herbivore. Herbivores that share a niche, but are of different guilds, may affect nutrient distribution to the niche and hence affect the other feeder. Temporal interactions between insect herbivores may be physiological in nature, but could incorporate plant morphological responses to herbivory; for example, a plant might be less apparent (smaller, fewer leaves) offering fewer niches for future phytophages, or a new flush of leaves may occur after the initial herbivory event, which may have a different physiology affecting later feeders (Leather 1993).

We propose that the effect of host-plant mediation of spatial and spatio-temporal interactions is dependent upon the life-history strategy of the plant. Short-lived plants are less likely to invest in chemical defences, so the responses of insects feeding on them are likely to be governed by the quantity and quality of nutrients. However, long-lived plants are more likely to invest in defensive compounds and so an induced mechanism might control the indirect interaction. Additionally, plant compensation for herbivory may be important. Herbaceous species compensate for herbivory by either mobilizing or producing nutrients enabling plant regrowth. Consequently, this relative increase in available nutrients might be important for insect herbivores sharing the host plant (spatial and/or temporal separation). We speculate that long-lived plants are more likely to compensate through mobilizing stored nutrients or redistributing photosynthate which may influence herbivores which are separated by niches and by time. However, due to continuing coevolution between insects and their host plants, any such hypothesis may not be entirely supported. Exceptions may result from insects adapting to overcome induced defences or to seasonal fluctuations in nutrients (McNeill 1973). Root architecture of the host plant may also be an important feature in determining the nature of the interaction. It is easy to envisage that plants with a large tap root might be more susceptible to below-ground insect herbivory than those with a fibrous root system. Moreover, a fibrous system is more efficient at scavenging for water and nutrients over a greater area (Fitter 1987) and thus the physiological consequences of root feeding might be reduced. Also, a plant with a fibrous root system may be likely to compensate for loss of root tissue. Thus, it is not difficult to see that the response of foliar feeders to such plants may be weaker.

We have suggested that pairwise, or direct, interactions (Fig. 12.1a) provide building blocks for ecological communities. Now that these direct interactions are well understood (having been studied for so long and in such detail), we should start considering the role of indirect interactions between spatially, temporally or spatio-temporally separated species in determining community structure and function. We argue that host-plant mediated interactions between herbivores, be they temporally or spatially separated (Fig. 12.2), are commonly at the core

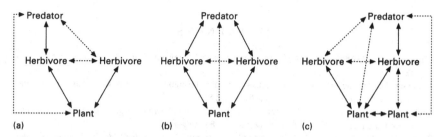

Fɪɢ. 12.7. Simple multitrophic communities illustrating the central position of herbivore–plant–herbivore interactions, direct (solid lines) and indirect (dotted lines). (a) Specialist predator, herbivore, plant interactions; (b) generalist predator, herbivore, plant interactions; (c) simple community, two plants, two herbivores, one specialist predator.

of multitrophic interactions in communities (Fig. 12.7). The simple interactions between predators, herbivores and plants can be substantially altered by the presence of another herbivore sharing the same host plant (Fig. 12.7a, b). Indeed, the core of a simple community, consisting of specialist predator, two specialist herbivores and two plants, is that of herbivore–plant–herbivore (Fig. 12.7).

Through moving away from direct interaction between two species and concentrating on three or more species, we gain a better understanding of community structure and function. By understanding how species interact and how these interactions are modified by other species within the community, we will be able to predict how communities will respond to major perturbations, such as climatic change. To develop this example: root herbivory induces a stress response within a host plant leading to an increase in the performance (and fecundity) of foliar-feeding insects (Fig. 12.3). Climate change models predict that there will be substantial periods of drought during European summers (Houghton *et al.* 1992). Consequently, the stress response caused by root herbivory might be enhanced, thus leading to substantial outbreaks of above-ground insect herbivores, some of which may be pest species. What then would be the knock-on effects for insect parasitoids and predators? Would they increase in numbers and, if so, could the outbreak be modified by root-feeding insects? Such questions provide some of the challenges facing community ecologists in the next decade. An understanding of multitrophic interactions, with cores such as herbivore–plant–herbivore, will be essential to predicting how communities will change with a changing environment.

ACKNOWLEDGEMENTS

The experimental work we describe was supported by the NERC. We are grateful to John Hollier for the data and analysis leading to Fig. 12.6 and to Laurel Fox for useful discussion.

REFERENCES

Abrams, P. A. (1987). On classifying interactions between populations. *Oecologia,* **73,** 272–281.

Andersen, D. C. (1987). Below-ground herbivory in natural communities: a review emphasizing fossorial animals. *Quarterly Review of Biology,* **62,** 261–286.

Bach, C. E. (1991). Direct and indirect interactions between ants (*Pheidole megacephala*), scales (*Coccus viridis*) and plants (*Pluchea indica*). *Oecologia,* **87,** 233–239.

Begon, M., Harper, J. L. & Townsend, C. R. (1990). *Ecology: Individuals, Populations and Communities.* Blackwell Scientific Publications, Oxford.

Bentley, S., Whittaker, J. B. & Malloch, A. J. C. (1980). Field experiments on the effects of grazing by a chrysomelid beetle (*Gastrophysa viridula*) on seed production and quality in *Rumex obtusifolius* and *Rumex crispus. Journal of Ecology,* **68,** 671–674.

Brown, V. K. & Gange, A. C. (1989). Differential effects of above- and below-ground insect herbivory during early plant succession. *Oikos,* **54,** 67–76.

Brown, V. K. & Gange, A. C. (1990). Insect herbivory below ground. *Advances in Ecological Research,* **20,** 1–58.

Brown, V. K., Gange, A. C., Evans, I. M. & Storr, A. L. (1987). The effect of insect herbivory on the growth and reproduction of two annual *Vicia* species at different stages in plant succession. *Journal of Ecology,* **75,** 1173–1189.

Caldwell, M. M., Richards, J. H., Johnson, D. A., Nowak, R. S. & Dzurec, R. S. (1981). Coping with herbivory: photosynthetic capacity and resource allocation in semiarid *Agropyron* bunchgrasses. *Oecologia,* **50,** 14–24.

Chongrattanameteekul, W., Foster, J. E. & Araya, J. E. (1991). Biological interactions between the cereal aphids *Rhopalosiphum padi* (L.) and *Sitobion avenae* (F.) (Hom., Aphididae) on wheat. *Journal of Applied Entomology,* **111,** 249–253.

Croxford, A. C. (1990). *Wound-induced changes in plants and their defensive role against insect herbivores.* PhD Thesis, University of Southampton.

Evans, E. W. (1991). Experimental manipulation of herbivores in native tallgrass prairie: responses of aboveground arthropods. *American Midland Naturalist,* **125,** 37–46.

Faeth, S. H. (1986). Indirect interactions between temporally separated herbivores mediated by the host plant. *Ecology,* **67,** 479–494.

Fitter, A. H. (1987). An architectural approach to the comparative ecology of plant root systems. *New Phytologist,* **106** (Suppl.), 61–77.

Gange, A. C. (1990). Effects of insect herbivory on herbaceous plants. *Pests, Pathogens and Plant Communities* (Ed. by J. J. Burdon & S. R. Leather), pp. 49–62. Blackwell Scientific Publications, Oxford.

Gange, A. C. & Brown, V. K. (1989). Effects of root herbivory by an insect on a foliar-feeding species, mediated through changes in the host plant. *Oecologia,* **81,** 38–42.

Goldson, S. L., Bourdot, G. W. & Proffitt, J. R. (1987). A study of the effects of *Sitona discoideus* (Coleoptera: Curculionidae) larval feeding on the growth and development of lucerne (*Medicago sativa*). *Journal of Applied Ecology,* **24,** 153–161.

Goldson, S. L., Frampton, E. R., Barratt, B. I. P. & Ferguson, C. M. (1984). The seasonal biology of *Sitona discoideus* Gyllenhal (Coleoptera: Curculionidae), an introduced pest of New Zealand lucerne. *Bulletin of Entomological Research,* **74,** 249–259.

Harborne, J. B. (1988). *Introduction to Ecological Biochemistry,* 3rd edn. Academic Press, London.

Hartley, S. E. & Lawton, J. H. (1987). Effects of different types of damage on the chemistry of birch foliage, and the responses of birch-feeding aphids. *Oecologia,* **74,** 432–437.

Haukioja, E. (1980). On the role of plant defences in the fluctuation of herbivore populations. *Oikos,* **35,** 202–213.

Hendrix, S. D. (1980). An evolutionary and ecological perspective of the insect fauna of ferns. *American Naturalist*, 115, 171–196.

Holland, E. A. & Detling, J. K. (1990). Plant response to herbivory and belowground nitrogen cycling. *Ecology*, 71, 1040–1049.

Holt, R. D. (1977). Predation, apparent competition, and the structure of prey communities. *Theoretical Population Biology*, 12, 197–229.

Holt, R. D. & Lawton, J. H. (1993). Apparent competition and enemy-free space in insect host-parasitoid communities. *American Naturalist*, 142, 623–645.

Houghton, J. T., Callandler, B. A. & Varney, S. K. (1992). *Climate Change 1992*. Cambridge University Press, Cambridge.

Hsiao, T. C. (1973). Plant responses to water stress. *Annual Review of Plant Physiology*, 24, 519–570.

Hunter, M. D. (1987). Opposing effects of spring defoliation on late season oak caterpillars. *Ecological Entomology*, 12, 373–382.

Karban, R. (1986). Interspecific competition between folivorous insects on *Erigeron glaucus*. *Ecology*, 67, 1063–1072.

Karban, R. & Myers, J. H. (1989). Induced plant responses to herbivory. *Annual Review of Ecology and Systematics*, 20, 331–348.

Kidd, N. A. C., Lewis, G. B. & Howell, C. A. (1985). An association between two species of pine aphid, *Schizolachnus pineti* and *Eulachnus agilis*. *Ecological Entomology*, 10, 427–432.

Lawton, J. H. & MacGarvin, M. (1986). The organization of herbivore communites. *Community Ecology: Pattern and Process* (Ed. by J. Kikkawa & D. J. Anderson), pp. 163–186. Blackwell Scientific Publications, Oxford.

Leather, S. R. (1993). Early season defoliation of bird cherry influences autumn colonization by the bird cherry aphid, *Rhopalosiphum padi*. *Oikos*, 66, 43–47.

Leather, S. R. & Dixon, A. F. G. (1984). Aphid growth and reproductive rates. *Entomologia Experimentalis et Applicata*, 35, 137–140.

Llewellyn, M. (1982). The energy economy of fluid-feeding herbivorous insects. *Proceedings of the 5th International Symposium on Insect–Plant Relationships* (Ed. by J. H. Visser & A. K. Minks), pp. 243–252. Wageningen, Netherlands.

Louda, S. M. (1984). Herbivore effect on stature, fruiting and leaf dynamics of a native crucifer. *Ecology*, 65, 1379–1386.

Louda, S. M. & Potvin, M. A. (1995). Effect of infloresence-feeding insects on the demography and lifetime fitness of a native plant. *Ecology*, 76, 229–245.

Louda, S. M., Keeler, K. H. & Holt, R. D. (1990). Herbivore influences on plant performance and competitive interactions. *Perspectives in Plant Competition* (Ed. by J. B. Grace & D. Tilman), pp. 413–444. Academic Press, New York.

Martin, M.-A., Cappuccino, N. & Ducharme, D. (1994). Performance of *Symydobius americanus* (Homoptera: Aphididae) on paper birch grazed by caterpillars. *Ecological Entomology*, 19, 6–10.

Masters, G. J. (1992). *Interactions between foliar- and root-feeding insects*. PhD Thesis, University of London.

Masters, G. J. (1995a). The effect of herbivore density on host plant mediated interactions between two insects. *Ecological Research*, 10, 125–133.

Masters, G. J. (1995b). The impact of root herbivory on aphid performance: field and laboratory evidence. *Acta Oecologica*, 16, 135–142.

Masters, G. J. & Brown, V. K. (1992). Plant-mediated interactions between two spatially separated insects. *Functional Ecology*, 6, 175–179.

Masters, G. J., Brown, V. K. & Gange, A. C. (1993). Plant mediated interactions between above- and below-ground insect herbivores. *Oikos*, 66, 148–151.

Mattson, W. J. & Haack, R. A. (1987). The role of drought stress in provoking outbreaks of phytophagous insects. *Insect Outbreaks* (Ed. by P. Barbosa & J. C. Schultz), pp. 365–407. Academic Press, New York.

May, R. M. (1988). How many species are there on earth? *Science*, **241**, 1441–1449.

McNeill, S. (1973). The dynamics of a population of *Leptopterna dolabrata* (Heteroptera: Miridae) in relation to its food resources. *Journal of Animal Ecology*, **42**, 495–507.

McQuate, G. T. & Connor, E. F. (1990). Insect responses to plant water deficits. II. Effect of water deficits in soybean plants on the growth and survival of Mexican bean beetle larvae. *Ecological Entomology*, **15**, 433–445.

Moran, N. A. & Whitham, T. G. (1990). Interspecific competition between root-feeding and leaf galling aphids mediated by host-plant resistance. *Ecology*, **71**, 1050–1058.

Müller-Scharer, H. (1991). The impact of root herbivory as a function of plant density and competition – survival, growth and fecundity of *Centaurea maculosa* in field plots. *Journal of Applied Ecology*, **28**, 759–776.

Müller, H., Stinson, C. S. A., Marquarot, K. & Schroeder, D. (1989). The entomofaunas of roots of *Centaurea maculosa* Lam., *C. diffusa* Lam. and *C. vallesiaca* Jordan in Europe. Niche separation in space and time. *Journal of Applied Entomology*, **107**, 83–95.

Powell, R. D. & Myers, J. H. (1988). The effect of *Sphenoptera jugoslavica* Obenb. (Col., Buprestidae) on its host plant *Centaurea diffusa* Lam. (Compositae). *Journal of Applied Entomology*, **106**, 25–45.

Putman, R. J. (1994). *Community Ecology*. Chapman & Hall, London.

Quiring, D. T. & McNeil, J. N. (1984). Adult–larval intraspecific competition in *Agromyza frontella* (Diptera: Agromyzidae). *Canadian Entomologist*, **116**, 1385–1391.

Reid, W. V. (1992). Conserving life's diversity: can the extinction crisis be stopped? *Environmental Science and Technology*, **26**, 1090–1095.

Ridsdill Smith, T. J. (1977). Effects of root-feeding by scarabaeid larvae on growth of perennial ryegrass plants. *Journal of Applied Ecology*, **14**, 73–80.

Roberts, R. J. & Morton, R. (1985). Biomass of larval scarabaeidae (Coleoptera) in relation to grazing pressures in temperate, sown pastures. *Journal of Applied Ecology*, **22**, 863–874.

Root, R. B. (1973). Organization of a plant–arthropod association in simple and diverse habitats: the fauna of collards (*Brassica oleracea*). *Ecological Monographs*, **43**, 95–124.

Schmitt, R. J. (1987). Indirect interactions between prey: apparent competition, predator aggregation, and habitat segregation. *Ecology*, **68**, 1887–1897.

Seastedt, T. R., Ramundo, R. A. & Hayes, D. C. (1988). Maximization of densities of soil animals by foliage herbivory: empirical evidence, graphical and conceptual models. *Oikos*, **51**, 243–248.

Silva-Bohorquez, I. (1987). *Interspecific interactions between insects on oak trees, with special reference to defoliators and the oak aphid.* PhD Thesis, University of Oxford.

Southwood, T. R. E. (1978). The components of diversity. *Diversity of Insect Faunas* (Ed. by L. A. Mound & N. Waloff), pp. 19–40; Blackwell Scientific Publications, Oxford.

Spike, B. P. & Tollefson, J. J. (1991). Response of western corn rootworm-infested corn to nitrogen fertilization and plant density. *Crop Science*, **31**, 776–785.

Strauss, S. Y. (1987). Direct and indirect effects of host-plant fertilization on an insect community. *Ecology*, **68**, 1670–1678.

Strauss, S. Y. (1991). Indirect effects in community ecology: their definition, study and importance. *Trends in Ecology and Evolution*, **6**, 206–210.

Sutherland, O. R. W. (1971). Feeding behaviour of the grass grub *Costelytra zealandica* (White) (Coleoptera: Melolonthinae) – 1. The influence of carbohydrates. *New Zealand Journal of Science*, **14**, 18–24.

Tomlin, E. S. & Sears, M. K. (1992). Indirect competition between the Colorado potato beetle (Coleoptera: Chrysomelidae) and the potato leafhopper (Homoptera: Cicadellidae) on potato: laboratory study. *Environmental Entomology*, **21**, 787–792.

Vaughton, G. (1990). Predation by insects limits seed production in *Banksia spinulosa* var. neoanglica (Proteaceae). *Australian Journal of Botany*, **38**, 335–340.

Vranjic, J. A. & Gullan, P. J. (1990). The effect of a sap-sucking herbivore, *Eriococcus coriaceus*

(Homoptera: Eriococcidae), on seedling growth and architecture in *Eucalyptus blakelyi*. *Oikos*, **59**, 157–162.

Wedderburn, M. E., Tucker, M. A., Pengelly, W. J. & Ledgard, S. F. (1990). Responses of a New Zealand North Island hill perennial ryegrass collection to nitrogen, moisture stress, and grass grub (*Costelytra zealandica*) infestation. *New Zealand Journal of Agricultural Research*, **33**, 405–411.

West, C. (1985). Factors underlying the late seasonal appearance of the lepidopterous leaf mining guild on oak. *Ecological Entomology*, **10**, 111–120.

Williams, K. S. & Myers, J. H. (1984). Previous herbivore attack of red alder may improve food quality for fall webworm larvae. *Oecologia*, **63**, 166–170.

13. GALL-INDUCING INSECT HERBIVORES IN MULTITROPHIC SYSTEMS

P. W. PRICE,* G. W. FERNANDES† AND
R. DÈCLERCK-FLOATE‡

*Department of Biological Sciences, Northern Arizona University, Flagstaff, AZ, USA; †Departamento Biologia Geral, Universidade Federal de Minas Gerais, Belo Horizonte, Minas Gerais, Brazil and ‡Agriculture and Agri-Food Canada, Research Centre, Lethbridge, Alberta, Canada

INTRODUCTION

The debate on the adaptive significance of gall formation for insect herbivores is of long standing and has viewed the gall inducer in a central position in multi-trophic interactions. Over 100 years ago the debate focused on whether the gall was a plant defence against the herbivore trophic level, whether the gall circumvented plant defences, or whether protection against the third trophic level of carnivores formed the selective advantage (e.g. Hollis 1889, 1890; McLachlan 1889; Mivart 1889; Romanes 1889, 1890; Wetterham 1889, reviewed in Price *et al.* 1987).

The galling habit is extremely common over a large range of vegetation types and latitudinal gradients. In many localities, gallers are the most conspicuous insect herbivores of the fauna and the most abundant. They attack a wide variety of plant species and commonly support rich communities of parasitoids and inquilines. The local and global richness of galling species, the large variety of plant species attacked, and the complexity of carnivore assemblages based on gallers makes these systems some of the most numerous multitrophic level systems in terrestrial environments and some of the richest in biodiversity. Hence, the galling habit provides us with a valuable opportunity for understanding multitrophic level interactions:

1   Galls are persistent, conspicuous and easily censused.

2   They contain evidence of the survivorship of the insect and important clues on the causes of death, such that the trophic level responsible can be determined; for example plant resistance, intraspecific competition, or carnivores.

3   Much informaton can be obtained without killing insects because galls can be collected after emergence.

4   These opportunities allow the population dynamics of rare and uncommon species to be studied effectively, providing an unusual perspective.

5   Galls are usually initiated where the ovipositing female chooses to lay an egg, whether the gall is induced by the female adult as in sawflies, or by larval feeding as in other taxa. Therefore, the linkage between decisions made by an ovipositing

female and success of her progeny can be evaluated accurately. Such preference–performance linkage permits a view of the selective forces of prime importance in a trophic system and the ecological consequences of female behaviour.

6   Galls are commonly initiated in rapidly developing, undifferentiated plant tissue, enabling redirection of differentiation into a gall vs. the usual plant structure. Variable phenology and growth rates of shoots are likely to be critical in determining the quality of galls for larval establishment and survival. Therefore, the bottom-up effects of resource variation in plants for galling herbivores can be assessed using studies on the preference-performance linkage.

7   Variation in gall size, toughness and structure, as defined by plant genotype, environment and module variation, are likely to influence strongly inquilines, parasitoids, and predators, providing a tangible link among three or more trophic levels which are readily evaluated.

We published a review on tritrophic interactions involving plants, insect herbivores and carnivores in 1980 (Price *et al.* 1980). Only 5% of the papers addressed galling systems, the other 95% focusing mainly on free-feeding insect species. Chemical interactions, especially up the trophic system, were illustrated in 80% of the papers and the remainder dealt with physical factors. Virtually all papers indicated strong effects from plants up through the trophic system. There was virtually no consideration of phenotypic or genotypic variation in plants that modified trophic relationships. This perspective from 15 years ago provides us with a comparative basis for evaluating the interaction involving gall-inducing insect species, their host plants and other members of the community.

Given the close association between galling insects and their host plants (e.g. Weis *et al.* 1988; Weis 1994), there is the clear expectation that bottom-up influences on multitrophic interactions involving plants, gallers and carnivores will be strong. We document first the evidence and the several kinds of interactions that flow up the trophic system. Then, we examine the evidence for top-down effects through systems and their strengths relative to bottom-up impacts.

BOTTOM-UP INTERACTIONS

Global pattern of local galling species richness

The strong influence of the abiotic environment and plant resources on the frequency of galling insects over the globe is very clear (Fig. 13.1). Although gallers can be found at any latitude, there is a very strong peak in richness in warm temperate latitudes, or their altitudinal equivalents in the tropics, on the predominant scleromorph vegetations typical of Mediterranean climates. The adaptive nature of the galling habit clearly results in the richest faunas where the climate is dry for a large part of the year, and soil nutrient status is low, with a flora adapted with

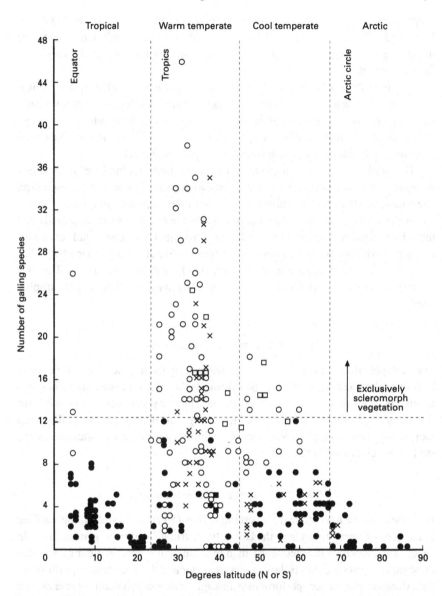

FIG. 13.1. Distribution of local species richness in galling insect assemblages on a latitudinal gradient, with altitude adjusted to the equivalent latitude at sea level. Samples were based on a 1-h local census or a 10-m wide transect including 45 trees, 100 shrubs and 1000 herbs. ○, samples in scleromorph vegetation; ●, samples in mesic non-scleromorph vegetation; x, samples on scleromorph vegetation in relatively mesic sites; □, samples from fynbos vegetation in South Africa.

tough, persistent leaves and commonly shrubby or low tree growth (Fernandes & Price 1991; Price *et al.* 1995). Chaparral, fynbos, desert scrub and cerrado are major vegetation types with members adapted in this way and to the inevitable fires in these dry climates.

An alternative scenario for this global pattern is not provided by a top-down perspective. While natural enemies of gallers typically cause higher mortality locally in mesic than xeric sites (Fernandes & Price 1991, 1992; Fernandes *et al.* 1994), galling richness is high on scleromorphic species even in such mesic sites, indicating a stronger association with plant form than carnivore impact.

This global view sets our perspective for the multitrophic level interactions involved in galling systems. There is a macroevolutionary link between scleromorph vegetation and the adaptive radiation of galling insects. As a consequence, there is an escalation of diversity at the carnivore trophic level among the parasitoids and inquilines. Seldom is the richness of faunas associated with insect galls exceeded on other herbivorous insect guilds (cf. Askew 1961; Hawkins & Goeden 1984; Waage & Greathead 1986; Hawkins 1994; Hawkins & Sheehan 1994). This is a bottom-up, resource-based system with increasing species richness up the trophic levels.

Plant architecture and related effects

The multiple effects of plant form and growth rate up the trophic system have yet to be fully understood, but some examples illustrate the diverse nature of these bottom-up effects. They include strong effects of plant phenotype, genotype and life form (i.e. trees vs. shrubs) on susceptibility to galling, life-history traits of gallers, gall morphology and consequent residence by inquilines and attack by parasitoids. Case studies are discussed in this order.

Plant phenotype: growth rate and age

In all the cases we have studied concerning willow (*Salix*) hosts and galling sawflies in the genus *Euura*, throughout the Holarctic distribution of these genera, juvenile shoots are attacked more frequently than reproductive shoots on older ramets. On these rapidly growing shoots, larvae survive better and these relationships illustrate the strongest preference–performance linkage between ovipositing females and larval survival discovered to date (e.g. Craig *et al.* 1989; Price 1994; Price *et al.* 1995). For *Euura lasiolepis*, a species we have studied in detail, rapid growth of ramets translates into larger galls and greater protection of sawfly larvae from parasitoids. A pattern of negative density dependence of decreasing per cent parasitoid attack with increasing density of galls per willow clone is generally observed for small chalcidoid parasitoids, such as *Pteromalus* species (Price &

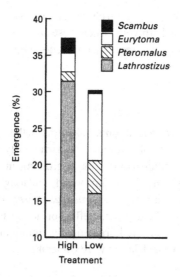

FIG. 13.2. Change in parasitoid communities in an experiment which provided high and low water levels to plants, using *Salix lasiolepis* and the galling sawfly *Euura lasiolepis* in a tritrophic system (from Craig 1994). Means are from galls on 20 potted willows in each treatment.

Clancy 1986a; Price 1988). This pattern results from the higher densities of gallers occurring on more favourable clones, producing larger galls on average and reducing parasitoid access.

Involving the same species interactions on willow, individual plants change in susceptibility with age and with water availability. Juveniles are more susceptible than reproductive individuals, a trait associated with their more rapid rate of growth. After heavy winter precipitation, which increases growth rate of willows, attack by sawflies and their survival of plant resistance factors increases, and parasitoid attack declines, resulting in strong increases in population density (Price & Clancy 1986a, b; Price *et al.* 1990). Experimental manipulation of young ramets with high- and low-water treatments has shown repeatedly the impact of rapid and slow growth on the preference–performance relationship (Price & Clancy 1986b; Preszler & Price 1988; Craig *et al.* 1989). Up the trophic system parasitoid communities also are affected, with each species being influenced in a different way depending on its size and attack pattern (Craig 1994) (Fig. 13.2). For example, the ichneumonid *Lathrostizus* with a long ovipositor benefits from well-watered plants, on which galls may toughen more slowly, while smaller parasitoids such as *Pteromalus* are more abundant on poorly watered plants, probably because galls are smaller and host larvae are more accessible. Craig (1994) reviews the impact of intraspecific plant variation on parasitoid communities for both phenotypic and genotypic plant characters.

Woody plants invariably change in character with age (e.g. Moorby & Wareing 1963) with clear effects on herbivores when these are investigated (e.g. Kearsley & Whitham 1989). However, more research on effects at the third trophic level would be valuable.

Plant genotype

Genotypes in a population of plants commonly differ in their susceptibility to galling insects, with effects up the trophic levels. One particularly interesting example involves the comparison among willow clones growing in close proximity which differed in the rate with which galls became woody and tough, which in turn affected access to an ichneumonid parasitoid, *Lathrostizus euurae*. Having a relatively long ovipositor, gall diameter had only a small effect on access to host larvae. However, the rate of gall toughening differed dramatically among clones, resulting in very different lengths of time in which larvae were vulnerable to attack: different 'windows

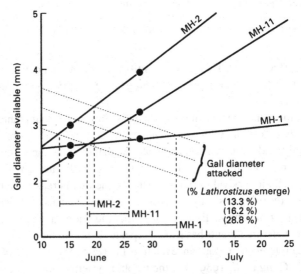

FIG. 13.3. Estimates of the window of vulnerability of *Euura lasiolepis* to the ichneumonid parasitoid *Lathrostizus euurae* on three willow clones growing together (from 989 galls on which *Lathrostizus* females were observed). As galls age, and expand in diameter, they toughen and provide more protection to larvae, but the rate differs among clones, MH-1, MH-11 and MH-2, resulting in different vulnerable periods and per cent parasitism. As galls age, the gall diameter attacked declines because galls toughen at a faster rate as the season progresses, and an estimate of the window of vulnerability is given by the intersection of mean gall diameter per clone and the mean gall diameter attacked, ± 1 SD, shown as dashed lines. (From Craig *et al.* 1990.) Gall toughness was measured with a Voland Texture Analyser, measuring the force needed to push a needle probe 2 mm into a gall at a speed of 1 mm s^{-1}.

of vulnerability' (Craig *et al.* 1990); for example, a ×2.8 increase in window length, from 6 to 17 days, resulted in a ×2.2 increase in percentage parasitoid emergence from galls, from 13.3 to 28.8% (Fig. 13.3). A different experiment using 15 willow genotypes of the same age in pots in the same location showed dramatic differences in attack by parasitoids, ranging from less than 5% on some genotypes to over 40% on others (Craig 1994).

Given that galling insects provide resources for carnivores other than parasitoids, such as avian predators (cf. Tscharntke, this volume), it is likely that multitrophic systems encompass a diverse community influenced by bottom-up effects through plant genotypic variation. The most compelling case we know of involves poplars, leaf-galling aphids, and other members of the community on the poplars: predatory birds and arthropods, other herbivores, and a mould (Dickson & Whitham 1996). Natural stands of hybrid poplars (*Populus augustifolia* × *P. fremontii*) show great variation in their genotypically-based resistance to the aphid *Pemphigus betae*. Susceptible genotypes support millions of aphids, while resistant trees close by have almost none. On susceptible trees there was a 1.3 times greater species richness of arthropods other than *P. betae* and these species, including predatory anthocorids and chamaemyiids, were 1.3 times more abundant than on resistant trees. Avian predators such as chickadees foraged more intensely on aphids in the susceptible trees. In a gall-removal experiment using susceptible genotypes, paired ramets of the same genotype were allocated to controls and a treatment of reducing aphid abundance to that observed in resistant trees. On the control ramets, there was a 1.5 times greater species richness, and a 1.3 times greater relative abundance per species in the arthropod community (Table 13.1). In the controls, bird predation

TABLE 13.1. Aspects of the food webs on poplars in the hybrid zone (*Populus angustifolia* × *P. fremontii*) in which plant genotypes influence the trophic interactions based on them. Genotypes include those susceptible to the galling aphid *Pemphigus betae* and those resistant to the aphid. (From Dickson & Whitham 1996.)

First trophic level	Aphid-susceptible genotype*	Aphid-resistant genotype or aphid removal treatment
Second trophic level		
Mean number of galls/1000 shoots	200	21
Number of foliage arthropod species	19	13
Number of individuals/1000 shoots	31	23
Third trophic level – per cent aphid galls attacked by natural enemies		
Birds	8	3
Insects	18	7
Fungi	13	8

* All values in the aphid-susceptible genotype column were significantly higher at $P < 0.05$.

was 2.7 times higher, insect predators attacked 2.6 times more galls, and fungal infection was 1.6 times higher. Inquilines increased on treatments two-fold, mostly involving free-feeding aphids that moved into galls after emergence of *Pemphigus*. 'Plant genetics affects diverse species from three trophic levels supporting a 'bottom-up' influence of plants on community structure' (Dickson & Whitham 1996).

Differences between trees and shrubs

Both trees and shrubs vary in their resistance to herbivores as they age, involving physiological ageing, ontogenetic ageing or both (Kearsley & Whitham 1989; Roininen *et al.* 1993). However, shrubs differ from trees in an important architectural way because they have reduced apical dominance and are composed of a cluster of more or less equivalent axes produced by frequent sprouting from the basal region (Steeves & Sussex 1989). Thus, a shrub generally maintains a persistent resource of juvenile shoots suitable for attack by gallers, while the ontogenetic and physiological age of most trees inexorably increase. This is because a tree forms a single main axis, or trunk, with strong dominance of a leading shoot or shoots, and buds closest to the apex sprout when conditions are suitable (Steeves & Sussex 1989) (i.e. after pruning or damage). These different growth forms have dramatic influences up the trophic levels.

Comparing members of the same genus *Salix*, one with the stature of a shrub, such as *S. lasiolepis*, and one a tree-forming species, *S. pentandra*, we can then examine the dynamics of two species of the galling sawfly genus, *Euura*, one on each host. Bottom-up impact becomes evident. A population of shrubs provides a renewable resource for sawflies over time resulting in a relatively stable population of *Euura* influenced mainly by winter precipitation (Price *et al.* 1995). However, a colonizing population of trees grows older uniformly and becomes resistant with age. *Euura* sawflies discover the newly established population, increase in abundance, and rapidly go locally extinct in about 7 years (Fig. 13.4), apparently because of some resistance factor associated with ontogenetic ageing, or loss of juvenility (Roininen *et al.* 1993). After this boom-and-bust phase of sawfly colonization, they are found only on older trees that are damaged or heavily pruned in managed landscapes, when regrowth is very rapid although in the tree crown.

The consequences for the third and higher trophic levels based on *Euura* sawflies on shrubs and trees are clear enough. On shrubs, a comparatively predictable resource persists over many years with local extinction and colonization episodes of low frequency and highly scattered over a landscape. On trees, extinction and colonization over a landscape is inevitable, with brief residence times of a few years, and a highly heterogeneous landscape of willows for sawflies driven by disturbance, with resources blinking on and off in small patches.

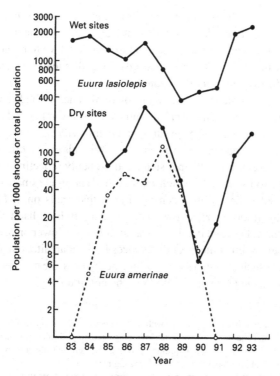

FIG. 13.4. Census data for two galling sawflies *Euura lasiolepis* on a shrubby willow and *Euura amerinae* on willow trees. (From Price *et al.* 1995 and Roininen *et al.* 1993.) In *E. lasiolepis*, populations per 1000 shoots were persistent for 11 years in both wet and dry sites. In *E. amerinae* the first gall to be found in the population appears in 1983, and no galls were present in 1991. For *E. amerinae*, total population reached a peak of 120 galls in 1988, with local extinction occurring 9 years after colonization.

Differences between shrub species

Even shrub species within the same genus can be sufficiently different in architecture to influence profoundly the herbivores, their gall structure, and the communities based on galls. One case involves two willow species and their cecidomyiid bud-gallers in the genus *Rabdophaga*. The upward flow of influences through the trophic system is driven by differences in terminal dominance between the willow species. *Salix lutea* has strong terminal growth with one shoot per ramet growing much more strongly than most of the others. Therefore, one bud is more favourable to galling, the female gallers have evolved to lay few eggs close to this bud, and larvae initiate usually one gall per shoot. This cone-shaped gall is generally large, but with scales derived from modified leaves and leaf stipules loosely adpressed.

The loose structure perhaps results from limitations on shoot compression during gall formation due to rapid shoot growth, or it may be a reflection of the underlying shoot architecture. Inquilines, at the second trophic level, were numerous per gall (Table 13.2), with almost 10 cm^{-3}, and 33% of galls yielded parasitoids of the galler or inquiline *Dasineura albovittata*.

At the other extreme of a continuum of dominance in bud development in shrubby willows, *S. exugua* has weak dominance and several lateral buds per stem can develop to a similar extent each summer. On this willow, another *Rabdophaga* species lays eggs in larger clutches because several buds can be effectively galled. Larvae initiate several bud galls per stem, and probably reflecting a shorter internode length on shoots, the galls which are modified shoots are small and composed of tightly packed scales, with a density higher than galls on *S. lutea* (0.36 vs 0.19 g cm^{-3}). Inquilines showed lower success than in the galls of *S. lutea*. Also, parasitism on the gallers and inquilines was significantly lower on *S. exigua* (Table 13.2). This example illustrates that differences in plant architecture can influence life-history traits such as clutch size of galling herbivores, the number and structure of galls, and the abundance of inquilines in the community.

TABLE 13.2. The characteristics of willow growth in the species *Salix lutea* and *S. exigua*, with responses in the second and third trophic levels involving cecidomyiid gallers in the genus *Rabdophaga*, and their parasitoids and inquilines. (From R. Dèclerck-Floate, unpublished, collected in 1991 at Henefer, Utah, from plants with interdigitating branches.)

	Salix lutea	*Salix exigua*
First trophic level		
Apical dominance	Strong	Weak
Mean number of buds/100 cm of shoot	4.2*	6.1
Per cent of undamaged stems producing lateral shoots	4*	52
Mean length of distal shoot on young ramets		
grown after apex is removed (cm)	117*	48
Mean gall density (g cm^{-3})	0.19*	0.36
Second trophic level		
Mean fecundity per *Rabdophaga* female	255*	530
Mean number of eggs/clutch	1.3*	4.4
Mean number of galls/shoot	1.0*	3.5
Mean number of inquiline species	3	3
Mean number of inquilines cm^{-3} (*Dasineura albovittata*)	9.8*	1.8
Third trophic level		
Number of parasitoid species on galler	2	4
Number of parasitoid species on inquilines	4	3
Mean number of parasitoid individuals per gall	0.59*	0.13
Per cent of galls with parasitoids emerged	33*	11

* Significantly different at $P < 0.001$.

From the examples provided, it should be apparent that bottom-up influences throughout trophic systems involving gall-inducing herbivores can be very strong. Such influences are probably ubiquitous. We made the point many years ago (Price *et al.* 1980) that plant effects on tritrophic interactions were general and important; for example, many cases exist where parasitoids are influenced by volatile chemicals in plants, indicating a strong bottom-up effect (see also Price 1981). In that review, we made little reference to gall-inducing insects, but clearly these kinds of species now provide a rich variety of new examples which extend the scope of our understanding on multitrophic interactions.

TOP-DOWN INTERACTIONS

Do top carnivores have cascading effects down trophic systems involving gall-inducing insect herbivores? Such trophic cascades are well researched and perhaps general in aquatic communities, and a balanced view of the relative roles of bottom-up and top-down forces in trophic systems has developed (e.g. Carpenter *et al.* 1985, Carpenter & Kitchell 1987, 1988; McQueen *et al.* 1986, 1989; Schindler 1978). However, such partitioning of effects up and down trophic levels is in its infancy in terrestrial systems (Hunter & Price 1992a, b), although recent analysis of the literature discussed below forms a promising start.

Conceptual perspective

Based on the different size structure of organisms in aquatic and terrestrial food webs, we may make some predictions. Top predators are large in the watery environment, with size declining with descent through the trophic levels. Phytoplankton are very much smaller than the large predatory fish such as pike and tuna. The immature stages of even top predators can be consumed by members lower in the food chain: eggs and young are vulnerable to intermediate and top predators. Hence, we should anticipate strong cascading effects down the food web, as have been documented.

The size structure of organisms among trophic levels is largely reversed in the majority of terrestrial food webs based on plants. Large plants are fed upon by small insect herbivores, which are in turn consumed by smaller parasitoids, or predators much smaller than the autotrophs. Herbivores are not usually predators as the equivalent zooplankton are in aquatic systems, and one plant individual can sustain a rich community in the higher trophic levels. Therefore, finding effects of top carnivores that organize plant communities in some way, can be expected to be rare in terrestrial systems. Insectivorous birds have been shown to increase growth of oak by eating insect herbivores (Marquis & Whelan 1994), but this says nothing about changing community organization as we see so frequently in aquatic systems.

Predictably, we have no cases to report of trophic systems involving gall inducers that demonstrate trophic cascades from top predators to strong effects on plant community structure. However, top predators have been studied to assess their impact on galling herbivore populations and plant damage.

Connection with Coccoloba

Small islands in the Bahamas may or may not have resident *Anolis* lizard populations, such that cascading effects of a top predator can be evaluated. Schoener (1988) found that in the presence of lizards, leaves of buttonwood *Conocarpus erectus* were much less damaged by insect herbivores than when lizards were absent. Herbivore damage was about 35% greater on lizard-free islands. Unfortunately, no galling herbivores were noted, but this and related studies stimulated an experimental approach using sea grape *Coccoloba uvifera* as the primary producer, a cecidomyiid galler, probably *Ctenodactylomyia watsoni* or *cocolobae*, web-spiders and *Anolis* lizards (Spiller & Schoener 1990, 1994). Lizards as the top predators had the potential to increase leaf damage by feeding on spiders, or to decrease galling by eating cecidomyiid adults directly. Parasitoids of the gall midge were not considered.

In fact, in the absence of lizards, total leaf damage was 3.3 times higher than with lizards at natural densities, although total damage was low at around a maximum of 3.5% of area at the end of the season. Lizards reduced spider densities, but spiders had no significant impact on leaf damage, and neither spiders nor lizards had significant effects on the gallers.

Parasitoid removal

Very few studies have involved experimental removal of all parasitoids in a direct attempt to evaluate their role in the population dynamics of galling insects and consequent impact on plant fitness. We know that galling sawflies in the genus *Euura* can have a significant impact on plant fitness in very local, highly favourable sites (Sacchi *et al.* 1988). However, experiments by Woodman (1990) lasting 3 years and three sawfly generations showed no impact of parasitoids on *Euura lasiolepis* populations and plant damage (see also Price 1990).

Of course, top-down effects are always present in galling-insect systems, but mortality is usually measured in the galling stage, after damage to the plant is inflicted. Parasitoids frequently cause high mortality on galling herbivores, but if this translates to lower populations in the following generation has not been established experimentally and convincingly. In addition, we have yet to see top-down influences on plant growth or fitness in galling-insect systems.

SYNTHESIS ON TERRESTRIAL MULTITROPHIC INTERACTIONS

As was the case 15 years ago, we are still collecting examples of multiple levels of interactions in terrestrial systems (Price *et al.* 1980; Barbosa & Letourneau 1988; Barbosa *et al.* 1991). Where is the synthesis and how can we synthesize hundreds of case studies reporting idiosyncratic interactions?

One approach is to attempt to partition the bottom-up effects and top-down effects on species at mid-trophic levels, concentrating on herbivores, for example. This would follow the effective examples in aquatic systems, and would similarly require extensive experimental work. Such an approach will take decades of research before general patterns can emerge, but it holds the greatest opportunity for the discovery of broad patterns in nature and the mechanistic explanation for these patterns. Hence, the approach would ultimately yield factually based theory on multitrophic-level systems. In our opinion, this should be a long-term goal.

In the shorter term, using perturbation experiments to study the population dynamics of herbivores provides a powerful mechanistic approach for understanding the relative roles of bottom-up and top-down forces in terrestrial systems. For example, Cappuccino (1992) studied the tephridid *Eurosta solidaginis*, which initiates galls on stems of the goldenrod *Solidago altissima*. She altered the density of galls in 20 subpopulations and examined the factors involved in post-treatment population change. Significant density dependence was observed only for the factor 'early larval death', involving the plant–insect interaction alone: the establishment of a larva in a feeding site. Predation by chickadees, woodpeckers, the stem-boring beetle *Mordellistena* and parasitism by *Eurytoma* were not density dependent, and could not account for the patterns of population convergence toward pretreatment densities. Evidence suggested that the plant population provided a heterogeneous resource that set limits on the carrying capacity of the local *Eurosta* population. Similar experiments using the gall sawfly *Euura lasiolepis* yielded the same conclusions (Price *et al.* 1995). Studying the free-feeding western tussock moth, *Orgyia vetusta*, Harrison (1994) showed that food-plant shortage resulted in density-dependent larval survival and adult mortality in density-perturbation experiments.

Although Murdoch (1970) called for density-perturbation experiments, Harrison and Cappuccino (1995) could find only 11 such studies on herbivorous insects 25 years later. They found 60 studies on animals in general in those 25 years in two major journals in ecology. Searching for general patterns in bottom-up, top-down and lateral effects within a trophic level, such as competition, they concluded that bottom-up forces were much more likely to produce population regulation than top-down or lateral forces. Among the 60 studies, 89% of studies found resource-based density dependence, while 39% found carnivores producing density-dependent effects. In the small number of studies on insect herbivores the pattern remained, with 68% showing bottom-up and 33% top-down regulatory forces.

Contributing to the perspective that bottom-up influences predominate in these kinds of terrestrial systems is the evidence we presented in our review (Price *et al.* 1980) and the studies on galling insects discussed here. Thus, a large body of data encompassing a diverse spectrum of herbivorous insects provides a clear picture of the dominance of plant-based effects up trophic systems.

With the paper by Harrison and Cappuccino (1995), a synthesis based on experimental studies has begun. For the study of multitrophic-level systems these patterns pose the question of what actually regulates the carnivore trophic levels. Does the bottom-up influence on herbivores continue with a resource-based, supply-side ecology up the food web? And what generalities can be developed on the population dynamics of terrestrial autotrophs? Are top-down forces significant, and what mechanisms are involved?

The details of complex interactions involving indirect effects in multitrophic-level systems clearly require more study in particular systems, but these should be balanced by greater concern with the discovery of broad patterns and ultimately, factually based theory. We may well espouse the position taken by Holt and Lawton (1994) that a start may utilize *contingent theories* relevant to particular systems, and end with the broader perspective of interwoven themes and theory. In a similar vein, Schoener (1987) recognized that the diversity of interactions in nature could not be captured realistically by a central monolithic theory. Rather, pluralistic theory will be required for the factually based discovery of patterns and their mechanistic explanations; for example, in multitrophic-level systems, we may discover that bottom-up forces and indirect effects are most relevant to some kinds of organisms and top-down effects predominate in others. Such partitioning may well be relevant to terrestrial versus aquatic systems. The debate on general patterns and mechanisms in multitrophic patterns is now engaged. Partitioning our efforts between experimental studies, conceptual developments and pattern detection will no doubt eventually yield pluralistic, factually based theory on these complex and fascinating systems.

ACKNOWLEDGEMENTS

Over the years we have been supported financially by grants from the National Science Foundation (current grant DEB-9318188).

REFERENCES

Askew, R. R. (1961). On the biology of the inhabitants of oak galls of Cynipidae (Hymenoptera) in Britain. *Transactions of the Society for British Entomology*, **14**, 237–268.

Barbosa, P. & Letourneau, D. K. (1988). *Novel Aspects of Insect–Plant Interactions.* John Wiley & Sons, New York.

Barbosa, P., Krischik, V. A. & Jones, C. G. (Eds) (1991). *Microbial Mediation of Plant–Herbivore Interactions.* John Wiley & Sons, New York.

Cappuccino, N. (1992). The nature of population stability in *Eurosta solidaginis*, a nonoutbreaking herbivore of goldenrod. *Ecology*, **73**, 1792–1801.

Carpenter, S. R. & Kitchell, J. F. (1987). The temporal scale of variance in lake productivity. *American Naturalist*, **129**, 417–433.

Carpenter, S. R. & Kitchell, J. F. (1988). Consumer control of lake productivity. *BioScience*, **38**, 764–769.

Carpenter, S. R., Kitchell, J. F. & Hodgson, J. R. (1985). Cascading trophic interactions and lake productivity. *BioScience*, **35**, 634–639.

Craig, T. P. (1994). Effects of intraspecific plant variation on parasitoid communities. *Parasitoid Community Ecology* (Ed. by B. A. Hawkins & W. Sheehan), pp. 205–227. Oxford University Press, Oxford.

Craig, T. P., Itami, J. K. & Price, P. W. (1989). A strong relationship between oviposition preference and larval performance in a shoot-galling sawfly. *Ecology*, **70**, 1691–1699.

Craig, T. P., Itami, J. K. & Price, P. W. (1990). The window of vulnerability of a shoot-galling sawfly to attack by a parasitoid. *Ecology*, **71**, 1471–1482.

Dickson, L. L. & Whitham, T. G. (1996). Genetically-based plant resistance traits affect arthropods, fungi and birds. *Oecologia* **106**, 400–406.

Fernandes, G. W. & Price, P. W. (1991). Comparison of tropical and temperate galling species richness: the roles of environmental harshness and plant nutrient status. *Plant–Animal Interactions: Evolutionary Ecology in Tropical and Temperate Regions* (Ed. by P. W. Price, T. M. Lewinsohn, G. W. Fernandes & W. W. Benson), pp. 91–115. John Wiley & Sons, New York.

Fernandes, G. W. & Price, P. W. (1992). The adaptive significance of insect gall distribution: survivorship of species in xeric and mesic habitats. *Oecologia*, **90**, 14–20.

Fernandes, G. W., Lara, A. C. F. & Price, P. W. (1994). The geography of galling insects and the mechanisms that result in patterns. *The Ecology and Evolution of Gall-Forming Insects* (Ed. by P. W. Price, W. J. Mattson & Y. N. Baranchikov), pp. 42–48. United States Department of Agriculture, Forest Service, North Central Experiment Station General Technical Report NC-174.

Harrison, S. (1994). Resources and dispersal as factors limiting a population of the tussock moth (*Orygia vetusta*), a flightless defoliator. *Oecologia*, **99**, 27–34.

Harrison, S. & Cappuccino, N. (1995). Experimental approaches to understanding population regulation. *Population Dynamics: New Approaches and Synthesis* (Ed. by N. Cappuccino & P. W. Price), pp. 131–147. Academic Press, San Diego.

Hawkins, B. A. (1994). *Pattern and Process in Host–Parasitoid Interactions.* Cambridge University Press, Cambridge.

Hawkins, B. A. & Goeden, R. D. (1984). Organization of a parasitoid community associated with a complex of galls on *Atriplex* spp. in southern California. *Ecological Entomology*, **9**, 271–292.

Hawkins, B. A. & Sheehan, W. (Eds) (1994). *Parasitoid Community Ecology.* Oxford University Press, Oxford.

Hollis, W. A. (1889). Galls. *Nature*, **41**, 131.

Hollis, W. A. (1890). Galls. *Nature*, **41**, 272.

Holt, R. D. & Lawton, J. H. (1994). The ecological consequences of shared natural enemies. *Annual Review of Ecology and Systematics*, **25**, 495–520.

Hunter, M. D. & Price, P. W. (1992a). Playing chutes and ladders: heterogeneity and the relative roles of bottom-up and top-down forces in natural communities. *Ecology*, **73**, 724–732.

Hunter, M. D. & Price, P. W. (1992b). Natural variability in plants and animals. *Effects of Resource Distribution on Animal–Plant Interactions* (Ed. by M. D. Hunter, T. Ohgushi & P. W. Price), pp. 1–12. Academic Press, San Diego.

Kearsley, M. J. C. & Whitham, T. G. (1989). Development changes in resistance to herbivory: implications for individuals and populations. *Ecology*, **70**, 422–434.

Marquis, R. J. & Whelan, C. J. (1994). Insectivorous birds increase growth of white oak through consumption of leaf-chewing insects. *Ecology*, **75**, 2007–2014.

McLachlan, R. (1889). Galls. *Nature*, **41**, 131.

McQueen, D. J., Post, J. R. & Mills, E. L. (1986). Trophic relationships in freshwater pelagic systems. *Canadian Journal of Fisheries and Aquatic Sciences*, **43**, 1571–1581.

McQueen, D. J., Johannes, M. R. S., Post, J. R., Stewart, T. J. & Lean, D. R. S. (1989). Bottom-up and top-down impacts on freshwater pelagic community structure. *Ecological Monographs*, **59**, 289–309.

Mivart, S. G. (1889). Galls. *Nature*, **41**, 174–175.

Moorby, T. & Wareing, P. F. (1963). Aging in woody plants. *Annals of Botany* (NS), **106**, 291–309.

Murdoch, W. W. (1970). Population regulation and population inertia. *Ecology*, **51**, 497–502.

Preszler, R. W. & Price, P. W. (1988). Host quality and sawfly populations: a new approach to life table analysis. *Ecology*, **69**, 2012–2020.

Price, P. W. (1981). Semiochemicals in evolutionary time. *Semiochemicals: Their Role in Pest Control* (Ed. by D. A. Nordlund, R. L. Jones & W. J. Lewis), pp. 251–279. John Wiley & Sons, New York.

Price, P. W. (1988). Inversely density-dependent parasitism: the role of plant refuges for hosts. *Journal of Animal Ecology*, **57**, 89–96.

Price, P. W. (1990). Evaluating the role of natural enemies in latent and eruptive species: new approaches in life table construction. *Population Dynamics of Forest Insects* (Ed. by A. D. Watt, S. R. Leather, M. D. Hunter & N. A. C. Kidd), pp. 221–232. Intercept, Andover.

Price, P. W. (1994). Phylogenetic constraints, adaptive syndromes, and emergent properties: from individuals to population dynamics. *Researches on Population Ecology*, **36**, 3–14.

Price, P. W. & Clancy, K. M. (1986a). Interactions among three trophic levels: gall size and parasitoid attack. *Ecology*, **67**, 1593–1600.

Price, P. W. & Clancy, K. M. (1986b). Multiple effects of precipitation on *Salix lasiolepis* and populations of the stem-galling sawfly. *Euura lasiolepis*. *Ecological Research*, **1**, 1–14.

Price, P. W., Bouton, C. E., Gross, P., McPheron, B. A., Thompson, J. N. & Weis, A. E. (1980). Interactions among three trophic levels: influence of plants on interactions between insect herbivores and natural enemies. *Annual Review of Ecology and Systematics*, **11**, 41–65.

Price, P. W., Fernandes, G. W. & Waring, G. L. (1987). Adaptive nature of insect galls. *Environmental Entomology*, **16**, 15–24.

Price, P. W., Cobb, N., Craig, T. P., Fernandes, G. W., Itami, J. K., Mopper, S. & Preszler, R. W. (1990). Insect herbivore population dynamics on trees and shrubs: new approaches relevant to latent and eruptive species and life table development. *Insect–Plant Interactions*, vol. 2 (Ed. by E. A. Bernays), pp. 1–38. CRC Press, Boca Raton, FL.

Price, P. W., Craig, T. P. & Roininen, H. (1995). Working toward theory on galling sawfly population dynamics. *Population Dynamics: New Approaches and Synthesis* (Ed. by N. Cappuccino & P. W. Price), pp. 321–338. Academic Press, San Diego.

Roininen, H., Price, P. W. & Tahvanainen, J. (1993). Colonization and extinction in a population of the shoot-galling sawfly, *Euura amerinae*. *Oikos*, **68**, 448–454.

Romanes, G. J. (1889). Galls. *Nature*, **41**, 80–174.

Romanes, G. J. (1890). Galls. *Nature*, **41**, 369.

Sacchi, C. P., Price, P. W., Craig, T. P. & Itami, J. K. (1988). Impact of shoot galler attack on sexual reproduction in the arroyo willow. *Ecology*, **69**, 2021–2030.

Schindler, D. W. (1978). Factors regulating phytoplankton production and standing crop in the world's freshwaters. *Limnology and Oceanography*, **23**, 478–486.

Schoener, T. W. (1987). Mechanistic approaches to community ecology: a new reductionism? *American Zoologist*, **26**, 81–106.

Schoener, T. W. (1988). Leaf damage in island buttonwood, *Conocarpus erectus*: correlations with pubescence, island area, isolation and the distribution of major carnivores. *Oikos*, **53**, 253–266.

Spiller, D. A. & Schoener, T. W. (1990). A terrestrial field experiment showing the impact of eliminating top predators on foliage damage. *Nature*, **347**, 469–472.

Spiller, D. A. & Schoener, T. W. (1994). Effects of top and intermediate predators in a terrestrial food web. *Ecology*, **75**, 182–196.

Steeves, T. A. & Sussex, I. M. (1989). *Patterns in Plant Development*. Cambridge University Press, Cambridge.

Waage, J. & Greathead, D. (Eds) (1986). *Insect Parasitoids*. Academic Press, London.

Weis, A. E. (1994). What can gallmakers tell us about natural selection on the components of plant defense? *The Ecology and Evolution of Gall-Forming Insects* (Ed. by P. W. Price, W. J. Mattson & Y. N. Baranchikov), pp. 157–171. United States Department of Agriculture, Forest Service, North Central Experiment Station General Technical Report NC-174.

Weis, A. E., Walton, R. & Crego, C. L. (1988). Reactive plant tissue sites and the population biology of gall makers. *Annual Review of Entomology*, **33**, 467–486.

Wetterham, D. (1889). Galls. *Nature*, **41**, 131.

Woodman, R. L. (1990). *Enemy impact and herbivore community structure: tests using parasitoid assemblages, predatory ants, and galling sawflies on arroyo willow*. PhD Thesis, Northern Arizona University, Flagstaff.

14. HOST–MULTIPARASITOID INTERACTIONS

T. H. JONES, M. P. HASSELL AND H. C. J. GODFRAY
Department of Biology and NERC Centre for Population Biology, Imperial College at Silwood Park, Ascot, Berkshire SL5 7PY, UK

INTRODUCTION

Parasitoids are insects whose larvae develop by feeding on the bodies of other insects, their ravages eventually causing the death of the host. They are abundant and ubiquitous members of nearly all terrestrial environments. Somewhere between 0.5 and 2 million species of parasitoids occur on earth (Gaston 1991; LaSalle & Gauld 1992), and few herbivorous insects escape from their attack. In forestry and agriculture, parasitoids have a major role in preventing pest population outbreaks, and they have been used successfully on numerous occasions as biological control agents. Their economic importance has led to intensive study of their evolutionary ecology and population dynamics (Godfray 1994). A great deal is now understood about the behavioural cues and stimuli used in host location and host acceptance, as well as the physiological processes underlying host–parasitoid interactions. There is a particularly well-developed, mechanistic theory of host–parasitoid population dynamics which has stimulated many empirical studies, both in the laboratory and in the field. Although much research on parasitoids was initially prompted by applied issues, they have proved valuable model systems for investigating general problems in population dynamics, as well as in several areas of evolutionary ecology, particularly the study of the sex ratio.

The study of parasitoid communities is not as well developed as the study of their population dynamics and evolutionary ecology. In one way, this is surprising because parasitoid communities have some definite advantages over other terrestrial communities in their ease of study (see Hawkins & Sheehan 1994). Unlike communities dominated by predators and saprophages, it is relatively straightforward to identify the major trophic interactions and even to quantify them. However, a major disadvantage of studying parasitoid communities is the taxonomic problems associated with identifying and even delimiting parasitoid species. Modern systematic research has gone a long way in eroding these problems, but parasitoids, especially the majority of hymenopterous parasitoids, can still claim to be one of the most taxonomically challenging large groups of metazoans.

Another problem with the development of parasitoid community ecology is that there are a number of rather distinct areas of research that bear directly on community-level problems, but which have largely developed in isolation. Fusing these different strands is important, both in framing the important conceptual questions to be addressed, as well as attempting to solve them. In this chapter, we review briefly these different approaches, and then suggest ways in which they may be combined. The different areas we have in mind are (i) the description of parasitoid communities by the construction of food webs; (ii) the identification of patterns in parasitoid communities using macroecological techniques; (iii) the study of indirect and higher-order interactions including apparent competition and tritrophic interactions; (iv) the extension of classical one-host–one-parasitoid population dynamics to more species rich communities.

PARASITOID WEBS

A parasitoid web is a subset of a traditional food web that contains hosts (and possibly food plants), parasitoids and hyperparasitoids (Memmott & Godfray 1992, 1994). As mentioned above, parasitoid webs are relatively easy to construct because it is more straightforward to associate a host and its parasitoids than, say, a prey species and its predators. Parasitoid webs can be classified in a number of ways. A distinction can be made between (i) connectance webs that provide only presence or absence information about individual trophic links; (ii) semi-quantitative webs that provide information about the relative numbers of different parasitoids on each host; and (iii) quantitative webs that in addition provide information about the relative abundances of different host species. An orthogonal classification is into (i) community webs that describe a circumscribed set of hosts, and all their parasitoids and hyperparasitoids (the set might, for example, be all aphids or all leafminers at a particular locality); and (ii) source webs that study the parasitoids and hyperparasitoids associated with a particular host (sink webs based on the hosts of one parasitoid or hyperparasitoid are theoretically possible, but we know of no example). In this section, we concentrate on community webs and discuss source webs in the following section. A final categorization can be made between webs constructed at one time and one place, and those that sum data over time and/ or space. Of these different types of webs, it is clear that community webs tell us most about how parasitoid communities are constructed, and quantitative community webs constructed at one time and place are particularly informative.

There are still relatively few examples of parasitoid webs in the literature. The first major webs were semi-quantitative community webs based around British cynipid gall wasps (Askew 1961) and British lepidopterous leafminers (Askew & Shaw 1979). Semi-quantitative webs have also been built for American gall wasp communities (Force 1974; Hawkins & Goeden 1984) and there are a few connectance webs containing many parasitoids (Rejmanek & Stary 1979; Hopkins 1984; Whittaker 1984). The only fully quantitative parasitoid web published so far is of leafminers in tropical dry forest in Costa Rica (Memmott *et al.* 1994), although two further webs based on British aphids and British leafmining moths in the genus *Phyllonorycter* are in preparation (Godfray *et al.*, unpublished data).

The importance of parasitoid webs is that they provide raw data about the complexity of natural host–parasitoid communities. An individual parasitoid web gives valuable information about the connectedness of species in a community, the possibility of indirect interactions mediated by shared natural enemies, and the degree to which individual host–parasitoid associations can be abstracted from the community in which they are embedded and studied in isolation; for example, consider Fig. 14.1 which shows the connectance and quantitative parasitoid webs for the Costa Rican dry forest leafminer community (Memmott *et al.* 1994). Although the hosts are taxonomically diverse (from three orders of insects), they inhabit an ecologically rather uniform niche (the mine) and are attacked by parasitoids that frequently kill their host at oviposition and hence do not need to be highly specialized to overcome the host immune system. In consequence, there is a large guild of rather polyphagous parasitoids and the connectance web is highly connected. Consideration of the connectance web alone suggests a very complex community, whose high-dimensional dynamics would be very difficult to understand and study. However, the quantitative web provides some suggestions about how the community might be structured. Certain host species are numerically dominant (in particular a group of hispine beetles attacking a forest floor bamboo) and may act as keystone species (Pimm 1982) whose dynamics have a disproportionate effect on other species in the community. Consider a parasitoid species that attacks the keystone species and another species of host. This parasitoid might be the major parasitoid of the non-keystone species and a minor parasitoid of the keystone species, yet because of the abundance of the latter, the keystone species may still be the parasitoid's most important host. In such a case, the dynamics of the non-keystone species would be driven, via a shared parasitoid, by fluctuations in the density of the keystone species.

Such arguments illustrate the potential value of quantitative parasitoid webs, but also their limitations. Parasitoid webs are important for generating hypotheses about species interactions, but these hypotheses have to be tested by field manipulation experiments (Paine 1980, 1988). Yet given the enormous effort required to plan and conduct experiments in nature, constructing parasitoid webs as an initial step is surely a better alternative to blind multifactorial experiments – death by ANOVA.

As more parasitoid webs become available, it will be possible to get further insights into parasitoid community structure using a comparative approach; for example, the initial results of our quantitative food web based on leafminers and aphids at a study site at Silwood Park in southern England (Müller, Rott & Godfray unpublished data) suggest that the leafminer web may have a similar structure to the Costa Rican web illustrated in Fig. 14.1, while the aphid community is very

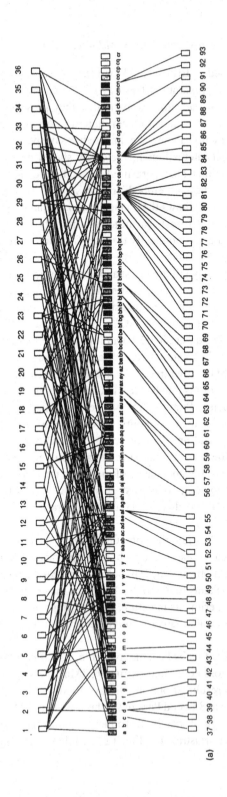

(a)

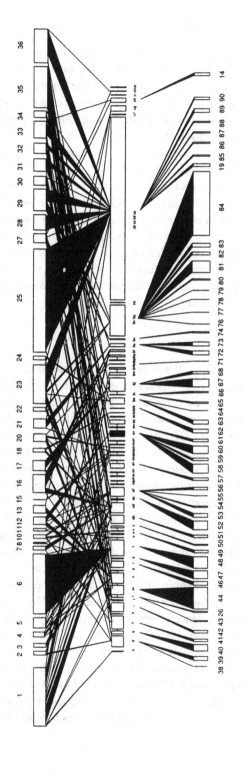

(b)

Scale ▭ 100 hosts ▭ 25 parasitoids

FIG. 14.1. Connectance (a) and fully quantitative (b) parasitoid web describing the parasitoids attacking a community of leaf miners in tropical dry forest, Costa Rica. In (a), the leafminers are represented by rectangles labelled by letters in the central row. The different shading of the rectangles represents leafminers in different insect orders (Diptera, black; Lepidoptera, grey; Coleoptera, white; uncertain, hatched) and species feeding on the same plant species are linked by horizontal lines. Parasitoids are represented by numbered rectangles in the top and bottom rows. For clarity, parasitoids reared from a single hosts are placed in the bottom row, and those reared from several hosts in the top row. (b) The quantitative web is arranged in a similar manner, except that the widths of the rectangles represent relative abundance (note parasitoid and host abundances are shown with different scales). (From Memmott & Godfray 1994.)

different. Askew and Shaw's (1979) semi-quantitative leafminer parasitoid web
also points to similarities in the tropical and temperate leafminer webs. Unlike the
two leafminer communities, the aphid community has a very large hyperparasitoid
trophic level while at the primary parasitoid trophic level the degree of connectedness
is substantially less. If parasitoid webs associated with hosts in different feeding
niches have consistent structural differences, then it may be possible to explain
some of the macroevolutionary patterns associated with the number of parasitoids
that attack different categories of herbivores (see next section).

A final caveat about the interpretation of parasitoid webs needs to be stated.
Parasitoids are not the only natural enemies attacking their hosts and frequently it
will be necessary to consider the role of predators and even diseases in structuring
communities. Moreover, a resource–consumer relationship is not the only way
two species can interact; for example, competition between members of the same
trophic level may also be important. Nevertheless, for many communities, we believe
the construction of parasitoid webs offers a unique insight into how a community
is structured and how it might be investigated.

THE MACROECOLOGICAL APPROACH

Although relatively few parasitoid communities have been investigated in the field,
numerous individual host species (especially pest species) have been studied in
depth and the number and identity of their parasitoids recorded. There are thus a
large number of source webs in the literature that might provide information about
patterns in parasitoid community ecology. Of course, there are fearsome statistical
problems in detecting patterns from literature data that were originally collected
for many different reasons. The number of species of parasitoid that are recorded
attacking a host species depends strongly on sample size – the numbers of host
reared – and it is essential that sample size should be factored out in any statistical
analysis. There are many sampling biases that might arise: species in certain habitats,
with wide as opposed to restricted range, of economic importance, with different
numbers of generations a year, and that cannot be sampled in certain life stages,
may all provide information of different reliability. Unreliable data lead to both
benign and non-benign problems: the benign problem is a simple increase in
statistical error that can be overcome by collecting a larger data set. However, a
much more insidious problem is covariance of error with explanatory variables
which can lead to spurious results: this cannot be overcome statistically and requires
new data collected in a manner that avoids the sample bias.

Over the past 10 years, B. A. Hawkins has amassed a huge data set of the number of species of parasitoids attacking many different species of host. As Hawkins
(1994) himself has reviewed this work in depth in a recent monograph, we discuss
this subject only briefly. The most robust conclusion that emerges is of the great

importance of the host-feeding niche in determining the number of parasitoids attacking a particular herbivorous insect. The greatest number of parasitoids are found on leafmining insects with a reduction as hosts feed in increasingly concealed sites (gallers, borers and root feeders) or in increasingly exposed sites (rollers/ webbers, external feeders).

How might this pattern arise? In the earliest paper reporting this result, Hawkins and Lawton (1987) suggested that over both evolutionary and ecological time, recruitment of parasitoids to hosts in different niches may vary due to differences in detectability. Leafminers, imprisoned between the two leaf lamellae, with their position frequently marked by a visually obvious mine, may be particularly apparent to parasitoids. Equilibrium parasitoid numbers will also be influenced by risk of mortality, and they suggested that the risk of mortality may be low in particularly safe feeding niches such as leafmines. Exactly how differences in recruitment and mortality determine the number of parasitoid species per host species depends, however, on the assumptions made about the population dynamics of the two classes of insects. More recently, Hochberg and Hawkins (1992) have described an explicitly dynamic model that predicts this pattern. We delay discussion of this work until the next section where we treat other approaches to the population dynamics of parasitoid communities. A final explanation of these patterns develops earlier ideas of Askew (1980) about the influence of taxonomic and ecological homogeneity on parasitoid host range. If hosts of a particular feeding niche tend to be similar, either (or both) in ecology or taxonomy, then host-range expansion in both ecological and evolutionary time will be facilitated. Thus, consider leafminers again; the leafmining habit is fairly uniform being constrained by leaf morphology. The ecology of leafminers is relatively uniform compared with species that feed externally or in galls, shoots or roots. In addition, relatively few taxa have evolved to become leafminers, but many have radiated widely giving rise to genera with many species. If leafminers are ecologically and phylogenetically relatively uniform, it is comparatively easy for a polyphagous species to have a large host range, and for a specialist species to make an evolutionary jump to a new host (Godfray 1994).

In addition to the major influence of feeding niche, more subtle effects were detected. Within gall midges (Cecidomyiidae), gall morphology influences parasitoid number, an effect that can probably be explained by the hypotheses discussed above. It had been thought that herbivores on plants of different successional stage were attacked by different numbers of parasitoid species, a result that would parallel the distribution of host species per host plant in different successional stages. However, Hawkins' (1994) most recent analyses suggest that the influence of successional stage is not of great importance.

The analyses discussed above are chiefly concerned with community statics. A different question is the speed with which communities are assembled. We can get some information on this by asking how fast parasitoid species numbers build up on host species that are introduced accidentally or deliberately into a new area. Cornell and Hawkins (1993) collected data from the literature on the number of parasitoids attacking hosts in areas where they were endemic and in areas where they had been introduced. As one would expect, the two figures were highly correlated, but interestingly hosts as invaders were attacked by significantly lower numbers of parasitoid species than hosts as residents. Cornell and Hawkins (1993) were also able to detect a weak but significant influence of the time since introduction on the number of parasitoid species attacking a host. These data suggest either that parasitoid recruitment takes some time to occur, or that hosts as invaders tend to attract fewer species of parasitoid, perhaps because they tend to be dissimilar to native hosts and hence are harder to locate by parasitoids.

Another way of addressing this question is to compare the number of parasitoids attacking an introduced host with the number of parasitoids attacking similar hosts in the same area. An opportunity to do this was recently presented by the invasion of the UK by two leafmining moths in the genus *Phyllonorycter* (Gracillariidae) which feed on the leaves of *Platanus* spp. and *Pyracantha* spp., two widely planted (but introduced) garden plants. Fortuitously, the parasitoid complex attacking native *Phyllonorycter* species has been extensively studied (Askew & Shaw 1979, and references therein). The numbers of parasitoid attacking the two new species

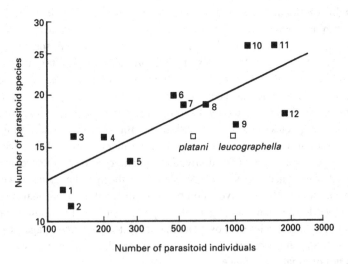

FIG. 14.2. Number of parasitoid species reared from different groups of leafmining *Phyllonorycter* as a function of sample size. The 12 numbered solid squares represent native *Phyllonorycter* that attack different species of host plant (from Godfray *et al.* 1995). The open squares are two species, *P. platani* and *P. leucographella*, that have invaded in the past 10 years.

of *Phyllonorycter* approximately 10 years after their invasion was compared with the numbers attacking the resident species (Godfray *et al.* 1995). Controlling for sample size, the data for the invading species (16 parasitoid species each) each fell well within the scatter of the resident species (Fig. 14.2). This result suggests that a novel host species, which is ecologically and taxonomically similar to native species, can accumulate a 'full' parasitoid complement in a very short period of time. One unexpected observation was that one of the parasitoid species attacking the *Phyllonorycter* on *Platanus* was previously unknown from the British Isles and had apparently colonized at the same time.

INDIRECT INTERACTIONS

As mentioned in the Introduction, extensive studies of parasitoid ecology and evolutionary biology have revealed a variety of ways that host–parasitoid interactions may be influenced by other species in a community. In this section, we review some of these interactions, but make no attempt at comprehensiveness.

Tritrophic interaction

The host plant may exert a variety of effects on the parasitoid–herbivore–plant interaction. At the behavioural level, many parasitoids use plant-derived cues in locating microhabitats where hosts may be present, or use volatiles released by damaged plants as more accurate clues to the location of the host. Turlings and his co-workers (Turlings *et al.* 1990, 1991) have shown that some plants produce specific volatile chemicals when fed on by herbivores and that the herbivores' parasitoids use these chemicals in host location. The production of these volatiles may be just a by-product of wound formation or a direct defence against predation, but it is also possible that it might have evolved as a trans-trophic level signalling system. Parasitoid searching efficiency is also known to be influenced by host-plant characters, for example plant hairiness (Hulspas-Jordan & Van Lenteren 1978), while the nutrient status of the plant, and the distribution of secondary metabolites, can have both direct effects on the parasitoid and indirect effects mediated by the host (Barbosa 1988). Finally, some plants have extrafloral nectaries which provide food for insects including parasitoids, which may be preferentially attracted to these species (Van Emden 1963; Leius 1967).

These considerations, and others reviewed by Price *et al.* (1980) and Price (1981), warn against ignoring the plant trophic level when studying host–parasitoid communities. The lower trophic level may also be crucially important in cases where herbivores are food limited and are largely regulated from below rather than by their natural enemies, a subject discussed more fully by Price *et al.* (this volume).

Apparent competition

Consider two species of host that share a common parasitoid. If the abundance of one host increases, then parasitoid numbers are also likely to rise. This may be because in the short term parasitoids are attracted into the area, or in the long term because the increased numbers of hosts offer greater opportunities for parasitoid reproduction leading to population growth. Now the increase in the number of parasitoids will have a negative effect on the population dynamics of the second species of host. We thus have a situation that in many ways is similar to competition: two species have reciprocally negative effects on each other's population growth rates; but in this case the species need never directly interact, the negative effects are mediated via a shared natural enemy. Because of the parallels with traditional competition, Holt (1977, 1984) christened this indirect interaction apparent competition, and further distinguished between short-term and long-term apparent competition (Holt & Kotler 1987), the former involving a behavioural response by the natural enemy, the latter population growth. Short-term apparent competition need not be the only immediate interaction between two species linked by a natural enemy. If densities of one species go up, leading to natural enemy satiation, the mortality experienced by the other species may actually drop, giving rise to 'short-term apparent mutualism' (Holt 1977; Abrams 1987).

The potential importance of apparent competition in community ecology has often been discussed (Holt & Lawton 1993, 1994, and references therein). In particular, it has been suggested that communities of herbivorous arthropods might be structured by apparent competition mediated by shared predators, pathogens and parasitoids (e.g. Jeffries & Lawton 1984; Lawton 1986; Godfray 1994). Some studies have found patterns in communities that are consistent with apparent competition mediated by parasitoids (reviewed by Godfray 1994), but experimental evidence is sparse. The most important study to date is that of Settle and Wilson (1990a, b) who studied two species of leafhopper that feed on the leaves of cultivated grape in California. During their study, one species (*Erythroneura variabilis*) invaded the San Joaquin Valley, an extension of range that was associated with a marked decline in the abundance of a second species, *E. elegantula*. The two insects did compete, but intra- and inter-specific competition were of similar magnitude and Settle and Wilson argued that competition was unlikely to explain the drop in numbers of *E. elegantula*. Instead, they suggested that *E. elegantula* declined because of attack by a shared mymarid egg parasitoid, *Anagrus epops*. *E. variabilis* eggs are better protected from attack by the parasitoid which gives this species an advantage in comparison with *E. elegantula*.

In our view, the paucity of experimental studies of apparent competition mediated by parasitoids probably does not reflect the importance of this type of interaction in insect communities. Indeed, many applied entomologists take it for granted that biological control of pests can be improved by creating habitats for the alternative hosts of the pest's natural enemies. We believe that further experimental studies in this area will be richly rewarding.

BUILDING UP: POPULATION DYNAMICS

A. J. Nicholson, one of the founders of modern population dynamics, was particularly interested in host–parasitoid population dynamics (Nicholson 1933; Nicholson & Bailey 1935). He asked how populations of host and parasitoid might interact if each had one synchronized generation a year, and if encounters of hosts by parasitoids were a random process. In collaboration with Bailey, a mathematician, he developed a coupled discrete time formulation for this problem that has since been called the Nicholson–Bailey model. This model predicts diverging oscillations: parasitoid densities increase if hosts are abundant leading to over-exploitation and a crash in host densities followed by a subsequent crash in parasitoid densities. In the absence of parasitoids, host densities rapidly increase leading to ever-increasing cycles of boom and bust. In nature, the interaction cannot persist (Hassell 1978).

That the null model of host–parasitoid interactions predicts unstable oscillations has led to intensive research into what features of host–parasitoid biology might lead to persistent interactions (Hassell 1978; Hassell & Pacala 1990). After much debate, the consensus today is that for a 'Nicholsonian' parasitoid to regulate the numbers of its host, there must be sufficient variance in the risk of parasitoid attack across host individuals (in many circumstances, a simple rule of thumb operates that coefficient of variance in the risk of parasitism must exceed one (Pacala *et al.* 1990)). Heterogeneity of risk is important because it implies some hosts are relatively protected from parasitoid attack and so can survive through periods when parasitoid densities are high. This, and the fact they also produce offspring that can act as hosts for the next parasitoid generation, acts to tame the diverging oscillations inherent in the system. There are a number of ways in which heterogeneity of risk can be generated biologically: (i) some hosts may exist in a physical refuge from parasitoid attack; (ii) some hosts may have the ability to defend themselves physically or physiologically from parasitism; (iii) hosts may be distributed heterogeneously across patches in the environment, and parasitoids may tend not to visit certain patches; and (iv) the host and parasitoid populations may not be perfectly synchronized in time so that some hosts enjoy a reduced risk of parasitism. In populations with overlapping generations, a long-lived invulnerable developmental stage is another means of avoiding over-exploitation during periods of high parasitoid population density. Finally, other mechanisms come into play if we consider spatially distributed populations. Originally, Nicholson speculated that locally unstable populations might still persist regionally if populations that went extinct locally could be rescued by immigration from other, out of phase, populations. This prescient suggestion anticipated a major theme of modern metapopulation theory and has been confirmed by explicit modelling. In continuous space (or on regularly connected lattices), locally unstable populations can persist regionally as ever-changing waves of (parasitoid) pursuit and (host) evasion that may be spatially well organized into discrete spirals or that exhibit more complex, apparently chaotic, spatial structures (Hassell *et al.* 1991, 1994; Comins *et al.* 1992).

The recent consensus on how parasitoids might regulate their hosts is important as it suggests means by which models of one-host–one-parasitoid can be extended to more complex communities. In the past, a number of studies have considered slightly more complex communities; for example, two parasitoids attacking one host (May & Hassell 1981; Hogarth & Diamond 1984; Kakehashi *et al.* 1984; Hassell & May 1986; Godfray & Waage 1991; Briggs 1993), two hosts attacked by a common parasitoid (Holt 1977; Holt & Lawton 1994), or a community composed of a host, parasitoid and hyperparasitoid (Beddington & Hammond 1977). Rather than describe each of these studies in turn, we move straight on to some recent work that analyses a community of a host attacked by a specialist and generalist parasitoid, and a theoretical exploration of the dynamics of a five-species community, referring to these earlier results where appropriate.

The case study concerns the cabbage root fly *Delia radicum* (Jones *et al.* 1993). For the specialist cynipid parasitoid *Trybliographa rapae*, there is evidence from field data of the existence of within-generation spatial patterns where parasitism is directly correlated with host density per plant (Jones & Hassell 1988). Exploration of the theoretical model describing the interaction showed that, were *T. rapae* to act alone, this spatial heterogeneity in parasitism would promote stability, but only within a very narrow range of the net host rate of increase. In contrast, *Aleochara bilineata*, the generalist staphylinid parasitoid, acts as a simple, between-generation density-dependent factor. Were it present alone, stable host populations would give way to locally unstable ones with limit cycles and higher-order behaviour as the net host rate of increase increases. It is, however, when the two natural enemies are combined that the most interesting patterns are found. Provided that the survivorship of *T. rapae* is sufficiently high, the interplay of the two natural enemies can lead to alternative stable states, although the population levels lie largely outside the range of observed densities from the field. Interestingly, these patterns are very similar to those found by Hassell and May (1986) in the treatment of a generalist predator and specialist parasitoid attacking an insect host/prey. In their particular case, the parameter combinations required for alternative three-species states require rather extreme values of the net host rate of increase.

The important conclusion from this type of study is that in many cases, instead of the relative straightforward dynamics exhibited by simple single host–single parasitoid interactions, the addition of just one other parasitoid to the system gives rise to a far wider range of dynamical behaviours.

Consider the five species community pictured in Fig. 14.3. Two species of hosts are each attacked by a specialist parasitoid while a third species of parasitoid, the generalist, can attack both hosts. We want to know under what circumstances the full five-species community can occur, and when it cannot occur, what sub-community we would expect (Wilson *et al.* in press).

To begin, we need to specify who wins in competition when two parasitoids attack the same host individual. Observations of natural communities provide some help here. Endoparasitoids are often specialist insects because they require complex adaptations to protect themselves against the defences of the host. Ectoparasitoids, on the other hand, are often more generalist because they kill their host at oviposition. In competition between endoparasitoids and ectoparasitoids, the latter usually win as they just consume the host and any parasitoid it contains. We thus assume that the generalist is competitively superior to the specialist. Next, we have to decide how hosts vary in their susceptibility to parasitism. A useful phenomenological approach is to assume risk is gamma-distributed across hosts; integrating risk over the season, the proportion of hosts escaping parasitism is the zero term of a negative binomial distribution with clumping parameter k. As first found by May (1978), stability requires $k < 1$.

Moving from one to two parasitoids requires another assumption that is crucial in building up from single to complex communities. We need to understand the correlation (ρ) in the risk of attack between the two species of parasitoid. Perhaps the simplest assumption is that the risk of attack is independent ($\rho = 0$). However, for many biological mechanisms generating heterogeneity of risk, independence is unlikely. If some hosts are in a physical refuge from one species of parasitoid, it is likely but not certain that they will also be protected from a second species. A second limiting case is thus to assume that hosts vary in risk, but that the risk

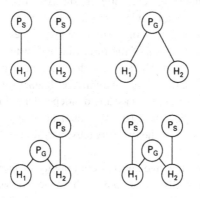

FIG. 14.3. Hypothetical communities containing two host species, two specialist parasitoids and a generalist parasitoid, and subsets thereof.

of attack from either parasitoid is the same ($\rho = 1$). If the two parasitoid species have very different searching strategies, or if there is a trade-off in the ability of a host to defend itself against the two species, then negative correlations are possible ($\rho < 0$).

The importance of the correlation in risk can be illustrated by earlier work that sought to answer the question of whether it was better to release one or two species of parasitoid to control an insect pest. May and Hassell (1981) and later Hogarth and Diamond (1984) and Godfray and Waage (1991), assumed independent parasitoid attack and concluded two parasitoids were better than one, while Kakehashi *et al.* (1984) assumed a perfect correlation in risk and concluded the reverse. It is also much easier for two species of parasitoid to coexist on the same species of host when the distribution of risk is independent. This is because the independent distribution leads to intra-specific competition being stronger than inter-specific competition and hence a species is at an advantage when rare.

A further assumption that has interesting ramifications is the comparative fecundity of the two species of host. Consider a sub-community consisting of the two hosts and the shared generalist. This simple interaction has been studied by Holt and colleagues as an example of apparent competition (see last section). Coexistence is possible only if the two hosts have the same fecundity and hence the same equilibrium population density (this assumes the parasitoid is equally efficient at locating both species of host). If one host has a larger fecundity than the other, it can support a population of parasitoids that is sufficient to drive the second species to extinction. Holt and Lawton (1993) call such an outcome dynamic monophagy and discuss its possible occurrence in nature.

The standard way to analyse a host–parasitoid model is to derive the equilibrium densities and then to investigate the stability properties of the system. For this more complex community, we also need to know whether a species can invade a community from which it is currently absent. As will become apparent, some invasible communities are not stable so that standard techniques are not applicable. Such situations were explored using numerical techniques.

An example of the analysis of the three-species system is given in Fig. 14.4 (from Wilson *et al.* in press). The axes are the heterogeneity of risk of parasitism among hosts, and the ratio of the searching efficiency of the specialist to the generalist. To obtain this figure, we assumed independence in the risk of parasitism by the two species of parasitoid ($\rho = 0$), equal fecundity of the two hosts, and equal searching efficiency of the two specialist parasitoids. From this analysis we can conclude the following:

1   The persistence of the five-species community is facilitated by increasing heterogeneity of risk. As discussed above, where the risks of parasitoid attack are uncorrelated, increasing heterogeneity leads to stronger intra-specific than interspecific competition increasing the likelihood of coexistence.

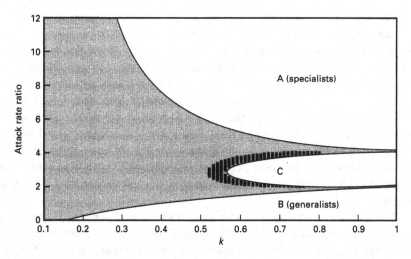

FIG. 14.4. Circumstances under which the different communities and subcommunities depicted in Fig. 14.3 are predicted to occur as a function of k, the amount of density dependence in the parasitoid attack function (small values imply greater density dependence), and the ratio of searching efficiencies of specialist and generalist parasitoids (values greater than 1 imply the specialists are superior). In the light-shaded region, all five species coexist at a stable equilibrium, while in the darker shaded region all five species coexist, but with limit cycle and chaotic dynamics. In region A, the specialists occur, but the generalist is absent while the reverse occurs in region B. In region C, no community is both stable and uninvasable and very complex dynamics, described in the text, are predicted. (From Wilson *et al.* in press).

2 As the searching efficiency of the specialist increases relative to the generalist, there is an increased likelihood that the generalist is outcompeted and excluded from the system. If the searching efficiency of the generalist increases, the reverse happens.

3 The final conclusion is perhaps the most interesting. There is a region of parameter space in which no community is both stable and uninvasible. In this region, all three parasitoid species can invade a community in which they are absent, yet the full five species community cannot persist: the densities of species show large amplitude fluctuations leading inevitably to the extinction of the two specialists and to the establishment of a community containing the two hosts and the generalist. However, once this community has been attained, either or both specialists are able to invade, leading to a four- or five-species community with complex dynamics that ultimately breaks down with the loss of the specialists. No one community can persist and the observed state of the system thus depends critically on the speed with which the five-species community collapses, and the frequency with which the specialists reinvade.

This last conclusion is interesting because it is a type of dynamic behaviour that can be shown only by communities more complex than the simple communities that have traditionally been studied. The appearance of complex dynamics is reasonably robust. If we relax any of the major assumptions discussed above, the detailed picture changes but regions of complex dynamic stability still remain; for example, allowing hosts to have different fecundities means that a community consisting of one generalist and two hosts cannot occur and in the region where this community occurs in Fig. 14.4 a single host and the generalist is found. However, a region of dynamic instability occurs as the searching efficiency of the specialist increases. One way that the complex dynamic stability can be lost is if the correlation in risk of attack is perfect ($\rho = 1$) but as long as ρ is less than about 0.9 it is still present.

The only other population dynamic study of a large host–parasitoid community is that of Hochberg and Hawkins (1992, 1993). They took a very different approach and aimed to see whether they could predict the relationship between feeding niche and parasitoid species diversity discussed above in the section on the macro-ecological approach. They assumed that a certain fraction of the host population is in a proportional refuge, completely protected from parasitism, while the remainder of the population is subject to attack by both generalist and specialist parasitoids. The dynamics of generalist parasitoids are uncoupled from that of the host, which distinguishes them from the monophagous and tightly coupled specialists. Starting with a large assemblage of specialists and generalists, the model was run to equilibrium and 1000 hosts 'sampled' to see how many were parasitized and by how many species.

The model predicts a humped relationship between species diversity and the extent of the proportional refuge. When the refuge is very large, the host cannot be controlled by the parasitoid, and in consequence the sample of 1000 hosts contains few parasitoids. As more and more hosts become susceptibile to parasitism, a greater number of parasitoids are picked up in the 1000 host sample and diversity rises. However, a threshold is reached when sufficient hosts are susceptible that the specialist parasitoids reduce host density to a very low level at which few generalist parasitoids are common enough to exceed a detection threshold. Thus diversity peaks for intermediate levels of the refuge. Hochberg and Hawkins (1992, 1993) suggest that the sequence: external feeders; leaf rollers; leafminers; gallers; stem borers and root feeders, represents ever increasing refuges and that their model offers an explanation for the pattern of parasitoid diversity and feeding niche described above. We are not convinced that the particular model they use is a good representation of a complex parasitoid community, nor that the relationship between feeding niche and the extent of a proportional refuge is real. Nevertheless, we applaud this work as an important step in moving towards more complex and more realistic host–parasitoid community models.

CONCLUSIONS

We are bullish about the possibilities of substantial progress being made in parasitoid community ecology in the near future. We believe we need more descriptive studies of parasitoid webs that, to quote Robert Paine in a related context, will provide 'road maps of interactions' and 'grist for theoretical development'. Such parasitoid webs may help to explain some of the important patterns established by macroecological studies, and also suggest how experiments can be designed to investigate the importance of indirect interactions. While experimental studies are essential, they can only be properly interpreted within the framework of a formal theory of population dynamics: the spoken word is just not good enough a tool to explore complex quantitative hypotheses. We believe that much of the controversy of bottom-up versus top-down effects in insect population dynamics will disappear when ecologists dismount their soapboxes and rephrase their arguments in a quantitative framework. Developing a population dynamic framework of parasitoid communities will not be easy – that much is clear from the discussion in the last section–but we have the advantages of a well-developed dynamic theory of simple interactions, and of working with a type of biological interaction that is relative easy to study quantitatively, and to manipulate, in the field.

REFERENCES

Abrams, P. A. (1987). On classifying interactions between populations. *Oecologia*, **73**, 272–281.

Askew, R. R. (1961). On the biology of the inhabitants of oak galls of Cynipidae (Hymenoptera) in Britain. *Transactions of the Society for British Entomology*, **14**, 237–268.

Askew, R. R. (1980). The diversity of insect communities in leaf-mines and plant galls. *Journal of Animal Ecology*, **49**, 817–829.

Askew, R. R. & Shaw, M. R. (1979). Mortality factors affecting the leaf-mining stages of *Phyllonorycter* (Lepidoptera: Gracilariidae) on oak and birch. 2. Biological of the parasite species. *Zoological Journal of the Linnaean Society*, **67**, 51–64.

Barbosa, B. (1988). Natural enemies and herbivore plant interactions: influence of plant allelochemicals and host specificity. *Novel Aspects of Insect-Plant Allelochemicals and Host Specificity* (Ed. by P. Barbosa & D. Letourneau), pp 201–229. John Wiley & Sons, New York.

Beddington, J. R. & Hammond, P. S. (1977). On the dynamics of host-parasite-hyperparasitoid interactions. *Journal of Animal Ecology*, **46**, 811–821.

Briggs, C. (1993). Competition among parasitoid species on a stage-structured host, and its effect on host suppression. *American Naturalist*, **141**, 372–396.

Comins, H. N., Hassell, M. P. & May, R. M. (1992). The spatial dynamics of host–parasitoid systems. *Journal of Animal Ecology*, **61**, 735–748.

Cornell, H. V. & Hawkins, B. A. (1993). Accumulation of native parasitoids on introduced hosts: a comparison of 'hosts-as-natives' and 'hosts-as-invaders'. *American Naturalist*, **141**, 847–865.

Force, D. C. (1974). Ecology of host–parasitoid communities. *Science*, **184**, 624–632.

Gaston, K. J. (1991). Body size and probability of description: the beetle fauna of Britain. *Ecological Entomology*, **16**, 505–508.

Godfray, H. C. J. (1994). *Parasitoids. Behavioural and Evolutionary Ecology.* Monographs in Behaviour and Ecology, Princeton University Press, Princeton.

Godfray, H. C. J. & Waage, J. K. (1991). Predictive modelling in biological control: the mango mealy bug (*Rastrococcus invadens*) and its parasitoids. *Journal of Applied Ecology,* **28,** 434–453.

Godfray, H. C. J., Agassiz, D. J. L., Nash, D. R. & Lawton, J. H. (1995). The recruitment of parasitoid species to two invading herbivores. *Journal of Animal Ecology,* **64,** 393–402.

Hassell, M. P. (1978). *The Dynamics of Arthropod Predator–Prey Systems.* Princeton University Press, Princeton.

Hassell, M. P. & May, R. M. (1986). Generalist and specialist natural enemies in insect predator-prey interactions. *Journal of Animal Ecology,* **55,** 923–940.

Hassell, M. P. & Pacala, S. W. (1990). Heterogeneity and the dynamics of host-parasitoid interactions. *Philosophical Transactions of the Royal Society of London, B,* **330,** 203–220.

Hassell, M. P., Comins, H. N. & May, R. M. (1991). Spatial structure and chaos in insect population dynamics. *Nature,* **353,** 255–258.

Hassell, M. P., Comins, H. N. & May, R. M. (1994). Species coexistence and self-organizing spatial dynamics. *Nature,* **370,** 290–292.

Hawkins, B. A. (1994). *Pattern and Process in Host–Parasitoid Interactions.* Cambridge University Press, Cambridge.

Hawkins, B. A. & Goeden, R. D. (1984). Organization of a parasitoid community associated with a complex of galls on *Atriplex* spp. in southern California. *Ecological Entomology,* **9,** 271–292.

Hawkins, B. A. & Lawton, J. H. (1987). Species richness for parasitoids of British phytophagous insects. *Nature,* **326,** 788–790.

Hawkins, B. A. & Sheehan, W. (Eds) (1994). *Parasitoid Community Ecology.* Oxford Science, Oxford.

Hochberg, M. E. & Hawkins, B. A. (1992). Refuges as a predictor of parasitoid diversity. *Science,* **255,** 973–976.

Hochberg, M. E. & Hawkins, B. A. (1993). Predicting parasitoid species richness. *American Naturalist,* **142,** 671–693.

Hogarth, W. L. & Diamond, P. (1984). Interspecific competition in larvae between entomophagous parasitoids. *American Naturalist,* **124,** 552–560.

Holt, R. D. (1977). Predation, apparent competition and the structure of prey communities. *Theoretical Population Biology,* **12,** 197–229.

Holt, R. D. (1984). Spatial heterogeneity, indirect interactions, and the coexistence of prey species. *American Naturalist,* **124,** 377–406.

Holt, R. D. & Kotler, B. P. (1987). Short-term apparent competition. *American Naturalist,* **130,** 412–430.

Holt, R. D. & Lawton, J. H. (1993). Apparent competition and enemy-free space in insect host-parasitoid communities. *American Naturalist,* **142,** 623–645.

Holt, R. D. & Lawton, J. H. (1994). The ecological consequences of indirect effects in ecological communities. *Annual Review of Ecology and Systematics,* **25,** 443–466.

Hopkins, M. J. G. (1984). The parasitoid complex associated with stem boring *Apion* (Coleoptera: Curculionidae) feeding on *Rumex* species (Polygonaceae). *Entomologist's Monthly Magazine,* **120,** 187–192.

Hulspas-Jordan, P. M. & Van Lenteren, J. C. (1978). The relationship between host-plant leaf structure and parasitization efficiency of the parasitic wasp *Encarsia formosa* Gahan (Hymenoptera: Aphelinidae). *Mededelingen van de Faculteit Landbouwwetenschappen Rijksuniversiteit Gent,* **43,** 431–440.

Jeffries, M. J. & Lawton, J. H. (1984). Enemy free space and the structure of ecological communities. *Biological Journal of the Linnean Society,* **23,** 269–286.

Jones, T. H. & Hassell, M. P. (1988). Patterns of parasitism by *Trybliographa rapae,* a cynipid parasitoid of the cabbage root fly, under laboratory and field conditions. *Ecological Entomology,* **13,** 309–317.

Jones, T. H., Hassell, M. P. & Pacala, S. W. (1993). Spatial heterogeneity and the population dynamics of a host-parasitoid system. *Journal of Animal Ecology,* **62,** 251–262.

Kakehashi, N., Seasick, Y. & Iwasa, Y. (1984). Niche overlap of parasitoids in host–parasitoid systems: its consequence to single versus multiple introduction controversy in biological control. *Journal of Applied Ecology*, **21**, 115–131.

LaSalle, J. & Gauld, I. D. (1992). Parasitic Hymenoptera and the biodiversity crisis. *Redia*, **74** (Appendice), 315–334.

Lawton, J. H. (1986). The effects of parasitoids on phytophagous insect communities. *Insect Parasitoids* (Ed. by J. K. Waage & D. Greathead), pp. 265–287. Academic Press, London.

Leius, K. (1967). Influence of wild flowers on parasitism of tent caterpillar and codling moth. *Canadian Entomologist*, **93**, 771–780.

May, R. M. (1978). Host–parasitoid systems in patchy environments: a phenomenological model. *Journal of Animal Ecology*, **47**, 833–843.

May, R. M. & Hassell, M. P. (1981). The dynamics of multiparasitoid–host interactions. *American Naturalist*, **117**, 234–261.

Memmott, J. & Godfray, H. C. J. (1992). Parasitoid webs. *Hymenoptera and Biodiversity* (Ed. by J. LaSalle & I. D. Gauld), pp. 217–234. CAB International, Wallingford.

Memmott, J. & Godfray, H. C. J. (1994). The use and construction of parasitoid webs. *Parasitoid Community Ecology* (Ed. by B. A. Hawkins & W. Sheehan), pp. 300–318. Oxford Science, Oxford.

Memmott, J., Godfray, H. C. J. & Gauld, I. D. (1994). The structure of a tropical host–parasitoid community. *Journal of Animal Ecology*, **63**, 521–540.

Nicholson, A. J. (1933). The balance of animal populations. *Journal of Animal Ecology*, **2**, 131–178.

Nicholson, A. J. & Bailey, V. A. (1935). The balance of animal populations. Part 1. *Proceedings of the Zoological Society of London*, **3**, 551–598.

Pacala, S., Hassell, M. P. & May, R. M. (1990). Host–parasitoid associations in patchy environments. *Nature*, **344**, 150–153.

Paine, R. T. (1980). Food webs: linkage, interaction strength and community infrastructure. *Journal of Animal Ecology*, **49**, 667–685.

Paine, R.T. (1988). Food webs: road maps of interactions or grist for theoretical development. *Ecology*, **69**, 1648–1654.

Pimm, S. L. (1982). *Food Webs*. Chapman & Hall, London.

Price, P. W. (1981). Semiochemicals in evolutionary time. *Semiochemicals, Their Role in Pest Control* (Ed. by D. A. Nordlund, R. L. Jones & W. J. Lewis), pp. 251–279. John Wiley & Sons, New York.

Price, P. W., Bouton, C. E., Gross, P., McPheron, B. A., Thompson, J. N. & Weis, A. E. (1980). Interactions among three trophic levels: influence of plants on interactions between insect herbivores and natural enemies. *Annual Review of Ecology and Systematics*, **11**, 41–65.

Rejmanek, M. & Stary, P. (1979). Connectance in real biotic communities and critical values for stability of model ecosystems. *Nature*, **280**, 211–313.

Settle, W. H. & Wilson, L. T. (1990a). Invasion by the variegated leafhopper and biotic interactions: parasitism, competition and apparent competition. *Ecology*, **71**, 1461–1470.

Settle, W. H. & Wilson, L. T. (1990b). Behavioural factors affecting differential parasitism by *Anagrus epos* (Hymenoptera: Mymaridae), of two species of erythroneuran leafhoppers (Homoptera: Cicadellidae). *Journal of Animal Ecology*, **59**, 877–891.

Turlings, T. C. J., Tumlinson, J. H. & Lewis, W. J. (1990). Exploitation of herbivore-induced plant odours by host seeking parasitic wasps. *Science*, **250**, 1251–1253.

Turlings, T. C. J., Tumlinson, J. H., Heath, R. H., Proveaux, A. T. & Doolittle, R. E. (1991). Isolation and identification of allelochemicals that attract the larval parasitoid, *Cotesia marginiventris* (Cresson), to the microhabitat of one of its hosts. *Journal of Chemical Ecology*, **17**, 2235–2251.

Van Emden, H. F. (1963). Observations on the effect of flowers on the activity of parasitic Hymenoptera. *Entomologist's Monthly Magazine*, **98**, 265–270.

Wilson, H. B., Hassell, M. P. & Godfray, H. C. J. (1996). Host–parasitoid food webs: dynamics, persistence and invasion. *American Naturalist*, **148**, 787–806.

Whittaker, P. L. (1984). The insect fauna of mistletoe (*Phoradendron tomentosum* Loranthaceae) in Southern Texas. *Southwestern Naturalist*, **29**, 435–444.

15. VERTEBRATE EFFECTS ON PLANT–INVERTEBRATE FOOD WEBS

T. TSCHARNTKE

FG Agrarökologie, Georg-August Universität, Waldweg 26, D-37073 Göttingen, Germany

INTRODUCTION

Population dynamics and community organization are greatly influenced by interactions within and between all trophic levels, and analyses of simple two-species interactions often fail to explain patterns of coexistence and abundance (Price 1985; Tscharntke 1992b; Jones *et al.* this volume). In this chapter, the complexity of food web interactions is shown by exploring vertebrate effects on invertebrate communities, as exemplified by insectivorous birds and grazing mammals.

Insectivorous birds are often regarded as biological control agents of pests. However, prey items (e.g. caterpillars) eaten by birds may have been parasitized, and birds preying on parasitized herbivores do not promote biocontrol. In such situations, avian predators are competitive rather than complementary with the mortality caused by parasitoids. Further examples illustrating multitrophic interactions that are usually unrecognized at first glance include grazing by mammals, which do not only consume plant material, but also the plant-inhabiting phytophagous and entomophagous insects associated with these food plants.

Cascade effects among trophic levels induced by vertebrates include several such direct and indirect interactions, which are poorly known but potentially important variables of food-web structure. However, multifactorial research covering several trophic levels is difficult or even impracticable, and thus such results are rarely found in the literature. Lack of quantitative and experimental data contrasts with the universal nature of vertebrate–invertebrate interactions, and makes an evaluation of the subject extremely difficult. Accordingly, this chapter is rather speculative. Accounts of general features derived from published food webs (e.g. Cohen *et al.* 1990; Pimm *et al.* 1991; Pimm 1992) lack the complexity typical of real food webs. Such generalities are often oversimplified, since they are based on food webs showing (i) only three (in terrestrial ecosystems) or four (in aquatic systems) trophic levels; and (ii) few links that can be easily measured, thereby largely ignoring interactions with the remainder of the community (Andrewartha & Birch 1984; Lawton 1989; Polis *et al.* 1989; Holt & Lawton 1994; Polis 1994).

Figure 15.1 illustrates the cascading effects up and down the trophic systems when both vertebrates and invertebrates are considered. Mean chain length of published food webs in grassland habitats is 2.1 (Briand & Cohen 1987; Hairston & Hairston 1993), but obviously food webs are much more complicated. Both bottom-up and top-down effects are known to include significant consequences for the populations concerned. Bottom-up effects are generally most important in energy transfer (Hunter & Price 1992; Price *et al.* this volume), since resource availability generally determines the success of species. Top-down trophic cascades are predominantly known in aquatic systems with algae as the primary producers (Strong 1992). However, the following examples illustrate that top-down effects also play a major role in terrestrial systems:

1 Increased insect mortality due to predation by birds reduced leaf damage and enhanced growth of oak saplings (Marquis & Whelan 1994).

2 Exclusion of parasitoids attacking seed-feeding weevils increased percentage of attacked seeds from 20 to 43% (Gomez & Zamora 1994).

3 Leaf damage by herbivorous insects is higher on islands without lizards than on islands with lizards (Spiller & Schoener 1990).

4 Syrphid flies are known to significantly reduce aphid populations (Chambers & Adams 1986), thereby reducing negative effects on cereal growth.

5 Density of herbivorous insect larvae on bilberry (*Vaccinium myrtillus*) was 63% lower (and larval damage to the annual shoots of bilberry was significantly less) where birds had access to larvae, compared to exclosures (Atlegrim 1989).

6 In a tropical agroecosystem, ants reduced populations of two pest species on

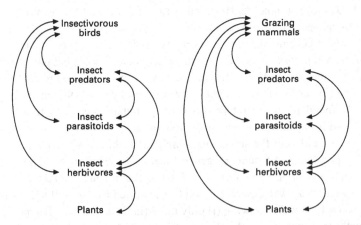

FIG. 15.1. Cascading top-down and bottom-up effects (indicated by arrows) in terrestrial trophic interactions. Predators of birds (e.g. raptors or cats) and grazing mammals (e.g. carnivorous mammals) represent a further trophic level not shown here, so six-level interactions are common in terrestrial food webs. Top-down and bottom-up arrows are given.

squash, thereby also reducing leaf damage and almost certainly increasing yields (Risch & Carroll 1982).

7 Rabbit control by myxomatosis resulted in a resurgence of many plant populations (Thomas 1960).

8 Many examples come from classical biological control (introductions of natural enemies), for example suppression of weed populations by herbivores, or suppression of forest pests (e.g. larch sawfly) by parasitoids, resulting in reduced tree defoliation (Huffaker 1971; Price 1987).

The five-trophic-level systems (Fig. 15.1) do not include top predators (raptors, carnivorous mammals), although such six-level systems are very common in terrestrial habitats. In addition, many micro-organisms and small Metazoa usually feed on and live within prey items of all trophic levels, making an evaluation of trophic effects even more complicated when further trophic levels are added (Janzen 1977; Hatcher & Ayres, Gange & Bower, Roncadori, Clay, Prestidge & Ball, this volume).

Vertebrates usually interact with plant–invertebrate food webs by feeding on several of the lower trophic levels simultaneously – that is, their feeding activity is level-overlapping, and links between species on adjacent levels are weak (Hairston & Hairston 1993; Polis 1994). They typically eat all trophic categories of insects (predatory, parasitized and unparasitized phyto- and saprophages). Not only do grazing mammals consume plants and plant-inhabiting insects, but insectivorous birds may feed temporarily on cereals or seeds, and many granivores such as tree sparrows switch to insects when feeding their nestlings. Thus, the trophic position of insectivorous and granivorous birds is as variable as that of grazers, and can only be arbitrarily assigned. Omnivory by these high-trophic-level organisms means that trophic cascades, in which populations alternate in importance as one descends the food chain, are mainly restricted to lower-trophic-level interactions.

For the rest of this chapter, I review effects of insectivorous birds and grazing mammals on invertebrate food webs. I direct particular attention to predator interference; namely vertebrates negatively affecting biocontrol abilities of invertebrate natural enemies.

BIRD PREDATION ON PHYTOPHAGOUS AND ENTOMOPHAGOUS INSECT POPULATIONS

Birds were used in biocontrol attempts much earlier than invertebrate predators or parasitoids (Otvos 1979, and references therein). In 1762, the mynah bird (*Acridotheres tristis*) was introduced from India to Mauritius to control the red locust (*Nomadacris septemtasciata*). This was the first successful importation of a predator, 112 years before the first transportation of an insect predator (*Coccinella unidecimpunctata*) along with other aphid natural enemies to New Zealand from

England. Despite several examples of bird introductions during the last century, opinions about the role of birds in insect control differ widely. Birds are opportunistic feeders and are considered to control insect populations at rather low endemic levels, but not when prey has reached outbreak levels (Holmes *et al.* 1979; Otvos 1979; Faeth 1987).

However, increasing local densities of populations of insectivorous birds, for example by installation of nesting boxes, cannot necessarily be judged to improve biocontrol, because birds may also prey on the natural enemies of pest insects. Few studies consider four-trophic-level effects by analysing whether birds prey on entomophagous, phytophagous, parasitized or unparasitized insect larvae, or facilitate or inhibit parasitization success indirectly. This lack of integration of entomological and ornithological research is evident in the literature. Life-table analysis, exemplified by Varley *et al.* (1973) for *Urophora* gall flies, is usually based on the arbitrary assumption that destruction of phytophagous insect larvae affects hosts and parasitoids equally. In contrast to this assumption, a literature compilation showed that, within the 12 systems found (see below), birds did not discriminate between parasitized and unparasitized prey in 25% of cases, preferred unparasitized larvae in 50%, and preferred parasitized larvae in 25%. Similarly, Hawkins and Cornell (in press) found, in a review of 82 life-table studies, that there is little consistency in the form of interactions between predators and parasitoids.

Bird predation: parasitized insects

A quarter of the 12 studies found birds primarily attacking parasitoid insects. Betts (1955) gives evidence that titmice (*Parus major* and *P. caeruleus*) consume more parasitized caterpillars of the winter moth (*Operophtera brumata*) than unparasitized ones, suggesting that the conspicuously sluggish habit of parasitized caterpillars enhances their selection as prey by birds.

Lipara lucens (Diptera: Chloropidae) induce cigar-like galls on the tops of shoots of common reed (*Phragmites australis*). During winter, galls are pecked open by blue tits (*Parus caeruleus*) (Mook 1971; De Bruyn 1994; T. Tscharntke, unpublished data). In four *Lipara* spp. studied, galls selected by birds had smaller than average diameter, with those forming the largest galls showing the greatest discrepancy in size between predated and non-predated galls. Smaller galls are less siliceous or softer, and can be opened more easily by tits. In both *Lipara lucens* and *L. rufitarsis*, parasitism is also highest in the smallest galls, since ovipositor length is sufficient for penetration of small galls only (Mook 1967, 1971; De Bruyn 1994). Accordingly, interference between birds and parasitoids is maximized, as both concentrate their feeding activity on galls of the same size. Such selection pressure on small galls suggests that induction of larger galls by *Lipara* should increase. However, thick reed shoots producing large galls are more resistant to gall induction than small shoots and cause high mortality of the very young gall-inducing larvae (Mook 1967, 1971; De Bruyn 1994). Intermediate gall size is a compromise between contrasting selection pressures of mortality due to plant resistance and predation or parasitism.

Selection pressure of vertebrate on invertebrate natural enemies should favour divergent prey specializations. Fritz (1982) argued that one evolutionary aspect of parasite–host interaction should be enhanced predator avoidance. Indeed, 50% of studies have found birds primarily attacking unparasitized larvae. In studies of *Lipara* spp., results confirmed expectations: percentage parasitism and percentage bird predation were negatively correlated (Tscharntke 1994; unpublished data). Four sibling species show species-specific patterns in the relative importance of mortality factors; for example, galls of *L. pullitarsis*, which are heavily pecked open by birds (78%), provide less chance for establishment of parasitoids (only 2% parasitism), whereas *L. lucens*, inducing particularly hard, complex and lignified galls, suffered from 36% parasitism, but less than 1% bird predation. In contrast to the review of Hawkins and Cornell (in press), comparative analyses of *Lipara* spp. show tradeoffs between predation and parasitism: avoidance of bird predation appeared to increase risk of parasitism. Summarizing results of these *Lipara* studies: (i) both parasitoids and birds attacked predominantly the small galls within each species, thereby increasing interference; while (ii) parasitoids attacked predominantly the *Lipara* species least predated by birds. This between-species differentiation appears to reduce the ecological importance of parasitoid suppression by birds.

Competitive interactions among natural enemies are also known in invertebrate predator–parasitoid systems; for example, ants kill more parasitized than unparasitized *Pieris* caterpillars (Jones 1987), and pentatomid predators cause significantly higher mortality of parasitized than unparasitized sawfly larvae (Tostowaryk 1971). Cocoons of pine sawflies eaten by mammals in the summer are almost exclusively parasitized, because the parasitoids delay adult emergence until after the start of mammal activity in the spring (Price & Tripp 1972).

Bird predation: unparasitized insects

Half of the 12 studies found birds primarily attacking unparasitized larvae (see also Otvos 1979; Fritz 1982). Such predation adds to the mortality caused by parasitoids and can be expected to increase control of phytophagous insects which may result in enhanced growth of the insects' host plants (Atlegrim 1989; Marquis & Whelan 1994).

Chipping sparrows (*Spizella passerina*) apparently recognized parasitized jack pine budworm (*Choristoneura pinus*) and avoided parasitized larvae and pupae. Parasitized prey seemed to be unpalatable and were dropped after selection (Sloan

& Simmons 1973). Abrahamson *et al.* (1989) analysed selection pressure on gall size of *Eurosta solidaginis* (Diptera; Tephritidae) on goldenrod (*Solidago altissima*) and found that percentage parasitism was highest in small galls, whereas birds preyed on large galls with greater frequency. Since parasitoids concentrated on fields where mean gall size was small, and bird attack was heavier in fields where mean gall size was large, interference between these mortality factors appears to be insignificant (Weis & Kapelinski 1994). In contrast to the *Lipara–Parus* interaction (see above), predation was higher in the largest galls, because of their high visibility or a learned preference for large prey items.

Examination of 305 gizzards of 54 bird species by Buckner and Turnock (1965) showed that most species fed on the larch sawfly (*Pristiphora erichsonii*) and only very few on parasitoids. According to MacLellan (1958), unparasitized larvae of the codling moth (*Laspeyresia pomonella*) were larger in size than parasitized larvae, and only 3% of larvae in the diet of woodpeckers (*Picoides* spp.) were parasitized vs. 14% of non-predated larvae. Similarly, Koroljkowa (1956) reports birds preferring non-parasitized caterpillars of gypsy moth (*Lymantria dispar*).

Feeding activity of woodpeckers (*Picoides* spp.) on western pine beetle larvae (*Dendroctonus brevicornis*) reduces bark thickness by puncturing, flaking and drilling the bark in search of food, so the remaining brood comes within the ovipositor range of parasitoids resulting in enhanced parasitism (Otvos 1979). In addition, birds may also facilitate the transfer of insect pathogens by consuming infected insects and then defecating in non-infected habitats (Otvos 1979, and references therein).

Bird predation: parasitized and unparasitized insects

A quarter of the 12 studies found insectivorous birds consuming more or less equal numbers of parasitized and unparasitized insects. In this case, only part of bird predation adds to the mortality exerted by parasitoids, the rest is competitive with the impact of parasitoids rather than complementary:

Herbivore mortality from bird predation of both parasitized and unparasitized herbivores (%) = (100 − parasitism (%)) × (total predation (%)/100).

Spring predation by birds caused a significant decrease in populations of the larch casebearer (*Coleophora laricella*), although 65% of the larvae eaten were parasitized. Birds (several species) did not discriminate between parasitized and unparasitized larvae (Sloan & Coppel 1968). Futura (1976, cited by Otvos 1979) studied low density gypsy moth (*Lymantria dispar*) populations and found that birds consumed larvae with and without endoparasitoids equally.

Analyses of the gall midge *Giraudiella inclusa* (Diptera: Cecidomyiidae) on *Phragmites australis* showed that during winter, blue tits (*Parus caeruleus*) peck

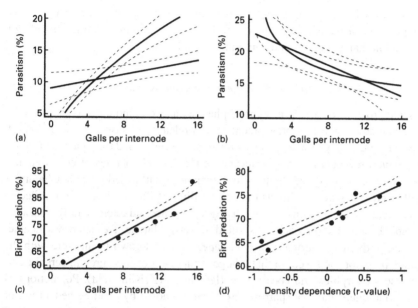

FIG. 15.2. Interference of bird predation and parasitism in *Giraudiella* midge galls. During winter, 64–80% of galls were pecked open by blue tits. (a) Spatial density dependence of *Aprostocetus calamarius* (Hymenoptera: Eulophidae): $r = 0.978$, $n = 8$, $P < 0.001$); and *Torymus arundinis* (Hymenoptera: Torymidae): $r = 0.715$, $n = 10$, $P = 0.02$. (b) Inverse spatial density dependence of *Aprostocetus gratus* (Hymenoptera: Eulophidae): $r = -0.877$, $n = 7$, $P < 0.01$); and *Platygaster* cf. *quadrifarius* (Hymenoptera: Platygasteridae): $r = -0.793$, $n = 8$, $P < 0.02$. (c) Per cent bird predation in relation to gall cluster size: $F = 75$, $r = 0.962$, $n = 8$, $P < 0.001$. (d) Mortality of parasitoid species due to bird predation, estimated by the parasitoids' density dependence response to *Giraudiella* galls, shown by the correlation coefficient: $F = 72.6$, $r = 0.955$, $n = 9$, $P < 0.001$. (From Tscharntke 1992b; arc-sine transformed percentages).

open midge galls, apparently unconcerned about parasitization. Tits interfere with the 14 parasitoid species inhabiting the midge galls; 56% of midge galls are parasitized, so the average of 70% gall predation by birds results in only 31% additional mortality (Tscharntke 1992b).

Of eight frequent parasitoid species attacking *Giraudiella* populations, two showed positive density-dependent (Fig. 15.2a), two negative density-dependent (Fig. 15.2b), and four density-independent responses. Blue tits preferentially pecked open the large clusters of midge galls from *Phragmites* habitats and thereby caused a particularly high mortality of parasitoid species concentrated in large gall clusters (Fig. 15.2c). Accordingly, blue tits caused maximum mortality against parasitoids with density-dependent responses (Fig. 15.2d), thereby suppressing the parasitoids' presumed ability to regulate host populations. Total parasitism, exerted by all parasitoid species, was density independent.

Bird predation on midge galls may even promote midge abundance, if birds' suppression of parasitoids turns out to be more important in midge population regulation than the direct midge mortality (see below).

Bird predation: entomophagous insects

Insectivorous birds attack both phytophagous and entomophagous prey. Predominant feeding on invertebrate predators may release pest insects from biocontrol. Recent studies on yellowhammers (*Emberiza citrinella*) and tree sparrows (*Passer montanus*) revealed a strong influence on the natural enemies of cereal aphids. Within a maximum range of 250 m from their nests within hedges, yellowhammers collect great numbers of insects for their nestlings (Lille 1992). The prey collected in cereal fields mainly consisted of (phytophagous) noctuid caterpillars, (predacious) carabid beetles and (predacious) syrphid (hoverfly) larvae. Tree sparrows, characteristic birds of the agricultural landscape, are also known to interfere with the aphid–predator food web, in that a large component of their prey in cereal fields consists of syrphids and coccinellids (ladybirds) (Kristin 1984). Populations of yellowhammers and tree sparrows are often promoted by planting hedgerows or by installing nesting boxes. However, they do not appear to contribute to the control of cereal aphids which are the main pest in cereals of central Europe. Insectivorous birds may even release populations of cereal aphids from control of these invertebrate predators. Hoverflies significantly reduce aphid populations, shown by experiments with predator exclusions (Chambers & Adams 1986).

In forest ecosystems, hoverflies but not aphids form a large proportion of the diet of at least three songbird species: nuthatch (*Sitta europaea*), great tit (*Parus major*), chiffchaff (*Phylloscopus collybita*) (Kristin 1991, 1992). Several birds feed their nestlings a very high proportion of spiders and insect predators (e.g. see Kristin 1992). Interference between avian and invertebrate predators does not meet expectations for optimal biocontrol of actual or potential insect pests by insectivorous birds, but appears to be a common factor in structuring food webs and in affecting the success of invertebrate biocontrol.

The predator interference hypothesis

Trophic interaction types, like predation, can only be fully evaluated in a broader context, when multitrophic direct and indirect effects are considered simultaneously. Predation by insectivorous birds can enhance pest populations, when the suppression of natural enemies is greater than the direct mortality exerted on phytophagous pests. Populations of invertebrate predators showing a strong numerical response may suffer more than their less effective competitors, since birds usually switch on to the most abundant prey items. Removal of higher-order predators allowing intermediate predators to increase is termed mesopredator release (Soule *et al.* 1988; Holt & Lawton 1994). Such top-down cascade effects are little known, but should regularly occur in insectivorous birds predating: (i) preferentially on the natural enemies of herbivores (e.g. in aphid–hoverfly systems); (ii) preferentially on parasitized larvae (e.g. in small *Lipara* galls, that are also preferred by parasitoids); or (iii) on both parasitized and unparasitized prey (e.g. in *Giraudiella* midge galls).

At least three examples support the predator interference hypothesis. Exclusion of *Anolis* lizards from experimental areas on Antillean islands resulted in a 20–30 times increase in the abundance of three large web-building spiders, which in turn, caused a 25% decrease in the abundance of aerial insect prey (Pacala & Roughgarden 1984; see also Spiller & Schoener 1988). Predation by anthocorid bugs on pyralid moths parasitized by braconid wasps indirectly caused an increase in the moth population, 180% over levels occurring with wasps only (Press *et al.* 1974). In classical biological control with parasitoids, confinement to only the most searching-efficient species can be better than the inclusion of all species; that is, also the less searching-efficient but competitively superior parasitoids, which may only reduce the success of their competitors (the single vs. multiple introduction controversy, see Zwölfer 1971; Ehler 1992).

The predator interference hypothesis, that biocontrol of invertebrate natural enemies can be prevented by vertebrate predation, is also based on theoretical grounds, since Lotka–Volterra equations imply an interesting conclusion (Wilson & Bossert 1971). When, for example, the mortality caused by insectivorous birds is of equal importance to phytophages and entomophages, then populations of phytophages increase and those of entomophages decrease. This differential effect is due to the different growth curves of prey and predator population:

(1) $dN_1/dt = B_1 N_1 N_2 - D_1 N_1$ for predator populations, and
(2) $dN_1/dt = B_2 N_2 - D_2 N_1 N_2$ for prey populations.

N_1, N_2 is the number of predators (1) and prey (2); B_1, B_2 is the birth rate, a constant; D_1, D_2 is the death rate, a constant.

Suppose, $N_1 N_2 = 100(100) = 10\ 000$, and birds consume 50% of both phytophage and entomophage populations: then, $N_1 N_2 = 50(50) = 2500$. The new $N_1 N_2$ value is only 25% of the original and will correspondingly reduce: (i) birth rate of predators; and (ii) mortality rate of prey, resulting in great advantages for the phytophagous prey. In general, such predator interference (or intraguild predation) should significantly affect predator–prey interactions in terrestrial ecosystems (but see the contrasting assumptions of Hairston & Hairston 1993).

Such top-down cascade effects promoting populations of lower trophic levels are well-known in lakes (the 'size efficiency hypothesis', e.g. Dodson 1974, Lampert 1987). Vertebrate predators (fish) select prey primarily on the basis of size, but not trophic level. They decrease populations of invertebrate predators, thereby causing increases in zooplankton populations that consume and suppress algal populations.

MAMMAL GRAZING AND THE PLANT–HERBIVORE–PARASITOID FOOD WEB

Effects of mammal grazing include profound and obvious changes in vegetation structure and, consequently, in the communities of associated insects. Such indirect effects can be separated from effects changing plant and insect populations directly (Harper 1977; Morris 1978; McNaughton 1984; Detling 1988; Huntly 1991; Tscharntke & Greiler 1995, and references therein).

With regard to indirect effects of grazing mammals, grassland composition is determined by selectivity of feeding, since fodder grasses, fodder legumes and several other species are nutritious food, while toxic or non-nutritious plants (e.g. *Rumex* spp., *Juncus* spp., *Veratrum album, Colchicum autumnale*) are rejected and therefore classified as pasture weeds. Selectivity of grazing depends on: (i) the kind of grazer species involved; and (ii) intensity of grazing. Fertilization, due to urine and faeces deposition, causes local changes in plant diversity. Floral composition and growth are also influenced by trampling which usually results in grazed pastures exhibiting more bare ground than ungrazed meadows. Such local elimination of vegetation causes mini-successions. Insects show a wide array of responses to these dramatic changes, depending on taxonomic and functional groups. Some insects benefit from grazing, while many others are negatively affected (e.g. Morris 1991).

Direct effects of grazing mammals include the often disregarded effect that grazing mammals consume plant material which may contain endophagous insect larvae. Such destructive foraging implies unknown effects on the plant–insect food webs of grasslands. In a way, insects and mammals compete for palatable forage (Watts *et al.* 1982); for example, caterpillar *Hemileuca oliviae* larvae commonly achieve a biomass that is four times that of the recommended stocking rate of cattle. Insects often consume much more grass than megaherbivores.

Grasses cover large proportions of grazed areas and support diverse, though often overlooked, insect communities (Tscharntke & Greiler 1995). Most studies on grass insects deal with pests or ectophagous insects, although endophages are generally specific grass inhabitants (e.g. Claridge & Dawah 1994). Analyses of the endophagous stem-borer community from 15 grass species showed a distinct pattern with regard to insect numbers per grass species (Fig. 15.3). Species richness of insects from 10 perennials increased with mean shoot length, which conforms with an increase in heterogeneity of shoots. Five annuals were almost unattacked, presumably due to their unpredictability in space and time (Greiler & Tscharntke

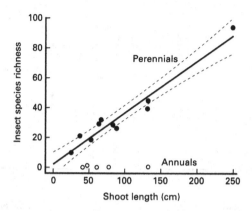

FIG. 15.3. Number of insect species per grass species in relation to mean shoot length of the grass. (T. Tscharntke, unpublished data.) Larvae from 70 000 shoots were reared, half of the shoots were dissected and measured in detail (1986–89, near Karlsruhe, each grass species with at least 12 different samples from at least four regions). $F = 168$, $r = 0.983$, $n = 15$, $P < 0.001$.

1992; Tscharntke & Greiler 1995). Accordingly, these two simple variables (shoot length, life-cycle dichotomy) were valid predictors of species richness, explaining 97% of the variance. On average, 34 species were found per perennial grass species. Two-thirds of these were parasitoids, and half were monophagous. The smallest grasses (*Corynepherus canescens, Agrostis capillaris*) were attacked by only one phytophagous chalcid wasp (Hymenoptera: Eurytomidae), one gall midge (Diptera: Cecidomyiidae), and one mealy bug (Homoptera: Pseudococcidae); which were common to all 10 perennials studied. The taller grasses were also attacked by chloropid flies (Diptera: Chloropidae), cephid wasps (Hymenoptera: Cephidae), moths, beetles and several small flies (Diptera: Schizophora). So, perennial grasses supported many specialized endophagous insects (and a multiple of the less-specialized ectophagous species), an observation that conflicts with the general assumption that simplicty of architecture and biochemistry (few secondary compounds) cause grasses to harbour species-poor and unspecialized communities.

Mammal grazing: direct effects on phytophagous insects

In a comparison of grazed and ungrazed habitats, flowering shoots of cocksfoot (*Dactylis glomerata*) were analysed with respect to endophagous insects (Fig. 15.4). Shoots in grazed habitats were only 50% as tall as those in ungrazed habitats (Fig. 15.4a), while those in exclosures in grazed habitats were intermediate, as expected from studies on ecotypic variation of pasture grasses with differential grazing histories (e.g. Painter *et al.* 1993).

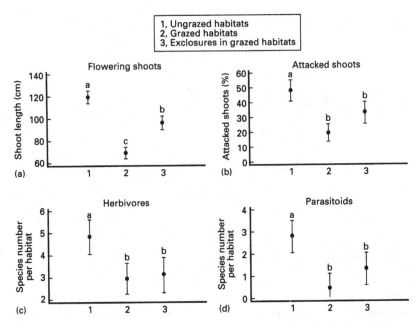

FIG. 15.4. Effects of sheep grazing on populations of cocksfoot (*Dactylis glomerata*), and endophagous insects. Four ungrazed habitats, five grazed habitats, and fenced areas (excluding sheep) in four of the grazed habitats were compared (1984 near Göttingen, based on the dissection and rearing of insect larvae from 975 *Dactylis* shoots (Dubbert & Tscharntke, unpublished data). Arithmetic means and 95% confidence limits are given. Homogenous groups separated by LSD-test have been labelled by different letters. (a) Mean length (cm) of flowering shoots ($F = 41$, $P < 0.001$). (b) Mean percentage of shoots attacked by endophagous insects ($F = 11.2$, $P = 0.002$). (c) Number of phytophagous insect species found per habitat ($F = 4.8$, $P = 0.035$). (d) Number of parasitoid species per habitat, dissected and reared from hosts ($F = 7.1$, $P = 0.012$).

The percentage of attacked shoots was significantly reduced in grazed areas (Fig. 15.4b), as was the number of herbivore species (Fig. 15.4c). Accordingly, shoot size and species number were not only positively correlated in a comparison of different grass species (Fig. 15.3), but also when long and short shoots of one species were compared (Fig. 15.4). Heterogeneity of large shoots not only translates to higher species numbers in evolutionary time scales (comparison of small and tall grass species), but also in ecological time scales. In addition, species richness of herbivores appeared to be less affected by grazing than that of parasitoids (Fig. 15.4c, d), suggesting a greater susceptibility to the disturbance regime of grazed areas for parasitoids than herbivores (see below).

Grazer-induced alterations in plant growth and architecture involve factors other than shoot length reduction, which further shape communities of associated insects. Simplification of grass architecture by consumption of apical plant parts should affect all species dependent on complex structures, for example flower or seed feeders, various tall-grassland species of *Auchenorrhyncha* and other sap feeders (Andrzejewska 1965; Morris 1981; Denno & Roderick 1990; Völkl *et al.* 1993). Grazing not only simplifies plant architecture, but also induces production of new tillers or lateral shoots. Such tiller regeneration promotes species which depend on meristematically active, soft, nutrient-rich tissue later in the season. In general, the nutritional status of grass as a feeding site declines with age of the shoot (Tscharntke 1988), and regeneration of tillering grasses makes nutritious food available later in the season. In *Phragmites australis*, the growing tips of side shoots were significantly more nutritious and less resistant than those of main shoots (with a similar shoot thickness): arithmetic means were 1.1 vs. 0.6 g dry weight, 3.5 vs. 2.5% nitrogen, and (as a measure of plant defence) 1.5 vs. 4% silicate (in all cases: $n = 10$, $P < 0.001$, T. Tscharntke, unpublished data).

Feeding commonly reduces subsequent food quality of the host plant, as re-growth leaves may be more toxic (Crawley 1983; but see Pullin 1987; Roininen *et al.* 1994). However, most grasses lack the variety of secondary defence compounds which occur in many dicots (Bernays & Barbehenn 1987). Young grass tillers of freshly growing fodder grasses do not show any induced response, but provide good-quality tissue. The higher tiller number in grazed swards, compared with ungrazed or cut swards, is associated with an increase in larval numbers of frit fly (*Oscinella frit*, Diptera: Chloropidae) (Moore & Clements 1984).

In the common reed (*Phragmites australis*), both stem-boring moths and grazing sheep kill the growing shoot tip, thereby causing fresh side shoots to grow from the nodes beneath the point of damage. Such destruction of apical meristems, activation of dormant buds and development of new tillers, is a characteristic feature of both *P. australis* and fodder grasses of grazed pastures. Analyses of grazing effects on *P. australis* may be particularly advantageous, the species harbours about 100 endophagous insect species (93% of which are monophagous), so influences on insect food webs should be pronounced.

In a field experiment, reed plots were manually cut to simulate sheep grazing or mass outbreaks of stem-boring moths, resulting in reduced shoot length and a seven to eight times increase in number of side shoots (Fig. 15.5a, b). It can be hypothesized that these changes in shoot growth will have two contrasting consequences on insect communities: (i) specialists of young side shoots should increase in abundance; and (ii) specialists of shoot tips should decrease in abundance. These hypotheses are supported by the results of Tscharntke (1989, 1990, 1994, unpublished data), in summary: (i) gall numbers of the so-called ricegrain midge *Giraudiella inclusa* doubled (Fig. 15.5c), since side shoots provided a more nutritious resource, and mortality of the gall-inducing larvae was subsequently greatly reduced; (ii) gall midges of *Lasioptera arundinis* can only induce galls on side shoots, so abundance of this specialist is even closely correlated with side

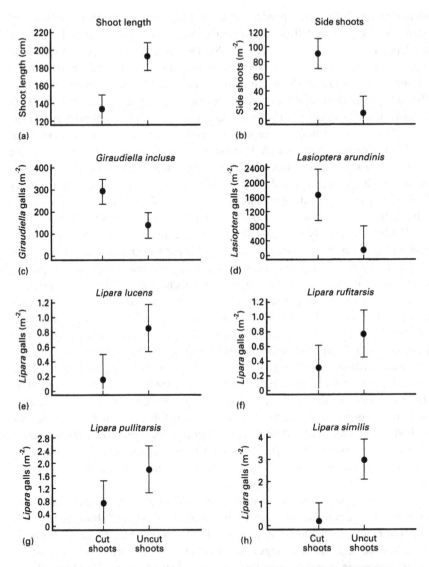

FIG. 15.5. Effects of cutting shoots of *Phragmites australis* on insect populations 1987–89, near Karlsruhe. (T. Tscharntke, unpublished data). There were 17 pairs of cut and uncut samples, each comprising three to four plots, each covering 0.6 m^{-2} and harbouring on average 45 shoots which were sampled, measured, and dissected (5850 shoots overall). Arithmetic means and 95% confidence limits are given ($\mu = 17$, paired t-test). (a) Shoot length (cm): $t = 12$, $P < 0.001$; (b) side shoots m^{-2}: $t = 11$, $P < 0.001$; (c) *Giraudiella inclusa* m^{-2}: $t = 6.3$, $P < 0.001$; (d) *Lasioptera arundinis* m^{-2}: $t = 7.2$, $P < 0.001$; (e) *Lipara lucens* m^{-2}: $t = 4$, $P < 0.001$; (f) *Lipara rufitarsis* m^{-2}: $t = 2.7$, $P < 0.02$; (g) *Lipara pullitarsis* m^{-2}: $t = 3.7$, $P = 0.002$; (h) *Lipara similis* m^{-2}: $t = 5.9$, $P < 0.001$.

shoot abundance. *L. Arundinis* even showed a disproportionate increase in abundance on cut plants, with a seven to eight times increase in side shoots resulting in a 16-fold increase in *Lasioptera* larvae (Fig. 15.5d); (iii) four sibling species of the chloropid genus *Lipara* all cause more or less conspicuous galls on the tips of shoots (see above). As expected, abundance of these gallers decreased (by more than five times), as a result of the destruction of their food resource (Fig. 15.5e–h).

Mammal grazing: direct effects on parasitoids

Very little is known about interactions between four or five trophic levels; for example how grazing by mammals may structure the food webs of phytophagous and entomophagous insects. The disturbance regime exerted by grazing megaherbivores should particularly affect species characterized by: (i) small populations; (ii) fluctuating populations; and (iii) high trophic positions (Kruess & Tscharntke 1994; Lawton 1995). High disturbance levels in grazed swards can be characterized by (i) small-scale patterns or mosaics of rapidly changing resource availability due to consumption and trampling; and (ii) large-scale patterns of fragmentation of those resources preferentially destroyed by grazing mammals (e.g. seeds). Both effects should result in changes in insect populations due to local extinction and recolonization. Since published information is lacking, three analyses of related situations may provide some insight:

1 Moore *et al.* (1986) studied parasitization of stem-boring Diptera (mainly *Oscinella frit*). Per cent parasitism of the larvae was reduced by ploughing and reseeding, compared with established grass. The main parasitoid was a wingless braconid wasp (*Chasmodon apterus*), which was greatly affected by the soil disturbance associated with ploughing and which could only slowly colonize from the edge of newly sown fields.

2 High disturbance levels in common reed (*P. australis*) led to reduced rates of parasitism (Table 15.1). All four herbivores analysed experienced less parasitism when reed shoots were heavily damaged, either by manual cutting or natural damage in the field, regardless of the size of host populations (Tscharntke 1992a, unpublished data).

3 Grazing by mammals enhances large-scale fragmentation of resources with insect species continually going extinct and then recolonizing local sites. As already mentioned, specialist insects depending on plant resources that become rare due to consumption by grazers, can be expected to suffer most. Since flowers, seeds and tall stems are regularly grazed in pastures, patches (e.g. of flowering clover) are generally only found in sites inaccessible to grazers. Insect communities of such clover islands can be expected to show characteristic features: analyses of both manually created and naturally occurring islands of red clover (*Trifolium pratense*)

TABLE 15.1. Effects of cutting *Phragmites* shoots on the parasitism of phytophagous insects.

	Mean per cent parasitism*	
	Cut shoots	Uncut controls
Diptera, Cecidomyiidae:		
Giraudiella inclusa†	50	56
	49	63
Lasioptera flexuosa†	1.0	14.3
	10.2	17.3
Lasioptera hungarica‡	0.5	1.3
	3.2	3.3
Hymenoptera, Eurytomidae:		
Tetramesa phragmites‡	80.7	80.9
	68.5	79
Mean difference in parasitism	4.3	
	7.9	

* The first pair of percentages is based on the cutting experiment (Fig. 15.5, but only $n = 12$ paired samples), the second pair on the comparison of 12 habitats with low damage levels (on average three side shoots m^{-2}) and 15 habitats with higher damage levels (on average 19 side shoots m^{-2}; 1987–1989, near Karlsruhe. (Tscharntke, unpublished data; estimation of *Giraudiella* parasitism is also based on Tscharntke 1992a (twice Wilcoxon signed-rank test: $n = 4$, $Z = 0.04$).)
†,‡ Host populations in plots with cut shoots are either significantly larger† or significantly smaller‡ than in control plots.

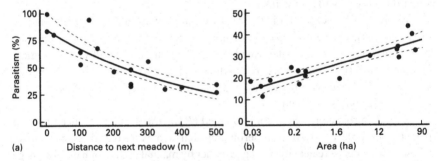

(a) Distance to next meadow (m)   (b) Area (ha)

FIG. 15.6. Per cent parasitism of stem-boring *Apion* species (Coleoptera: Apionidae) (a) decreases with isolation of manually created islands of red clover ($F = 54.6$, $r = 0.879$, $n = 18$ islands, $P < 0.001$), and (b) increases with size of old meadows characterized by wild red clover populations ($F = 66.7$, $r = 0.906$, $n = 17$ meadows, $P < 0.001$). (From Kruess & Tscharntke 1994; Kruess 1996 arc-sine transformed percentages.)

showed that numbers of phytophagous clover insects were not reduced by habitat fragmentation, while numbers of parasitoids were, resulting in a decrease of both parasitoid species richness and percentage parasitism (Kruess & Tscharntke 1994; Kruess 1996; Fig. 15.6). Both increased isolation and decreased area of such islands contributed to the release of phytophagous insects from parasitoid control.

In conclusion, disturbance effects of megaherbivores resulting in fluctuations of plant resources shortens chain length of the plant–invertebrate food web or decreases entomophage : phytophage ratios. Changing availability, rarity, and isolation of plant resources may contribute to the reduced control of actual or potential pest insects.

Compensation for extinct megaherbivores

Effects of grazing megaherbivores such as cows are mostly considered to be an external factor changing the natural, undisturbed establishment of so-called 'normal' plant–insect communities. However, the open landscape in central Europe was not only created by human activities such as clearing and ploughing. The large populations of pre-Pleistocene and Pleistocene megaherbivores should also have persistently influenced terrestrial communities: (i) shaping evolution of plants and insects; and (ii) enhancing open landscapes by stopping succession to beechwood forests. Accordingly, cows or sheep, determining structure of insect communities of temperate grasslands, are the modern successors of megaherbivores that are now extinct.

Contemporary patterns of insect communities on flowering plants evolved during the last 70 million years and in consequence, were paralleled by the diversification of modern birds and mammals. The abundant and diverse megafauna over the last 20 million years undoubtedly shaped the evolution of plant and insect characteristics. Mass extinctions of the megafauna (adult body mass >1000 kg) at the end of the Pleistocene (*c.* 11 000 years BP) have been explained by: (i) the environmental deterioration hypothesis (climatic changes during the ice ages); and (ii) the human overkill hypothesis (predation of a naive prey community by immigrant human hunters (Howe & Westley 1988; Owen-Smith 1989)). Large numbers of mammalian herbivores were dominant features of temperate landscapes, resembling the situation in today's eastern Africa, but these disappeared between Pleistocene and Holocene. Africa has kept its Pleistocene diversity, with some 35 genera of megaherbivores, whereas temperate regions lost over 60% of the Pleistocene diversity of about 40 genera (North America) or 20 genera (Europe), leaving behind plants and invertebrates characterized by adaptations that should have become anachronistic during the Pleistocene habitat changes (see Janzen & Martin 1982; Martin & Klein 1984; May 1993). Anachronistic adaptations to former selection pressures of mammalian megaherbivores may include the following: nettle hairs (see Pullin 1987); spines of cynipid galls (*Diplolepis nervosa*) on oak trees; secondary plant metabolites in noxious weeds of cattle ranges (e.g. *Colchicum autumnale* or *Veratrum album*); grazing tolerance of grasses (e.g. due to high silicate content) (Tscharntke & Greiler 1995); or other mechanical attributes (thorns of roses, hawthorn, blackthorn). Accordingly, both evolutionary and ecological arguments emphasize the need to consider vertebrate and invertebrate food webs simultaneously.

ACKNOWLEDGEMENTS

I am grateful to Roland Brandl, Bradford A. Hawkins, Rick Karban, Peter W. Price, Stefan Vidal, and an anonymous referee for useful comments on the manuscript. Many thanks to Mark Dubbert and Andreas Kruess for leaving me unpublished results of their thesis, Minh Hang Vu for careful grass shoot dissections and the drawings, and Susanne Jahn for initial preparation of figures.

REFERENCES

Abrahamson, W. G., Sattler, J. F., McCrea, K. D. & Weis, A. (1989). Variation in selection pressures on the goldenrod gall fly and the competitive interactions of its natural enemies. *Oecologia*, 79, 15–22.

Andrewartha, H. G. & Birch, L. C. (1984). *The Ecological Web: More on the Distribution and Abundance of Animals*. University of Chicago Press, Chicago.

Andrzejewska, L. (1965). Stratification and its dynamics in meadow communities of Auchenorrhyncha (Homoptera). *Ekologia Polska* Series A13, 685–715.

Atlegrim, O. (1989). Exclusion of birds from bilberry stands: impact on insect larval density and damage to bilberry. *Oecologia*, 79, 136–139.

Bernays, E. A. & Barbehenn, R. (1987). Nutritional ecology of grass foliage-chewing insects. *Nutritional Ecology of Insects, Mites, Spiders and Related Invertebrates* (Ed. by F. Slansky & J. G. Rodriguez), pp. 147–175. John Wiley & Sons, New York.

Betts, M. M. (1955). The food of titmice in oak woodland. *Journal of Animal Ecology*, 24, 282–323.

Briand, F. & Cohen, J. E. (1987). Environmental correlates of food chain length. *Science*, 238, 956–960.

Buckner, C. H. & Turnock, W. J. (1965). Avian predation on the larch sawfly *Pristiphora erichsonii* (Htg.) (Hymenoptera: Tenthredinidae). *Ecology*, 46, 223–236.

Chambers, R. J. & Adams, T. H. L. (1986). Quantification of the impact of hoverflies (Diptera: Syrphidae) on the cereal aphids in winter wheat: an analysis of field populations. *Journal of Applied Ecology*, 23, 895–904.

Claridge, M. F. & Dawah, H. A. (1994). Assemblages of herbivorous chalcid wasps and their parasitoids associated with grasses – problems of species and specifity. *Plant Galls* (Ed. by M. A. J. Williams), pp. 313–329. Clarendon Press, Oxford.

Cohen, J. E., Briand, F. & Newman, C. M. (1990). *Community Food Webs*. Springer, New York.

Crawley, M. J. (1983). *Herbivory*. Blackwell Scientific Publications, Oxford.

De Bruyn, L. (1994). Life cycle strategies in a guild of dipteran gall formers on the common reed. *Plant Galls* (Ed. by M. A. J. Williams), pp. 253–274. Oxford University Press, Oxford.

Denno, R. F. & Roderick, G. K. (1990). Population biology of planthoppers. *Annual Review of Entomology*, 35, 489–520.

Detling, J. K. (1988). Grasslands and savannas: regulation on energy flow and nutrient cycling by herbivores. *Concepts of Entomology* (Ed. by L. R. Pomeroy & J. J. Alberts), pp. 131–148. Springer, New York.

Dodson, S. (1974). Zooplankton competition and predation: an experimental test of the size-efficiency hypothesis. *Ecology*, 55, 605–613.

Ehler, L. E. (1992). Guild analysis in biological control. *Environmental Entomology*, 21, 26–40.

Faeth, S. H. (1987). Community structure and folivorous insect outbreaks: the roles of vertical and horizontal interactions. *Insect Outbreaks* (Ed. by P. Barbosa & J. C. Schultz), pp. 135–171. Academic Press, San Diego.

Fritz, R. S. (1982). Selection for host modification by insect parasitoids. *Evolution*, 36, 283–288.

Gomez, J. M. & Zamora, R. (1994). Top-down effects in a tritrophic system: parasitoids enhance plant fitness. *Ecology*, **75**, 1023–1030.

Greiler, H. J. & Tscharntke, T. (1992). Habitat impact on insect communities of annual and perennial grasses. *Proceedings of the 8th International Symposium on Insect–Plant Relationships* (Ed. by S. B. J. Menken, J. H. Visser & P. Harrewijn). *Series Entomologica*, **49**, 27–29.

Hairston, N. G. & Hairston, N. G. (1993). Cause–effect relationships in energy flow, trophic structure, and interspecific interactions. *American Naturalist*, **142**, 379–411.

Harper, J. L. (1977). *Population Biology of Plants*. Academic Press, London.

Hawkins, B. A. & Cornell, H. V. (in press). Predators, parasitoids and pathogens as mortality agents in phytophagous insect populations. *American Naturalist*.

Holmes, R. T., Schultz, J. C. & Nothnagle, P. (1979). Bird predation on forest insects: an exclosure experiment. *Science*, **206**, 462–463.

Holt, R. D. & Lawton, J. H. (1994). The ecological consequences of shared natural enemies. *Annual Review of Ecology and Systematics*, **25**, 495–520.

Howe, H. F. & Westley, L. C. (1988). *Ecological Relationships of Plants and Animals*. Oxford University Press, New York.

Huffaker, C. B. (1971). *Biological Control*. Plenum, New York.

Hunter, M. D. & Price, P. W. (1992). Playing chutes and ladders – heterogeneity and the relative roles of bottom-up and top-down forces in natural communities. *Ecology*, **73**, 724–732.

Huntly, N. (1991). Herbivores and the dynamics of communities and ecosystems. *Annual Review of Ecology and Systematics*, **22**, 477–504.

Janzen, D. (1977). Why fruits rot, seeds mold, and meat spoils. *American Naturalist*, **111**, 691–713.

Janzen, D. H. & Martin, P. (1982). Neotropical anachronisms: what the gomphotheres ate. *Science*, **215**, 19–27.

Jones, R. E. (1987). Ants, parasitoids, and the cabbage butterfly *Pieris rapae*. *Journal of Animal Ecology*, **56**, 739–749.

Koroljkowa, C. E. (1956). Die Bedeutung der Vögel bei der Vernichtung in Massen auftretender Schadinsekten. *Soobshcheniya Instituta Lesa Akademii Nauk SSSR*, **2**, 65–106.

Kristin, A. (1984). Ernährung und Ernährungsökologie des Feldsperlings *Passer montanus* in der Umgebung von Bratislava. *Folia Zoologica*, **33**, 143–157.

Kristin, A. (1991). Feeding of some polyphagous songbirds on Syrphidae, Coccinellidae and aphids in beech-oak forests. *Behavior and Impact of Aphidophaga* (Ed. by L. Polgar, R. J. Chambers, A. F. G. Dixon & I. Hodek), pp. 183–186. SPB Academic, The Hague.

Kristin, A. (1992). Trophische Beziehungen zwischen Singvögeln und Wirbellosen im Eichen-Buchenwald zur Brutzeit. *Der Ornithologische Beobachter*, **89**, 157–169.

Kruess, A. (1996). *Folgen der Lebensraum-Fragmentierung für Pflanze-Herbivor-Parasitoid-Gesellschaften: Artendiversität und Interaktionen*. Verlag Paul Haupt, Bern.

Kruess, A. & Tscharntke, T. (1994). Habitat fragmentation, species loss, and biological control. *Science*, **264**, 1581–1584.

Lampert, W. (1987). Predictability in lake ecosystems: the role of biotic interactions. *Ecological Studies*, **61**, 333–346.

Lawton, J. H. (1989). Food webs. *Ecological Concepts* (Ed. by J. M. Cherrett), pp. 43–78. Blackwell Scientific Publications, Oxford.

Lawton, J. H. (1995). Population dynamic principle. *Extinction Rates* (Ed. by J. H. Lawton & R. M. May), pp. 147–163. Oxford University Press, Oxford.

Lille, R. (1992). Auswirkungen von Brachflächen auf die Vogelwelt der Knicklandschaft: Die Goldammer als Anzeiger der Lebensraumqualität. *Bauernblatt-Landpost*, **31**, 15–16.

MacLellan, C. R. (1958). Role of woodpeckers on control of the codling moth in Nova Scotia. *Canadian Entomologist*, **90**, 18–22.

McNaughton, S. J. (1984). Grazing lawns: animals in herds, plant form, and coevolution. *American Naturalist*, **124**, 863–886.

Marquis, R. J. & Whelan, C. J. (1994). Insectivorous birds increase growth of white oak through consumption of leaf-chewing insects. *Ecology*, **75**, 2007–2014.

Martin, P. S. & Klein, R. G. (Eds) (1984). *Quaternary Extinctions. A Prehistoric Revolution.* University of Arizona Press, Tuczon.

May, T. (1993). Beeinflußten Großsäuger die Waldvegetation der pleistozänen Warmzeiten Mitteleuropas? Ein Diskussionsbeitrag. *Natur und Museum*, **123**, 157–170.

Mook, J. H. (1967). Habitat selection by *Lipara lucens* Mg. (Diptera, Chloropidae) and its survival value. *Archives Neerlandaises de Zoologie*, **17**, 469–549.

Mook, J. H. (1971). Influence of environment on some insects attacking common reed (*Phragmites communis* Trin.). *Hidrobiologia*, **12**, 305–312.

Moore, D. & Clements, R. O. (1984). Stem-borer larval infestation of ryegrass swards under rotationally grazed and cut conditions. *Journal of Applied Ecology*, **21**, 581–590.

Moore, D., Clements, R. O., Ridout, M. S. (1986). Effects of pasture establishment and renovation techniques on the hymenopterous parasitoids of *Oscinella frit* L. and other stem-boring Diptera in ryegrass. *Journal of Applied Ecology*, **23**, 871–881.

Morris, M. G. (1978). Grassland management and invertebrate animals – a selective review. *Scientific Proceedings of the Royal Dublin Society*, **A6**, 247–258.

Morris, M. G. (1981). Responses of grassland invertebrates to management by cutting. 3. Adverse effects on *Auchenorrhyncha*. *Journal of Applied Ecology*, **18**, 107–123.

Morris, M. G. (1991). The management of reserves and protected areas. *The Scientific Management of Temperate Communities for Conservation* (Ed. by I. F. Spellerberg, F. B. Goldsmith & M. G. Morris), pp. 323–348. Blackwell Scientific Publications, Oxford.

Otvos, I. S. (1979). The effect of insectivorous bird activities in forest ecosystems: an evaluation. *The Role of Insectivorous Birds in Forest Ecosystems* (Ed. by J. G. Dickson, R. N. Conner, R. R. Fleet, J. C. Kroll & J. A. Jackson), pp. 341–374. Academic Press, New York.

Owen-Smith, N. (1989). Megafaunal extinctions. The conservation message from 11 000 years BP. *Conservation Biology*, **3**, 405–412.

Pacala, S. J. & Roughgarden, J. (1984). Control of arthropod abundance by *Anolis* lizards on St Eustatius (Neth. Antilles). *Oecologia*, **64**, 160–162.

Painter, E. L., Detling, J. K. & Steingraeber, D. A. (1993). Plant morphology and grazing history-relationships between native grasses and herbivores. *Vegetatio*, **106**, 37–62.

Pimm, S. L. (1992). *The Balance of Nature.* University of Chicago Press, Chicago.

Pimm, S. L., Lawton, J. H. & Cohen, J. E. (1991). Food web patterns and their consequences. *Nature* **350**, 669–674.

Polis, G. A. (1994). Food webs, trophic cascades and community structure. *Australian Journal of Ecology*, **19**, 121–136.

Polis, G. A., Myers, C. A. & Holt, R. D. (1989). The ecology and evolution of intraguild predation: potential competitors that eat each other. *Annual Review of Ecology and Systematics*, **20**, 297–330.

Press, J., Flaherty, R. & Arbogast, R. (1974). Interactions among *Plodia interpunctella*, *Bracon hebetor*, and *Xylocoris flavipes*. *Environmental Entomology*, **3**, 183–184.

Price, P. W. (1985). Research questions in ecology relating to community ecology, plant-herbivore interactions, and insect ecology in general. *Trends in Ecological Research for the 1980s* (Ed. by J. H. Cooley & F. B. Golley), pp. 75–88. Plenum, New York.

Price, P. W. (1987). The role of natural enemies in insect populations. *Insect Outbreaks* (Ed. by P. Barbosa & J. C. Schultz), pp. 287–313. Academic Press, San Diego.

Price, P. W. & Tripp, H. A. (1972). Activity patterns of parasitoids on the Swaine jack pine sawfly, *Neodiprion swainei*, and parasitoid impact on the host. *Canadian Entomologist*, **104**, 1003–1016.

Pullin, A. S. (1987). Changes in leaf quality following clipping and regrowth of *Urtica dioica*, and consequences for a specialist insect herbivore, *Aglais urticae*. *Oikos*, **49**, 39–45.

Risch, S. J. & Carroll, C. R. (1982). Effect of a keystone predacious ant, *Solenopsis geminata*, on arthropods in a tropical agroecosystem. *Ecology*, **63**, 1979–1983.

Roininen, H., Price, P. W. & Tahvanainen, Y. (1994). Does the willow bud galler, *Euura mucronata*, benefit from hare browsing on its host plant? *The Ecology and Evolution of Gall-forming Insects* (Ed. by P. W. Price, W. J. Mattson & Y. N. Baranchikov), pp. 12–26. USDA, North Central Forest Experiment Station, St Paul, MN, General Technical Report NC-174.

Sloan, N. F. & Coppel, H. C. (1968). Ecological implications of bird predators on the larch casebearer in Wisconsin. *Journal of Economic Entomology*, **61**, 1067–1070.

Sloan, N. F. & Simmons, G. A. (1973). Foraging behavior of the chipping sparrow in response to high populations of Jack pine budworm. *American Midland Naturalist*, **90**, 210–215.

Soule, M. E., Bolger, D. T., Alberts, A. C., Wright, J., Sorice, M. & Hill, S. (1988). Reconstructed dynamics of rapid extinctions of chaparral-requiring birds in urban habitat islands. *Conservation Biology*, **2**, 75–92.

Spiller, D. A. & Schoener, T. W. (1988). An experimental study of the effect of lizards on web-spider communities. *Ecological Monographs*, **58**, 57–77.

Spiller, D. A. & Schoener, T. W. (1990). A terrestrial field experiment showing the impact of eliminating top predators on foliage damage. *Nature*, **347**, 469–471.

Strong, D. R. (1992). Are trophic cascades all wet? Differentiation and donor-control in speciose ecosystems. *Ecology*, **73**, 747–754.

Thomas, A. S. (1960). Changes in vegetation since the advent of myxomatosis. *Journal of Ecology*, **48**, 287–306.

Tostowaryk, W. (1971). Relationship between parasitism and predation of diprionid sawflies. *Annals of the Entomological Society of America*, **64**, 1424–1427.

Tscharntke, T. (1988). Variability of the grass *Phragmites australis* in relation to the behaviour and mortality of the gall-inducing midge *Giraudiella inclusa* (Diptera, Cecidomyiidae). *Oecologia*, **76**, 504–512.

Tscharntke, T. (1989). Attack by a stem-boring moth increases susceptibility of *Phragmites australis* to gall-making by a midge: mechanisms and effects on midge population dynamics. *Oikos*, **55**, 93–100.

Tscharntke, T. (1990). Fluctuations in abundance of a stem-boring moth damaging shoots of *Phragmites australis*: causes and effects of overexploitation of food in a late-successional grass monoculture. *Journal of Applied Ecology*, **27**, 679–692.

Tscharntke, T. (1992a). Coexistence, tritrophic interactions and density dependence in a species-rich parasitoid community. *Journal of Animal Ecology*, **61**, 59–67.

Tscharntke, T. (1992b). Cascade effects among four trophic levels: bird predation on galls affects density-dependent parasitism. *Ecology*, **73**, 1689–1698.

Tscharntke, T. (1994). Tritrophic interactions in gallmaker communities on *Phragmites australis*: testing ecological hypotheses. *The Ecology and Evolution of Gall-Forming Insects* (Ed. by P. W. Price, W. J. Mattson & Y. N. Baranchikov), pp. 73–92. USDA, North Central Forest Experiment Station, St Paul, MN, General Technical Report NC-174.

Tscharntke, T. & Greiler, H. J. (1995). Insect communities, grasses, and grasslands. *Annual Review of Entomology*, **40**, 535–558.

Varley, G. C., Gradwell, G. A. & Hassell, M. P. (1973). *Insect Population Ecology*. Blackwell Scientific Publications, Oxford.

Völkl, W., Zwölfer, H., Romstock-Völkl, M. & Schmelzer, C. (1993). Habitat management in calcareous grasslands – effects on the insect community developing in flower heads of *Cynarea*. *Journal of Applied Ecology*, **30**, 307–315.

Watts, J. G., Huddleson, E. W. & Owens, J. C. (1982). Rangeland entomology. *Annual Review of Entomology*, **27**, 283–311.

Weis, A. & Kapelinski, A. (1994). Variable selection on *Eurosta's* gall size. II. A path analysis of the ecological factors behind selection. *Evolution*, **48**, 734–745.

Wilson, E. O. & Bossert, W. H. (1971). *A Primer of Population Biology*. Sinauer, Stamford.

Zwölfer, H. (1971). The structure and effect of parasitoid complexes attacking phytophagous host insects. *Dynamics of Populations* (Ed. by P. J. Den Boer & G. R. Gradwell), pp. 405–418. Centre for Agricultural Publishing and Documentation (Pudoc), Wageningen.

CONCLUDING REMARKS

R. KARBAN

It is very clear where some of the big holes lie in our ability to construct conditional hypotheses. Some of these have been highlighted in this symposium and further study of these areas would surely repay the effort; for example, we know very little about below-ground herbivores and their interactions with above-ground organisms (Masters & Brown). Similarly, because of their small size, the overwhelming importance of microbes in terrestrial community processes has not been appreciated (Price 1988; Faeth & Wilson). Many induced plant responses to herbivory may actually be responses to microbes that gain entrance following herbivores (Trojan horses) that violate the plant's outer defences (Faeth & Wilson; Karban & Baldwin 1997). 'Decomposers' have been largely ignored by ecologists studying multitrophic interactions. Since each trophic level can use only a small amount of the energy and biomass of the level below it, much of the rest is channelled into the detrital shunt which probably consists of two to five functional trophic levels (Polis 1994). This unstudied subweb can have potentially enormous effects on classical subwebs (plant–herbivore–predator subwebs) by returning energy and nutrients to the classical webs and greatly influencing their dynamics; for example, some of the biomass lost from dying plants re-enters the classical web when detritivores are eaten by predators, completely by-passing the herbivore link.

Although these large gaps in our knowledge exist, we can begin to see generalities about the nature of important multitrophic interactions:

1 Indirect interactions between herbivores that are mediated through their host plants are ubiquitous and important (e.g. Faeth & Wilson; Masters & Brown; Tscharntke). Similarly, indirect interactions between herbivores mediated through shared predators and parasites are common (e.g. Price *et al.*; Jones *et al.*; Tscharntke).

2 Herbivore species using the same host need not compete, as interactions mediated by the host plant may take many forms and have diverse outcomes.

3 Species need not be taxonomically related to interact (e.g. Faeth & Wilson; Masters & Brown; Jones *et al.*; Tscharntke).

4 Species that do not overlap in space or in time may still affect one another.

5 These indirect interactions mediated through the host plant, or through shared predators, may produce very complicated dynamics (e.g. Jones *et al.*).

We must also continue developing conditional hypotheses about when and where communities are likely to be influenced by interactions from higher or lower trophic levels. The relative importance of top-down and bottom-up forces will be influenced by specific attributes of the community: later successional communities have more parasitoids per host and are more likely to exhibit top-down control of herbivores. This trend is echoed in comparisons between orchards and field crops and the patterns in both natural and agroecosystems may be related to more stable communities having more effective predation and parasitism (Varley 1959; Price 1991, 1992; Hawkins & Gross 1992). Aquatic systems seem to be more likely to exhibit strong top-down effects than do terrestrial systems. Lakes have been particularly well studied with regard to these questions; top-down and bottom-up forces each explained about 50% of the variance of biomass (Carpenter *et al.* 1991). Terrestrial systems have been less well studied, but early reviews suggest that top-down forces are not as frequent or as strong (Hunter & Price 1992; Strong 1992; Denno *et al.* 1995; Harrison & Cappuccino 1995). One hypothesis to explain this difference is that aquatic systems (and particularly those showing strong top-down control) are more simple than terrestrial systems (Strong 1992).

For herbivores, the way in which each species makes a living, and the relationship it has with its host plants, may influence the likelihood that top-down or bottom-up forces will predominate. Herbivores that live within their host plants may be more likely to be strongly affected by subtle changes in resource quality and quantity (Karban 1989; Denno *et al.* 1995; Price *et al.*). Herbivores with less mobility may be more influenced by bottom-up forces than more mobile insects that do not form intimate relationships with their host plants (Karban 1989; Denno *et al.* 1995). Herbivores with haustellate (sucking) mouthparts appear to be more affected by bottom-up influences than do herbivores with mandibular (chewing) mouthparts, although the causes of this trend are unclear (Karban 1989; Denno *et al.* 1995). The tissue fed upon by herbivores may modify the outcome of multitrophic interactions; for example, Masters & Brown have argued that root herbivory often benefits folivores, whereas leaf herbivory generally harms root feeders. Plants appear to respond differently to herbivory to different tissues and these responses produce different multitrophic interactions. Damage to leaves was found to induce resistance to subsequent herbivores, whereas damage to buds was found to make the plants more susceptible (Haukioja *et al.* 1990; Karban & Niiho 1995).

In summary, plants, herbivores, and predators and parasites of herbivores make up a large fraction of the Earth's biomass. A pluralistic view reveals clearly that herbivores are influenced by multitrophic interactions from both above and below. We must now begin to specify conditions that make different multitrophic interactions more or less important; for example, later successional communities and those in aquatic habitats are more likely to be influenced by top-down forces. Herbivores that live within plant tissue and less mobile herbivores are more likely to be influenced by bottom-up forces. In the future, we should continue to compare the conditions, be they life histories, community attributes, climatic or other environmental parameters, that tend to influence the relative importance of multitrophic interactions.

REFERENCES

Carpenter, S. R., Frost, T. M., Kitchell, J. F., Kratz, T. K., Schindler, D. W., Shearer, J., Sprules, W. G., Vanni, M. J. & Zimmerman, A. P. (1991). Patterns of primary production and herbivory in 25 North American lake ecosystems. *Comparative Analyses of Ecosystems: Patterns, Mechanisms, and Theories* (Ed. by J. Cole, G. Lovett & S. Findlay), pp. 67–96. Springer, New York.

Denno, R. F., McClure, M. S. & Ott, J. R. (1995). Interspecific interactions in phytophagous insects: competition re-examined and resurrected. *Annual Review of Entomology*, 40, 297–331.

Hairston, N. G., Smith, F. E. & Slobodkin, L. B. (1960). Community structure, population control, and competition. *American Naturalist*, 94, 421–425.

Harrison, S. & Cappuccino, N. (1995). Using density-manipulation experiments to study population regulation. *Population Dynamics, New Approaches and Synthesis* (Ed. by N. Cappuccino & P. W. Price), pp. 131–147. Academic Press, San Diego.

Haukioja, E., Ruohomaki, K., Senn, J., Suomela, J. & Walls, M. (1990). Consequences of herbivory in the mountain birch (*Betula pubescens* spp. *tortuosa*): importance of functional organization of the tree. *Oecologia*, 82, 238–247.

Hawkins, B. A. & Gross, P. (1992). Species richness and population limitation in insect parasitoid-host systems. *American Naturalist*, 139, 417–423.

Hunter, M. D. & Price, P. W. (1992). Playing chutes and ladders: bottom-up and top-down forces in natural communities. *Ecology*, 73, 1134–1136.

Karban, R. (1989). Community organization of *Erigeron glaucus* folivores: effects of competition, predation, and host plant. *Ecology*, 70, 1028–1039.

Karban, R. & Baldwin, I. T. (1997). *Induced Responses to Herbivory*. University of Chicago Press, Chicago.

Karban, R. & Niiho, C. (1995). Induced resistance and susceptibility to herbivory: plant memory and altered plant development. *Ecology*, 76, 1220–1225.

Lindeman, R. L. (1942). The trophic–dynamic aspect of ecology. *Ecology*, 23, 399–418.

Polis, G. A. (1994). Food webs, trophic cascades and community structure. *Australian Journal of Ecology*, 19, 121–136.

Price, P. W. (1988). An overview of organismal interactions in ecosystems in evolutionary and ecological time. *Agriculture, Ecosystems and Environment*, 24, 369–377.

Price, P. W. (1991). Evolutionary theory of host and parasitoid interactions. *Biological Control*, 1, 83–93.

Price, P. W. (1992). Plant resources as the mechanistic basis for insect herbivore population dynamics. *Effects of Resource Distribution on Animal–Plant Interactions* (Ed. by M. D. Hunter, T. Ohgushi & P. W. Price), pp. 139–173. Academic Press, San Diego.

Quinn, J. F. & Dunham, A. E. (1983). On hypothesis testing in ecology and evolution. *American Naturalist*, 122, 602–617.

Schoener, T. W. (1986). Overview: kinds of ecological communities – ecology becomes pluralistic. *Community Ecology* (Ed. by J. Diamond & T. J. Case), pp. 467–479. Harper & Row, New York.

Strong, D. R. (1992). Are trophic cascades all wet? Differentiation and donor-control in speciose ecosystems. *Ecology*, 73, 747–754.

Underwood, A. J. & Petroaitis, P. S. (1993). Structure of intertidal assemblages in different locations: how can local processes be compared? *Species Diversity in Ecological Communities* (Ed. by R. E. Ricklefs & D. Schluter), pp. 39–51. University of Chicago Press, Chicago.

Varley, G. C. (1959). The biological control of agricultural pests. *Journal of the Royal Society. Arts*. 107, 475–490.

Weldon, C. W. & Slauson, W. L. (1986). The intensity of competition versus its importance: an overlooked distinction and some implications. *Quarterly Review of Biology*, 61, 23–44.

White, T. C. R. (1978). The importance of relative shortage of food in animal ecology. *Oecologia*, 63, 90–105.

PART 4
COMPLEX ANIMAL
INTERACTIONS

INTRODUCTORY REMARKS

ROBERT M. MAY

Department of Zoology, University of Oxford, South Parks Road, Oxford OX1 3PS, UK

These comments on the session on 'Complex Animal Interactions' do not attempt to summarize the session. Rather, I try to focus on a few themes which encompass several of the talks, and raise interesting, unresolved questions.

ANIMAL MULTITROPHIC INTERACTIONS?

The title of the final session itself raises a question. Does it make sense to talk of animal multitrophic interactions, as such? Or must any meaningful discussion necessarily involve the resource base of primary producers? A survey of the five papers constituting the session suggests that animal multitrophic interactions cannot be discussed in isolation. For example, we find that Hall and Raffaelli survey a synoptic collection of multitrophic systems, which in almost all cases include the primary producer base. The paper by Moore and de Ruiter, to which I shall return in more detail later, is largely focused on the role of detritus and decomposers in the soil. Oksanen *et al.* deal with a system in which the vegetation is very much part of the overall dynamics. So all three are dealing with multitrophic systems which combine the 'plant' or primary producer base with the animal interactions ultimately sustained by it.

At first glance, the paper by Begon *et al.* may seem to be dealing with purely animal systems (host, pathogen, parasitoids). A quick look at the data makes it abundantly clear that the sustaining resource for the host is very much part of the dynamics in every case. Beginning with the data for the host population alone, with the observed single-host-generation cycles, one must immediately ask what limits host growth. The answer is surely the resource base, and presumably what one is seeing is the simple dynamics of the host–resource system. The data for the host–pathogen system, and for the host–parasitoid system, each individually, again show single-host-generation cycles, and again it seems reasonable to assume that the resource base is involved. When pathogens and parasitoids are both present, Begon *et al.* interestingly find more fragile dynamics, but even here it could be that the resource base is involved to a degree. In short, this exceedingly interesting study is one in which the three animal components are implicitly engaged with the non-animal resource component. The paper by Holt is the only one which can be said to deal purely with animal multitrophic interactions. Another difference it has from the other four is that it is purely theoretical, whereas the others are largely empirical.

This brisk overview suggests that it is hard to disentangle multitrophic animal interactions from the primary producers that sustain them. However, one circumstance where it may be possible to discuss animal interactions could arise when the resource level is included within the description of a 'host' or 'herbivore'. This assumes that the host or herbivore population has some constant or fluctuating level, as part of its own description, without explicitly looking at the interaction between the resource and the host or herbivore population (such as in human host–pathogen systems).

DECOMPOSERS AND DETRITUS

It is not only the primary producers that are essential to the discussion of multitrophic interactions. Also, as Moore and de Ruiter and Oksanen *et al.* both emphasize, the decomposer element of multitrophic systems is essential to their functioning. Yet this aspect of community structure has been largely ignored. You will usually search in vain for discussion of the role of decomposers when reading the chapter on ecosystem structure and multitrophic interactions in the standard ecology texts. This neglect of the decomposers and detritus element of multitrophic systems might be largely a phenomenon of terrestrial ecosystem studies. But I think the myopia extends equally into the sea.

As Moore and de Ruiter have emphasized, fitting a full discussion of the role of decomposition into consideration of multitrophic interactions is more difficult than adding in the primary producers. Decomposition does not fit tidily as just one more trophic level (setting aside the fact that nothing really fits tidily as a simple 'trophic level'!). Decomposers essentially sit beside the conventional vertically structured trophic hierarchy. The 'decomposer' trophic box is fed from every level of the vertical hierarchy, and feeds back into the base of the primary producers (also, of course, the decomposer box can feed back into any one, or even all, higher trophic levels). It is thus an awkward element to shoe-horn into the vertical structure. But decomposers, in all their complexity, must be dealt with; in the absence of full consideration of their role, one ultimately has a nonsense, in which the recycling of the constituent elements of the plants and animals is ignored.

Indeed, I would put it even more strongly. A full understanding of the causes and consequences of biological diversity, in all its richness, probably cannot be had until the contribution made by decomposers to the structure and functioning of ecosystems is fully understood.

16. TWO'S COMPANY, THREE'S A CROWD: HOST–PATHOGEN–PARASITOID DYNAMICS

M. BEGON, S. M. SAIT AND D. J. THOMPSON

Population Biology Research Group, School of Biological Sciences, The University of Liverpool, PO Box 147, Liverpool L69 3BX, UK

INTRODUCTION

Population ecology (both empirical and theoretical) has been preoccupied very largely with studies of single species and interactions between species pairs. This has been so, not presumably because of a widespread belief that species or species pairs exist in isolation, but rather in recognition of the practical difficulties of extending frontiers to encompass three or more species, and in the belief (or hope) that the narrower focus none the less captures satisfactorily the essence of population ecology. The advantages, however, of extending attention even to three species are clear. First, an incremental step is taken towards the reality of a web of interactions between species, holding out the hope of progress in our understanding of the ecology of populations. Second, the various three-species combinations (what Robert Holt has called 'community modules'; Fig. 16.1) may themselves be seen as simple but realistic building blocks from which ecological communities are constructed, forging new links between population and community ecology. And third, a deeper understanding can be developed of two-species interactions, since such interactions often occur not directly between the two species concerned but indirectly, through a third species.

Progress in the ecology of populations beyond two species has gained increasing momentum over the past two decades. Theoretical studies have examined dynamics when single 'prey' species are attacked by two or more predators, parasites or pathogens (e.g. May & Hassell 1981; Dobson 1985; Hassell & May 1986; Hochberg *et al.* 1990; Hochberg & Holt 1990), and when a predator, parasite or pathogen attacks two or more species of prey (e.g. Roughgarden & Feldman 1975; Comins & Hassell 1976; Holt 1977; Holt & Pickering 1985; Bowers & Begon 1991; Begon *et al.* 1992; Holt & Lawton 1993; Begon & Bowers 1994; Norman *et al.*, unpublished data). Further studies, especially related to plants, have examined the dynamics of inter-specific competition when the resource that is being competed for also has explicit dynamics (see Tilman 1990). These can be viewed as exercises in introducing 'mechanism' into the interactions between exploitative competitors (Tilman 1990),

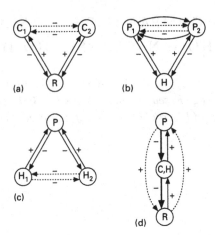

FIG. 16.1. Examples of community modules comprising three species. Solid lines indicate direct interactions, dotted lines indirect interactions. (a) Exploitative competition between two consumers, C_1, C_2, for a shared resource, R; (b) a combination of interference and exploitative competition between two predators, pathogens, parasites or parasitoids, P_1, P_2, for a shared prey or host, H; (c) two hosts or prey, H_1, H_2, sharing a common predator, P, and hence subject to apparent competition; (d) a resource, R, supporting a consumer, C or H, supporting a predator, P.

but they can also be seen as part of a more general movement into three-species ('predator–predator–prey') dynamics.

Empirical studies of population ecology beyond two species have been rare, especially if one excludes investigations which, say, monitor the co-occurrences of large numbers of potential competitors, but take little or no interest in the dynamics of their shared resources, or of the competitors themselves, and hence fail to address questions at the heart of population ecology – those concerning the dynamics of populations. Examples include the early work of Park (1948), who examined the dynamics when a shared sporozoan parasite, *Adelina triboli*, altered the outcome of competition between two species of flour beetle, *Tribolium confusum* and *T. castaneum*. Similar results – though more concerned with the eventual outcome than the dynamics of the interacting species – were obtained by Sibma *et al.* (1964) using oats, barley and a root-feeding nematode *Heterodera avenae*, and by Bentley and Whittaker (1979) and Cottam *et al.* (1986) using dock plants and the chrysomelid beetle *Gastrophysa viridula* – work extended by Hatcher *et al.* (1994) to include a rust fungus pathogen of the plants. Bess *et al.* (1961) examined the successive displacement amongst three species of braconid parasitoid from the genus *Opius* on Hawaii and the percentage of hosts (*Dacus dorsalis*) parasitized by each. The importance of apparent competition has been supported by Schmitt (1987), with the co-occurrence of gastropods and bivalves being determined by pressure from their shared invertebrate predators, and by Grosholz (1992), with the co-occurrence of two species of terrestrial isopod being determined by a shared viral infection. Tilman and coworkers have been able to support the predictions of consumer–consumer–resource models both in diatoms (Tilman *et al.* 1981) and higher plants (Tilman & Wedin 1991a, b).

Attention here is focused on host–pathogen–parasitoid systems – one specific example of a consumer–consumer–recourse 'community module'. The massive importance of insect host–parasitoid interactions in terrestrial communities, in both numerical and functional terms, has frequently been noted (see Godfray 1994). Similarly frequent have been complaints about the unwarranted neglect shown to host–pathogen interactions in studies of the dynamics of both populations and communities. Hence, the host–pathogen–parasitoid module can readily lay claim to importance, both in its own right and as an exemplar for the study of community modules generally.

A series of general points is illustrated by examining empirical results from a system comprising the Indian meal moth *Plodia interpunctella* (the host), the parasitoid *Venturia canescens*, and the *P. interpunctella* granulosis virus (PiGV), maintained in the laboratory. Our main focus of attention is the dynamics of populations. This contrasts with the vast majority of previous, related studies, and of studies reviewed in this symposium, which have focused on the end-results of population interactions, or on interactions occurring between individuals. None the less, our data have been collected at a range of scales. The population dynamics of replicated host, host–pathogen, host–parasitoid and host–pathogen–parasitoid systems have been monitored over extended periods. In addition, the competitive and co-operative interactions between pathogens and parasitoids within individual host larvae have been quantified. The ability to inter-relate ecological interactions occurring at population and individual scales, and thereby provide mechanistic rather than phenomenological explanations for population-level processes, is itself one of the attractions of study at the community-module level.

Particular attention is also drawn to the underlying inseparability, despite convention, of predatory and competitive interactions, to the unpredictability of even three-species systems in terms of their component two-species interactions, and especially to the dangers of misinterpreting three-species population dynamics as the dynamics of a two-species system, and thus failing to understand the basis of the dynamics.

POPULATION DYNAMICS: BACKGROUND

We begin by describing aspects of the population dynamics of the host, its granulosis virus (GV) and parasitoid, maintained in various combinations in population cages. Methodological details of the experimental system can be found in Sait *et al.* (1994a) and Begon *et al.* (1995). The moth is a pyralid and a pest of stored products world-wide. With five larval instars and an adult reproductive life of around 4 days, its generation length under the conditions of the present work is around 41 days. The parasitoid is an effectively parthenogenic ichneumonid wasp (males were never seen in the present study) that lays eggs singly in the later instars of its host (there is a transition from invulnerability to vulnerability over instars II and III – Sait *et al.* (1995)). Significant parasitoid development, however, does not start until the host enters its final instar (Harvey *et al.* 1994). The GV infects larval hosts only, when consumed by them with their food. The lethal dose increases with larval instar, making instar V effectively invulnerable to infection (Sait *et al.* 1994b). Infection itself proceeds from the midgut to most of the rest of the body, giving rise (typically 7–14 days later) to a cadaver almost full of infective viral particles, which may itself be eaten or may contaminate nearby food. Infective particles are not shed prior to host death.

HOST, HOST–PATHOGEN AND HOST–PARASITOID DYNAMICS

The dynamics of the host alone are illustrated in Fig. 16.2. Here, and throughout, the data are in the form of weekly counts of dead adults. The consistency between replicates was high. After an initial period of growth, populations fluctuated with marked regularity around an average density of roughly 120 moths, with typical maxima and minima of around 230 and 60. The period of these fluctuations was 6 or 7 weeks, approximately one host generation. Details of the strength of this periodicity for each replicate population (measured by the value of the autocorrelation function (ACF) at its peak), and of the length of the period (measured by the lag at which the peak ACF occurred and, where data are sufficient, by the peak period in a spectral analysis) are given throughout in the relevant figure legends. In each case, the fluctuations in the ACF were strongly damped with increasing period ('phase-forgetting' rather than 'phase-remembering' – Turchin and Taylor (1992)). This suggests an endogenous cause for the fluctuations provided by the populations themselves, rather than an exogenous cause provided by the environment in which they were maintained.

Host dynamics in host–pathogen systems differed slightly from those of the host alone (Fig. 16.3). The regularity and period of the fluctuations were effectively unaltered, but average host density (around 150) was slightly, though not significantly, higher ($t = 1.58$, $P = 0.19$). The amplitudes of the fluctuations were somewhat increased (with typical maxima and minima of around 320 and 40), and coefficients of variation of abundance significantly higher ($t = 2.90$, $P < 0.05$). Note that since the pathogen affects only the host larvae, pathogen dynamics cannot be portrayed on a scale directly comparable with that for the host (or the parasitoid, below). Instead, the number of infected larvae in a subsection of the food is shown. Their

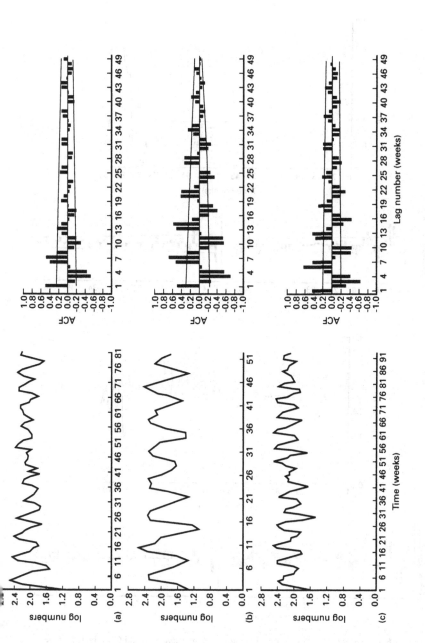

FIG. 16.2. Population dynamics of *P. interpunctella* alone (three replicates, a–c). Mean population densities (±SE) calculated after an initial settling down period (in brackets), were (a) 122.3 ± 6.7 (189 days); (b) 108.6 ± 11.0 (126 days); (c) 147.5 ± 7.6 (259 days). Autocorrelation functions (ACFs), were calculated to establish the period, strength and consistency of population cycles in the logged data, and spectral analysis carried out to determine the cycle period more precisely. In each case, the cycle period (the lag, in weeks, at which the ACF and spectral analysis reached their peaks) and the ACF (*P* < 0.05) of that period (a measure of the strength of the periodicity) are given: (a) 7, 7.2, 0.47; (b) 7, 5.7, 0.68; (c) 6, 6.1, 0.61.

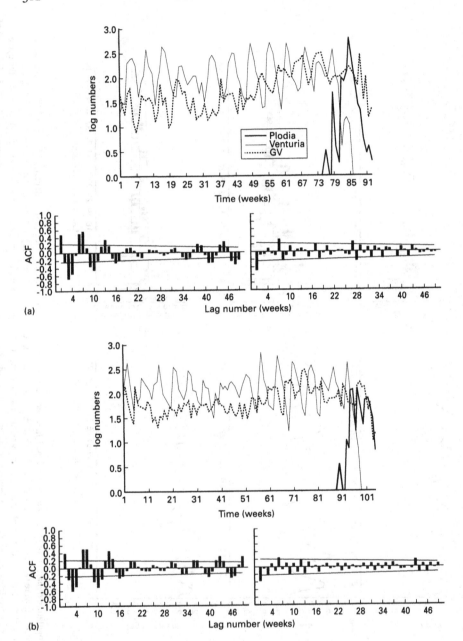

FIG. 16.3. *Above and opposite.* Population dynamics of *P. interpunctella* and GV-infected larvae (three replicates, a–c) with their corresponding autocorrelation functions (ACFs) for the host (left) and the infected larvae (right). ACFs were carried out on logged data, and, for the infected larvae, in order to detrend a significant increase, on differences between successive data points rather than on the data points themselves. Also shown, at the end of each data set, are the dynamics of the parasitoid, *V. canescens* after it had been allowed to invade the population cages.

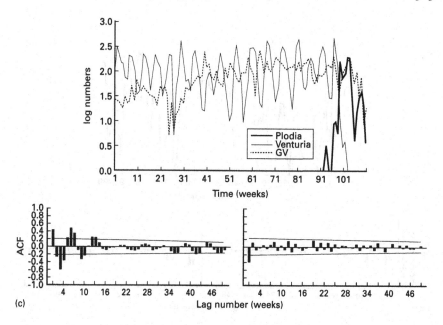

FIG. 16.3. *Continued.* Prior to the introduction of the parasitoid, mean population densities (±SE) following a settling-down period (brackets) were: (a) 153.5 ± 14.4 (301 days); (b) 151.1 ± 12.6 (217 days); (c) 134.7 ± 11.6 (182 days). The cycle periods (from ACF and spectral analysis) and the ACFs (*P* < 0.05) of those periods for the host (H) and infected larval (GV) populations were: (a) H: 7, 6.7, 0.58; GV: 7, 6.2, 0.32; (b) H: 6, 6.8, 0.51; GV: 6, 6.1, 0.24; (c) H: 6, 6.4, 0.52; GV: no significant peak.

dynamics showed less clear evidence of regular periodicity (a consequence, perhaps, of their being relatively long-lived within the system), though some signs of cycles comparable with those of the host are apparent in two of the three replicates once an overall increasing trend in pathogen abundance is taken into account. The numbers of infected larvae, combined with a typical host fecundity of around 200 eggs (Sait *et al.* 1994c) suggest a prevalence of infection of around 1% and certainly less than 5%.

Host–parasitoid dynamics (Fig. 16.4) displayed regular fluctuations with periods again similar to those in the host-alone and host–pathogen systems. Moreover, they were exhibited clearly not only by the host but also by the parasitoid. The parasitoid cycles were typically slightly behind those of the host. The parasitoid also significantly reduced host abundance (to a mean value of around 10) and substantially increased the amplitudes of the fluctuations (on a log scale), with typical maxima and minima of around 100 and zero.

Generation cycles

Note that a pattern of one-host-generation parasitoid–host cycles, as opposed to

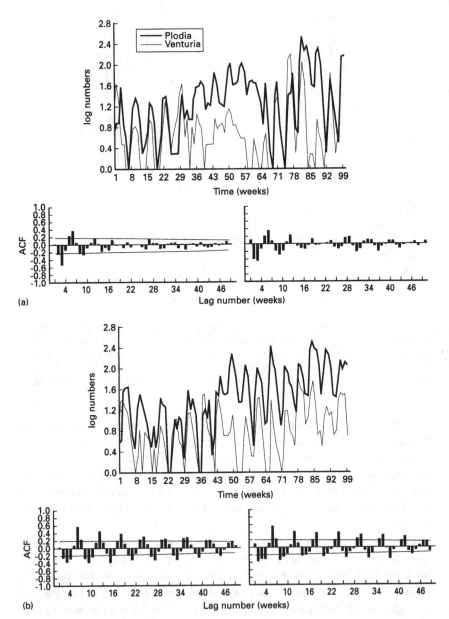

FIG. 16.4. *Above and opposite.* Population dynamics of *P. interpunctella* and *V. canescens* (three replicates, a–c) with their corresponding autocorrelation functions (ACFs) for the host (left) and parasitoid (right) (on logged data). For each host (H) and each parasitoid (P) population, mean population densities (±SE) following a settling down period (brackets) were: (a) H: 14.0 ± 3.1, P: 38.5 ± 5.5 (56 days); (b) H: 9.6 ± 1.2, P: 55.4 ± 6.5 (49 days); (c) H: 7.6 ± 1.4, P: 27.1 ± 4.1 (49 days). The cycle periods (from ACF and spectral analysis) and the ACFs (*P* < 0.05) of those periods were: (a) H: 6, 5.8, 0.39; P: 6, 6.0, 0.39; (b) H: 6, 6.3, 0.53; P: 6, 6.5, 0.56; (c) H: 6, 6.5, 0.66; P: 6, 6.4 0.55.

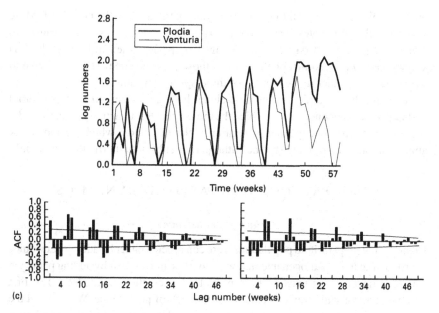

FIG. 16.4. *Continued.*

the more conventional multigeneration cycles (Begon *et al.* 1996), has been predicted to arise (Godfray & Hassell 1989; Gordon *et al.* 1991) under the conditions pertaining in the *Plodia–Venturia* system (Begon *et al.* 1995) – a prediction apparently confirmed here in replicated, long runs of data in which the cycles appear to have an endogenous cause. Previous empirical support for host–parasitoid generation cycles, by contrast, has come from short, unreplicated data sets (Godfray & Hassell 1989; Gordon *et al.* 1991; Reeve *et al.* 1994), and has been based either only on host–parasitoid (but not host-alone) dynamics, or on host dynamics alone supported by evidence of a significant involvement of a parasitoid in those dynamics. A similar prediction of generation cycles has been made for host–pathogen interactions when the host has an explicit stage-structure (Briggs & Godfray 1995). This too seems to be confirmed here, not apparently having been observed in any other system (but see Sait *et al.* 1994a).

On the other hand, generation cycles have also been predicted in single species populations in which larval competition acts immediately (rather than with a delay), and especially where competition is markedly asymmetric (Gurney & Nisbet 1985), as it is in *P. interpunctella* (Sait *et al.* 1994a) – and as we have seen, generation cycles were demonstrable here when the host was maintained alone. Clearly, therefore, it would be wrong to rush too readily into claiming that the present data support models predicting host–parasitoid or host–pathogen generation cycles. No proper explanation can fail to ignore the cycles generated by the host itself. More generally, this illustrates that caution may be necessary in ascribing causes to dynamical patterns in data sets involving two species when little or nothing is known about the dynamical behaviour of those species alone. This is a theme to which we return later in the context of three-species dynamics.

Overall, the dynamics of the host-alone, host–pathogen and host–parasitoid systems – the components from which the three-species interaction is constructed – suggest a pattern of endogenous generation cycles in the host, which are apparently amplified either marginally (by the pathogen) or substantially (by the parasitoid).

HOST–PATHOGEN–PARASITOID DYNAMICS

A lack of persistence

Host–pathogen–parasitoid populations were constructed either by adding parasitoids to an existing host–pathogen interaction (Figs 16.3 and 16.5) or by adding pathogen (infected hosts) to an existing host–parasitoid interaction (Fig. 16.6). The first notable contrast with previous patterns is the lack of persistence. Whereas all the previously described single- and two-species populations persisted either until they were abandoned or were used for other purposes, of 14 host–pathogen–parasitoid populations, 12 failed to persist and the remaining two show patterns (see below) that engender no confidence in their long-term persistence. Moreover, in each extinction, the host disappeared first, followed by its predators, rather than the population collapsing from a three- to a two-species, predator–prey system.

Three of the populations were constructed as a host–pathogen system maintained for almost 800 days to which the parasitoid was then added (Fig. 16.3). In these, the number of infected larvae found each week had risen to around 100. The mean time to extinction of the host once all three species had been combined was 82 ± 12 days (SE).

In the other 11 three-species populations, the third species was added earlier – as soon as the two-species system had become established. By this time, there were typically around 35 infected larvae each week in the host–pathogen populations. One host–parasitoid + pathogen population was still extant after around 370 days as a three-species system, and one host–pathogen + parasitoid population was unavoidably terminated after around 250 days. If the times to extinction of these are conservatively estimated as 370 and 250 days, respectively, then the mean times to extinction ($\pm$SE) are host–pathogen + parasitoid: 286 ± 67 days (Fig. 16.5), and host–parasitoid + pathogen: 259 ± 36 days (Fig. 16.6). Thus, amongst these 11 three-species populations, there appears to be both a qualitative consistency irrespective of the manner in which they were constructed (all went extinct or were apparently heading towards extinction) and an overall quantitative consistency

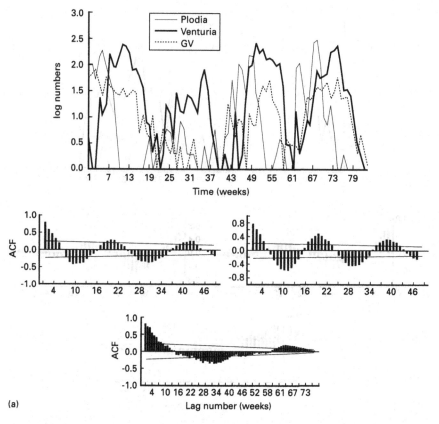

(a)

FIG. 16.5. *Above and on pp. 318–320.* Population dynamics of *P. interpunctella*, *V. canescens* and GV-infected larvae, when *V. canescens* was added to host/PiGV populations (five replicates, a–e), with their corresponding autocorrelation functions (ACFs) for the host (above left), the parasitoid (above right) and infected larvae (below) (on logged data). The cycle periods (at which ACFs reached their peaks) and the ACFs ($P < 0.05$ unless stated) for those periods for the host (H), the parasitoid (P) and infected larvae (GV) were: (a) H: 20, 0.24; P: 20, 0.53; GV: 64, 0.15; (b) H: 26, 0.27; P: 25, 0.31; GV: no significant peak; (c) H: 25, 0.17 ($P = 0.1$); P: 25, 0.24; GV: no significant peak; (d) H: 21, 0.24; P: 17, 0.30; GV: no significant peak; (e) no significant peaks.

(average time to extinction was approximately the same in the two groups). On the other hand, extinction was clearly less rapid in these host–pathogen + parasitoid populations, on average (Fig. 16.5) than when the initial prevalence of infection was higher (Fig. 16.3).

The three-species population dynamics here are therefore far less stable than those of their component two-species systems; diversity here does not promote stability. This is a striking result given the consistency of behaviour of the host-alone, host–pathogen and host–parasitoid populations. It emphasizes that the

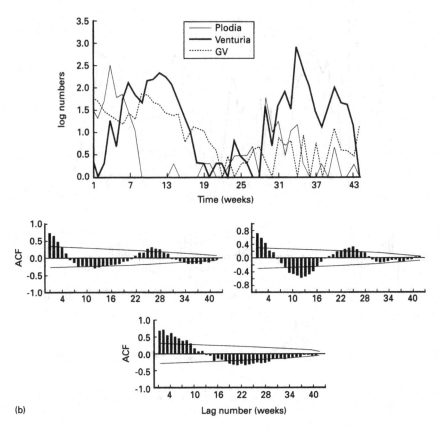

Fig. 16.5. *Continued.*

population dynamics of three-species systems may be difficult to predict even when the dynamics of their component subsystems are known in detail. In the present case, this is true both of dynamical patterns (the loss of generation cycles), and of persistence (the host was consistently driven to extinction in the three-species system, even though the host–parasitoid system would apparently cycle indefinitely, and the host–pathogen system would do so too, with no apparent effect of the pathogen on host abundance).

Multigeneration cycles

The dynamics of the various three-species systems were consistent in their lack of persistence and loss of generation cycles. However, it is notable that their detailed patterns differed according to how they were constructed. The host–parasitoid +

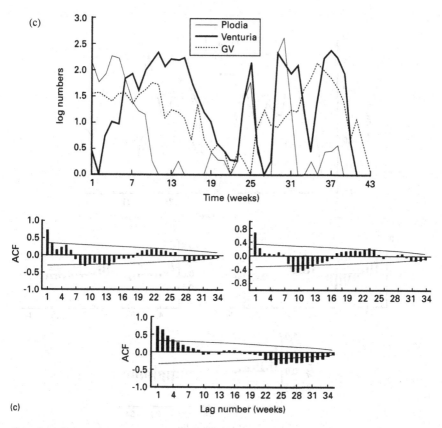

FIG. 16.5. *Continued.*

pathogen populations (Fig. 16.6) displayed no consistent pattern. It is striking, however, that the host–pathogen + parasitoid populations (Fig. 16.5) showed clear evidence of multigeneration cycles. This is seen most clearly in Fig. 16.5a, the longest run of data, where both host and parasitoid ran through four cycles of abundance, with significant peak ACFs at 20 weeks, prior to the extinction of the host. Three further replicates passed through two population cycles prior to host extinction (Fig. 16.5b, c) or termination (Fig. 16.5d). These had peak periods, respectively, for host and parasitoid, at 26 and 25 weeks, both significant (Fig. 16.5b), 25 and 25 weeks, the former not quite significant (Fig. 16.5c), and 21 and 17 weeks, both significant (Fig. 16.5d). On the other hand, the dynamics of virus-infected larvae in these populations showed no comparable cycles or regularities. In addition, a fifth replicate (Fig. 16.5e) and also the three populations to which the parasitoid was added later at higher disease prevalence (Fig. 16.3), all went extinct

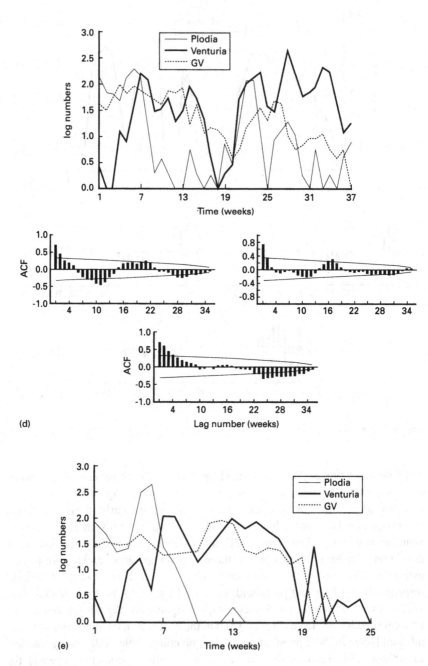

(d)

(e)

FIG. 16.5. *Continued.*

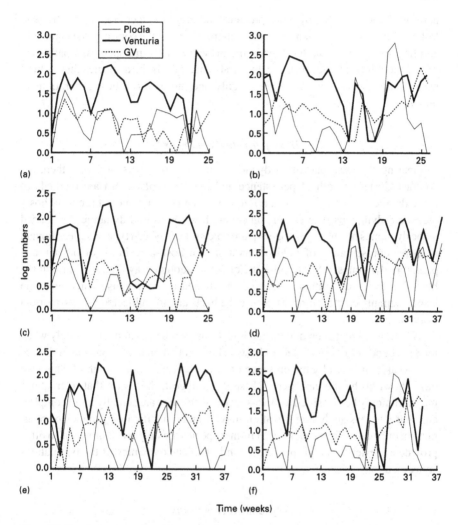

FIG. 16.6. Population dynamics of *P. interpunctella*, *V. canescens* and GV-infected larvae when PiGV was added to host–parasitoid populations (six replicates, a–f).

after a single rise and fall in parasitoid and host abundance, lasting a similar length of time to the cycles in the previous replicates (17–25 weeks in the case of the parasitoid).

Overall, therefore, a clear and remarkable picture emerges from the eight host–pathogen + parasitoid populations: cycles of abundance in both host and parasitoid, roughly three to four host generations in length, and of such amplitude that host extinction is likely following any one of them. The 'classical' expectation when a parasitoid and host (or any other predator and prey) interact alone is for 'Lotka–Volterra' cycles, several generations in length. Here, on the contrary, the parasitoid and host only exhibit such cycles in the presence of a third species, a pathogen, with which they interact strongly. In the absence of this third species they exhibit even clearer cycles – but these are roughly one (not several) host generations in length.

Host–parasitoid dynamics?

Comparing the host–parasitoid dynamics in Figs 16.4, 16.5 and 16.6 then, the contrasts in terms both of persistence and regularity of pattern are marked. To some degree, of course, this is a reflection of the fact that one set are truly host–parasitoid dynamics whereas the others are host–parasitoid dynamics embedded in the dynamics of a host–pathogen–parasitoid system. Which type of series, though, is more representative of field observations on host–parasitoid (or other prey–predator) dynamics? Or – to pose the question another way – how often will the host species in such interactions also be affected by pathogens, which none the less remain undetected simply because the hosts are not subjected to post-mortem examination?

We contend that such undetected infections may be common, especially when the prevalence of overtly infected hosts is low, and when host-alone dynamics are only weakly affected by the presence or absence of infection – both of which we have shown to be the case here. These data suggest, therefore, that interactions may commonly be classified as 'two-species' when they are actually more complex, and that whereas such simplification may often be justified on grounds of pragmatism or because interactions with further species are only weak, it may also often provide a barrier to the proper understanding of the dynamics of the populations concerned.

Dynamical complexity

The data also emphasize that the complexities of systems' population dynamics may vary in a very non-linear manner with the complexities of the systems themselves. Here, the complexities of host, parasitoid and pathogen dynamics were essentially the same in the host-alone, host–pathogen and host–parasitoid systems (host-generation-length cycles in each case) – but the move from two to three species was often accompanied by an immense leap forward in dynamical complexity. An alternative view would be that the complexity of an ecological system is only poorly measured by the number of species that comprise it. In either case, though, it is clear that the relationship between the number of interacting species in a system and the complexity of their dynamics can be a subtle one.

INTERACTIONS IN INDIVIDUAL HOSTS

One of the great attractions of the three-species perspective is that it holds out the prospect of understanding population dynamics in terms of overt individual dynamics – in the present case, understanding competition between populations of pathogens and parasitoids in terms of the competitive (and possibly co-operative) interactions between them within individual hosts.

A large number of the experiments carried out on early host instars can be summarized (Table 16.1) by noting that if hosts are dosed first with the GV such that the appearance of symptoms effectively prevents the host from entering its fifth instar, then the parasitoid (which does not develop significantly until the host's fifth instar, whenever the original attack occurred) cannot develop and is thus 'out-competed' by the pathogen. Clearly, the GV is the superior competitor in individual hosts in their early instars.

Further experiments were therefore carried out when viral attack preceded parasitoid attack, but viral symptoms appeared during the fifth instar. Denoting the day on which viral symptoms first appear as day 0, the outcomes were investigated of parasitoid attacks on days 4 to 6 inclusive, and day 11 (Table 16.2). For parasitoid attacks after day 1 (the second day of visible symptoms), no parasitoids survived to their adult stage. Indeed, the later in disease development their attack occurred, the more their development was stunted. And when attacks occurred around the day of first symptoms or earlier, even though parasitoid success rate was greater the earlier its attack came in disease development, the greatest success rate was only 33%, compared to around 90% in the absence of disease. Parasitoid success

TABLE 16.1. The outcome of mixed virus ingestion (PiGV) and parasitoid attack (*V. canescens*) within individual host larvae (*P. interpunctella*) of various instars. –, The parasitoid cannot attack first instar hosts; GV, the virus (assuming a lethal dose) prevents host development to the fifth instar, prevents parasitoid development and thus 'wins'; (GV), the virus again probably wins but may occasionally allow host development to the fifth instar and hence parasitoid development (see Table 16.2); Fig. 16.7, details of the equivocal outcome given in Fig. 16.7; ?, probably further equivocal outcomes similar to those in Fig. 16.7; P, viral infection, initiated in the fifth instar, has insufficient time to develop and so the parasitoid 'wins'.

Instar of parasitization	Instar of GV infection				
	I	II	III	IV	V
I	–	–	–	–	–
II	GV	GV	(GV)	?	P
III	GV	GV	(GV)	?	P
IV	GV	GV	(GV)	Fig. 16.7	P
V	GV	GV	(GV)	(GV)	P

TABLE 16.2. Fifth instar hosts were parasitized before and after the first appearance of disease symptoms. They were monitored until death or adult emergence. If the host died in the larval stage it was dissected and the stage of development reached by the parasitoid was recorded.

Time of parasitism in relation to first disease symptoms (days)	n	Numbers of parasitoids reaching this stage at host death							
		Egg	L1	L2	L3	L4	L5	Pupa	Adult
−4	7	0	0	0	0	0	0	7	0
−3	12	1	0	0	0	0	0	6	4
−2	31	2	2	1	2	0	5	13	6
−1	40	3	3	5	1	1	10	9	8
0	25	0	0	0	1	1	18	3	2
1	25	0	0	3	1	1	15	4	1
2	25	6	3	2	2	5	7	0	0
3	25	7	7	5	0	2	4	0	0
4	25	7	6	7	2	1	2	0	0
5	10	9	1	0	0	0	0	0	0
6	15	14	1	0	0	0	0	0	0
11	10	10	0	0	0	0	0	0	0

clearly varies with disease-stage, but the virus seems none the less to maintain its competitive edge whenever viral ingestion precedes parasitoid attack.

The outcomes were less one-sided when hosts were attacked by the parasitoid in the fourth instar and acquired a particular viral dose (1.95×10^7 infectious units) shortly afterwards in the fourth instar (Fig. 16.7; related results for other viral doses are omitted here for brevity). In the first place, parasitized larvae were less likely (57%) than unparasitized larvae (71%) to become diseased, even though significant parasitoid development would not have been initiated. However, amongst the 57% of parasitized hosts that did become overtly diseased, three different outcomes were possible.

1   In only a small proportion (22%) the GV effectively prevented parasitoid development, giving rise to a virus-filled host cadaver. Virus productivity was as high here as in unparasitized hosts, but these cases represented only 12% (22% × 57%) of the original cohort compared to 71% of unparasitized hosts.

2   In 39% of hosts, a parasitoid larva – not an adult – emerged from the host, having been prevented by the virus from completing its development. The quantities of infectious virus particles present in the doomed parasitoid larva and the remains of the host were substantially smaller than those in a normally infected host.

3   Finally, from a further 39% of diseased hosts, an adult parasitoid emerged and thus 'won' – compared to 90% adult parasitoids from non-diseased hosts. However, the adult parasitoids emerged at both a significantly smaller size and after a significantly longer period of development than parasitoids developing in uninfected

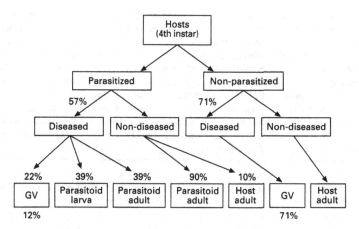

FIG. 16.7. Flow diagram documenting the fates of individual *P. interpunctella* (in percentage terms) when either parasitized by *V. canescens* and subsequently infected with granulosis virus in the fourth instar, or when only infected with virus. The figures of 12% and 71% at the bottom of the diagram refer to percentages of the original cohorts succumbing to GV infection alone when parasitized or not parasitized, respectively. The full figures for parasitoid adults, emerging from diseased and non-diseased hosts, respectively, were: weights 0.74 mg ± 0.07, 1.31 mg ± 0.03 (means ± SE), $t = 7.5$, $P < 0.01$; development times 26.2 days ± 1.3, 24.7 days ± 2.1 (means ± SE), $t = 2.4$, $P < 0.01$.

hosts. Note, too, that the host survival rates under these conditions were 10% after parasitoid attack, 29% after pathogen infection, but only 4% after both pathogen and parasitoid attack.

A comparable experiment was also carried out with host larvae parasitized in their third instar and acquiring a viral dose in their fourth. The susceptibility of parasitized hosts to infection was dramatically reduced: only 13% became diseased, compared to 57% (parasitized in the fourth instar, above) or 71% (unparasitized).

Overall, therefore, in these later host instars, competitive outcomes are more evenly balanced. More striking in these instars, perhaps, is the extent to which the pathogen, the parasitoid and the host are all adversely affected by the interaction.

Given these adverse effects, particularly of the virus on the parasitoid, it is surprising that experiments to determine the behavioural response of parasitoids to virally infected host larvae (Table 16.3) suggest that there is little discrimination between uninfected and infected hosts, regardless of host age or severity of infection. The only apparent exception is a slight, but significant, tendency for parasitoids to avoid ovipositing in infected hosts in the latest stages of overt infection. In lightly infected fifth instar hosts, there is still a small chance that the parasitoid can develop successfully, but eggs laid in infected fourth instars have absolutely no chance of survival, and constitute a waste of both eggs and time.

TABLE 16.3. A lack of discrimination against virus-infected hosts by *V. canescens*. One-day-old parasitoids were presented sequentially with uninfected and then infected hosts of different instars. In one part of the experiment, heavily infected fourth instar (L4) hosts were parasitized 10 days after the first appearance of disease symptoms (approximately half of the time from dosing to death). In the second, L5 hosts were presented to parasitoids on successive days after the appearance of disease symptoms. Parasitoids were observed with each host and two behavioural events recorded: 'probing' (insertion of the ovipositor in the host) to determine if discrimination occurred externally; oviposition (characterized by 'cocking' of the wasp abdomen) to determine if discrimination occurred internally. All parasitoids had previously probed and oviposited in their uninfected hosts. Only the number of ovipositions in the sequences L4, L4 and L5, L5 (day 10) were significantly fewer than expected ($\chi^2 = 5.8$, 6.0, respectively, $P < 0.05$).

Uninfected hosts	Infected hosts	Number of hosts presented	Probes	Ovipositions
L3	L4	15	15	14
L4	L4	19	19	14
L5	L5 (day 1 of symptoms)	10	10	10
L5	L5 (day 2 of symptoms)	10	10	10
L5	L5 (day 3 of symptoms)	10	10	10
L5	L5 (day 4 of symptoms)	10	10	8
L5	L5 (day 5 of symptoms)	10	10	9
L5	L5 (day 6 of symptoms)	10	10	10
L5	L5 (day 10 of symptoms)	15	15	10

Finally, aside from the competitive interactions, and because the parasitoids readily attack infected hosts, their role in transmitting virus is worth investigating and can be summarized simply (Table 16.4). No viral infections occur through direct transfer of virus on the ovipositing apparatus of a parasitoid when she first deposits an egg in even a heavily virus-infected host larva and then shortly afterwards lays an egg in an uninfected host. However, when parasitoids that had parasitized just one heavily infected host spent extended periods (until death) on food medium, contamination of that food and its subsequent ingestion by developing susceptible hosts led to a small but significant proportion (8%) becoming infected. Of course, these infected individuals would then become a new source of infection to other hosts. Thus, *Venturia* suffers badly in cases of mixed infection with the GV, incurs severe costs in eggs and time when it attacks infected hosts, and even acts as an agent in spreading the virus to other hosts.

TABLE 16.4. Vectoring of PiGV by *V. canescens*. To determine if oviposition was a route of pathogen transmission, wasps parasitized uninfected hosts once as controls, then heavily infected L4s once (see Table 16.3), and immediately afterwards uninfected hosts once (either L3 or L4). Hosts were then allowed to develop until death by disease or pupation. To determine if contamination of the parasitoid was a mechanism for pathogen transmission, five wasps each parasitized a heavily infected L4 once, were placed together with food and left until death. Twenty five eggs were then added to the food and the larvae allowed to develop until death by disease or pupation. Control treatments consisted of either host eggs but no parasitoids, or eggs and parasitoids that had previously parasitized uninfected hosts. Each treatment was replicated 20 times.

Transmission route	Treatment	*n*	Number of infecteds
Oviposition	Uninfected L3 (control)	15	0
	Uninfected L3 following parasitism of infecteds	40	0
	Uninfected L4 (control)	15	0
	Uninfected L4 following parasitism of infecteds	40	0
Contamination	Control (eggs, no parasitoids)	500	0
	Control parasitoids (no previous exposure to GV)	500	0
	Parasitoids (previous exposure to GV)	500	42

DISCUSSION

All ecologists know that the species with which they are primarily concerned exist in a web of interactions with other species. Most population ecologists, most of the time, isolate one or two species from this web. They do so not only as a practical necessity, but also because they indulge in an act of faith, which states that the isolation retains enough reality to forge an acceptable understanding of the species' dynamics. The present data allow us, perhaps for the first time, to put such acts of faith to the test – at least in the context of one particular system. Faith seems to have been misplaced – which is perhaps not surprising: acts of faith, after all, are designed for moral support, not as scientific hypotheses.

When generation cycles were discussed, the dangers were apparent of accepting proposed explanations for patterns in two-species population dynamics without first knowing the patterns exhibited by one or both of those species alone. The message was a reductionist one: beware of explanations for higher-level phenomena, however plausible, when they are not grounded in an understanding of the lower-level components. For the host–pathogen–parasitoid population dynamics, no a priori explanations were on offer, and there was therefore no danger of making the same mistake. What is more, any reliance on the two-species population dynamics as the key to understanding the dynamics of all three species would have been wholly unjustified. The message from the three-species dynamics, however, may, if anything, be an even more reductionist one: explanations for the patterns of host–pathogen–parasitoid population dynamics must make reference not only to the population dynamics of the component interactions, but also to the interactions of pathogens alone, parasitoids alone and pathogens and parasitoids together within individual hosts.

It is questionable, none the less, whether such knowledge could have prepared us for the remarkable switch from single-generation cycles in the one- and two-species systems to multigeneration cycles in the three-species system. Apparent empirical demonstrations of Lotka–Volterra, predator–prey, multigeneration cycles have been rare enough. With the largest body of such data, from rodents and other mammals (e.g. Hanski et al. 1991), and more generally, the implication has been that where such patterns are observed, the intensity of interaction has been sufficient to generate cycles in the abundances of predators and prey, in spite of their web of interactions with other species. From this perspective, the predator–prey interaction is of the essence, further interactions serving only to muddy the waters and possibly obscuring cycles altogether. This applies even to the recent suggestion that cycles in the snowshoe hare reflect both its role as predator (of vegetation) and prey (of mammalian predators) (Krebs et al. 1995).

The present results, however, raise the possibility that this may be exactly contrary to the truth, since here, multigeneration cycles in a predator (the parasitoid) and its prey are apparent only when there are further interactions with another 'predator' (the pathogen). From this alternative perspective, the predator–prey interaction, far from being of the essence in generating observed cycles, is actually irrelevant in its own right, assuming importance only as an integral part of a more complex interaction. Indeed, if results like this were repeated, or if theoretical studies suggested that they might be widespread, then population ecologists might be forced to reconsider the whole practice of isolating species-pairs, and assuming that those two species alone are responsible for their dynamics, even when those dynamics show marked regularities.

Other studies of pathogen–parasitoid interactions within individual hosts (see, for instance, Hochberg 1991) conform to the pattern described here in highlighting the importance of priority in access to hosts in determining the outcome of competition and also the likelihood of both pathogen and parasitoid being adversely affected by the interaction. What has not previously been clear, however, is first, the potential for a shifting balance between pathogen and parasitoid as the host grows and develops, and also that the superiority of one or other predator on a particular host stage may override any priority effects. The P. interpunctella/ PiGV/V. canescens system is prone to these stage-structure effects because the host becomes markedly less susceptible to the pathogen as it ages (Sait et al.

1994b) and because the parasitoid shows a marked preference for larger hosts (Sait *et al.* 1995) and delays the start of significant development, in any case, until the host enters its final instar – but both of these patterns are widespread in other systems. Hence, here, as elsewhere (Gurney & Nisbet 1985; Godfray & Hassell 1989; Gordon *et al.* 1991; Briggs & Godfray 1995), it may prove impossible to forge a proper understanding of the population dynamics of a system without an explicit recognition of the stage-structures of its interacting elements.

Overall, in spite of this shifting balance, the pathogen appears more usually to be the winner in competition with the parasitoid for those larval instars that they are both intrinsically able to exploit. On the other hand, to judge by the relative effects of the two on host abundance, the parasitoid seems to be superior in its exploitation of healthy hosts (that is, when infections are not 'mixed'), possibly as a result of actively searching for hosts rather than relying on chance encounters. Hochberg *et al.* (1990) modelled host–pathogen–parasitoid interactions, but in doing so imagined a highly seasonal environment that is not really applicable to the present system. None the less, it is interesting to note that they found that coexistence of parasitoid and pathogen required one enemy to be superior at exploiting healthy hosts and the other to be competitively superior within jointly infected and parasitized individual hosts (and also that coexistence could occur at constant, cyclic or chaotic densities). Why has this clear prediction failed so signally to find support from the present system – its first proper test?

It may simply reflect the contrast between a seasonal (in the model) and an aseasonal (this study) environment, but this is unlikely to be the reason, since coexistence based on a balance between intrinsic (non-competitive) and within-body competitive abilities is a characteristic of a wider range of related models (e.g. Hochberg & Holt 1990). Alternatively, parameter values for the present system might give rise to 'coexistence', but with cyclic or apparently chaotic dynamics of such amplitude, and/or such low host densities, that, in practice, stochastic extinction is effectively inevitable, even in the short term. Low densities are predicted, for example, when pathogen and parasitoid are relatively evenly matched (as it could be argued they are here, taking all host instars together), and unstable equilibria are predicted, for example, where the host has a high intrinsic rate of increase (as here) (Hochberg *et al.* 1990).

On the other hand, the failure of the prediction may simply be a consequence of the omission from the model (as from so many others) of host age-structure, which lies at the heart of the interaction here between pathogens and parasitoids within individual hosts. To put it crudely, whereas with both the host–pathogen and the host–parasitoid interaction, the host's larval life can be split into a vulnerable and an invulnerable phase, with the three-species interaction there is no escape. Naively, it might have been supposed that since the pathogen is superior on early instars and the parasitoid superior on late, the two might coexist as a consequence of a temporal niche-differentiation. When the niches occur as sequential phases in a host's life, however, the resources of the second phase are themselves the residue (and hence a part) of the first phase. The effects are cumulative – and hence ultimately fatal to all three species.

Is the investigation of a host–pathogen–parasitoid interaction a study of competition or predation? The answer, of course, is that when two predators compete for a prey which itself has explicit dynamics, the interaction is both competitive and predatory and they are inseparable within it. Thus, the traditional sharp distinction within population ecology between competition and predation can be seen to stem simply from the pragmatic tendency to study species-pairs in isolation. Three-species modules, by contrast (Fig. 16.1), present exploitative competition in terms of two linked predator–prey interactions (as here), and portray two prey species being attacked by a predator as the basis for 'apparent competition' between them (Holt 1977). Thus, as the frontiers of population ecology are extended beyond two species, the artificial barriers between competition and predation may dissolve, and they may not, and probably should not, survive as separate chapter headings in ecological texts.

In conclusion, therefore, extending the frontiers of population ecology to encompass three-species interactions not only adds reality to population ecology and reaches out to the ecology of communities – it also highlights the needs both for more studies incorporating age- and stage-structure and for others forming links between the dynamics of populations and the mechanisms operating amongst individuals.

REFERENCES

Begon, M. & Bowers, R. G. (1994). Host–host–pathogen models and microbial pest control: the effect of host self-regulation. *Journal of Theoretical Biology*, **169**, 275–287.

Begon, M., Bowers, R. G., Kadianakis, N. & Hodgkinson, D. E. (1992). Disease and community structure: the importance of host self-regulation in a host–host–pathogen model. *American Naturalist*, **139**, 1131–1150.

Begon, M., Harper, J. L. & Townsend, C. R. (1996). *Ecology: Individuals, Populations and Communities*, 3rd edn. Blackwell Science, Oxford.

Begon, M., Sait, S. M. & Thompson, D. J. (1995). Persistence of a parasitoid–host system: refuges and generation cycles? *Proceedings of the Royal Society of London, B*, **260**, 131–137.

Bentley, S. & Whittaker, J. B. (1979). Effects of grazing by a chrysomelid beetle, *Gastrophysa viridula*, on competition between *Rumex obtusifolius* and *Rumex crispus*. *Journal of Ecology*, **67**, 79–90.

Bess, H. A., Van den Bosch, R. & Haramoto, F. H. (1961). Fruit fly parasites and their activity in Hawaii. *Proceedings of the Hawaiian Entomological Society*, **17**, 367–378.

Bowers, R. G. & Begon, M. (1991). A host–host–pathogen model with free-living infective stages, applicable to microbial pest control. *Journal of Theoretical Biology*, **148**, 305–329.

Briggs, C. J. & Godfray, H. C. J. (1995). The dynamics of insect–pathogen interactions in stage-structured populations. *American Naturalist*, **145**, 855–887.

Comins, H. & Hassell, M. P. (1976). Predation in multi-prey communities. *Journal of Theoretical Biology*, 62, 93–114.

Cottam, D. A., Whittaker, J. B. & Malloch, A. J. C. (1986). The effects of chrysomelid beetle grazing and plant competition on the growth of *Rumex obtusifolius*. *Oecologia*, 70, 452–456.

Dobson, A. P. (1985). The population dynamics of competition between parasites. *Parasitology*, 91, 317–347.

Godfray, H. C. J. (1994). *Parasitoids: Behavioral and Evolutionary Ecology*. Princeton University Press, Princeton.

Godfray, H. C. J. & Hassell, M. P. (1989). Discrete and continuous insect populations in tropical environments. *Journal of Animal Ecology*, 58, 153–174.

Gordon, D. M., Nisbet, R. M., De Roos, A., Gurney, W. S. C. & Stewart, R. R. (1991). Discrete generations in host–parasitoid models with contrasting life cycles. *Journal of Animal Ecology*, 60, 295–308.

Grosholz, E. D. (1992). Interactions of intraspecific, interspecific, and apparent competition with host-pathogen population dynamics. *Ecology*, 73, 507–514.

Gurney, W. S. C. & Nisbet, R. M. (1985). Fluctuation periodicity, generation separation, and the expression of larval competition. *Theoretical Population Biology*, 28, 150–180.

Hanski, I., Hansson, L. & Henttonen, H. (1991). Specialist predators, generalist predators, and the microtine rodent cycle. *Journal of Animal Ecology*, 60, 353–367.

Hassell, M. P. & May, R. M. (1986). Generalist and specialist natural enemies in insect predator-prey interactions. *Journal of Animal Ecology*, 55, 923–940.

Harvey, J. A., Harvey, I. F. & Thompson, D. J. (1994). Flexible larval growth allows use of a range of host sizes by a parasitoid wasp. *Ecology*, 75, 1420–1428.

Hatcher, N. E., Paul, N. D., Ayres, P. G. & Whittaker, J. B. (1994). The effect of an insect herbivore and a rust fungus individually, and combined in sequence, on the growth of two *Rumex* species. *New Phytologist*, 128, 71–78.

Hochberg, M. E. (1991). Intra-host interactions between a braconid endoparasitoid, *Apanteles glomeratus*, and a baculovirus for larvae of *Pieris brassicae*. *Journal of Animal Ecology*, 60, 51–63.

Hochberg, M. E. & Holt, R. D. (1990). The coexistence of competing parasites. I. The role of cross-species infection. *American Naturalist*, 136, 517–541.

Hochberg, M. E., Hassell, M. P. & May, R. M. (1990). The dynamics of host–parasitoid–pathogen interactions. *American Naturalist*, 135, 74–94.

Holt, R. D. (1977). Predation, apparent competition and the structure of prey communities. *Theoretical Population Biology*, 12, 197–229.

Holt, R. D. & Lawton, J. H. (1993). Apparent competition and enemy-free space in insect host-parasitoid communities. *American Naturalist*, 142, 623–645.

Holt, R. D. & Pickering, J. (1985). Infectious disease and species coexistence: a model in Lotka–Volterra form. *American Naturalist*, 126, 196–211.

Krebs, C. J., Boutin, S., Boonstra, R., Sinclair, A. R. E., Smith, J. N. M., Dale, M. R. T., Martin, K. & Turkington, R. (1995). Impact of food and predation on the snowshoe hare cycle. *Science*, 269, 1112–1115.

May, R. M. & Hassell, M. P. (1981). The dynamics of multiparasitoid–host interactions. *American Naturalist*, 117, 234–261.

Park, T. (1948). Experimental studies of interspecific competition. I. Competition between populations of the flour beetles *Tribolium confusum* Duval and *Tribolium castaneum* Herbst. *Ecological Monographs*, 18, 267–307.

Reeve, J. D., Cronin, J. T. & Strong, D. R. (1994). Parasitism and generation cycles in a salt-marsh planthopper. *Journal of Animal Ecology*, 63, 912–920.

Roughgarden, J. & Feldman, M. (1975). Species packing and predation pressure. *Ecology*, 56, 489–492.

Sait, S. M., Begon, M. & Thompson, D. J. (1994a). Long-term population dynamics of the Indian meal moth *Plodia interpunctella* and its granulosis virus. *Journal of Animal Ecology*, 63, 861–870.

Sait, S. M., Begon, M. & Thompson, D. J. (1994b). The influence of larval age on the response of *Plodia interpunctella* to a granulosis virus. *Journal of Invertebrate Pathology*, 63, 107–110.

Sait, S. M., Begon, M. & Thompson, D. J. (1994c). The effects of a sublethal baculovirus infection in the Indian meal moth, *Plodia interpunctella*. *Journal of Animal Ecology*, 63, 541–550.

Sait, S. M., Andreev, R. A., Begon, M., Thompson, D. J., Harvey, J. A. & Swain, R. D. (1995). *Venturia canescens* parasitizing *Plodia interpunctella*: host vulnerability – a matter of degree. *Ecological Entomology*, 20, 199–201.

Schmitt, R. J. (1987). Indirect interactions between prey: apparent competition, predator aggregation, and habitat segregation. *Ecology*, 68, 1887–1897.

Sibma, L., Kort, J. & de Wit, C. T. (1964). Experiments on competition as a means of detecting possible damage by nematodes. *Jaarboek, Instituut voor Biologischen Scheikundig Onderzoek van Landbouwgewassen, 1964*, 119–124.

Tilman, D. (1990). Mechanisms of plant competition for nutrients: the elements of a predictive theory of competition. *Perspectives on Plant Competition* (Ed. by J. B. Grace & D. Tilman), pp. 117–141. Academic Press, New York.

Tilman, D. & Wedin, D. (1991a). Plant traits and resource reduction for five grasses growing on a nitrogen gradient. *Ecology*, 72, 685–700.

Tilman, D. & Wedin, D. (1991b). Dynamics of nitrogen competition between successional grasses. *Ecology*, 72, 1038–1049.

Tilman, D., Mattson, M. & Langer, S. (1981). Competition and nutrient kinetics along a temperature gradient: an experimental test of a mechanistic approach to niche theory. *Limnology and Oceanography*, 26, 1020–1033.

Turchin, P. & Taylor, A. D. (1992). Complex dynamics in ecological time series. *Ecology*, 73, 289–305.

17. COMMUNITY MODULES

R. D. HOLT

Department of Systematics and Ecology, Museum of Natural History, Dyche Hall, University of Kansas, Lawrence, Kansas 66045, USA

INTRODUCTION

Ecological communities are among the most complex entities studied by scientists, not least because they harbour thousands (at least) of species interacting in all sorts of idiosyncratic ways. There is no single, best approach to understanding communities. In this chapter, I argue that a useful approach between the baroque complexity of entire communities, and the bare bones of single and pair-wise population dynamics, is provided by close analyses of models of 'community modules' – small numbers of species (e.g. three to six) linked in a specified structure of interactions.

In the food-web literature, the word 'module' at times refers to discrete blocks and interactions within more complex webs. Here, the term 'community module' simply denotes multispecies extensions of pair-wise interactions, such as basic predator–prey, host–pathogen, and resource–consumer interactions. Familiar modules include (Fig. 17.1): (i) shared resources (potentially leading to exploitative competition, Tilman 1982); (ii) food chains (Oksanen *et al.* 1981); (iii) shared predation (potentially leading to apparent competition, Holt 1977); (iv) predation upon competing prey (e.g. keystone predation, Holt *et al.* 1994; Leibold 1996); and, (v) intraguild predation (Polis *et al.* 1989). After a few remarks on rationales for studying modules, I examine the implications of models of community modules for three issues in community ecology: the determinants of food–chain length; the potential for community saturation to result from shared predation; and the puzzle of species coexistence with strong intra-guild predation.

In some instances, a system may closely resemble a particular module. This can be ensured in model laboratory systems (Lawton 1995), and can also apply to some subwebs of natural communities, if a few species strongly interact (Paine 1992), or multiple species cluster into well-defined functional groups (e.g. Morin 1995). Host–pathogen and host–parasitoid systems often nicely match particular module structures (for useful reviews see Jones *et al.* 1994; Begon & Bowers 1995; Begon *et al.* this volume).

Modules also provide bite-size conceptual units that build up towards fuller communities. The hope is that analyses of modules may, at the very least, illuminate general processes and qualitative features of complex communities. For the purpose at hand, two examples suffice. First, theoretical analyses of community modules

COMMUNITY MODULES

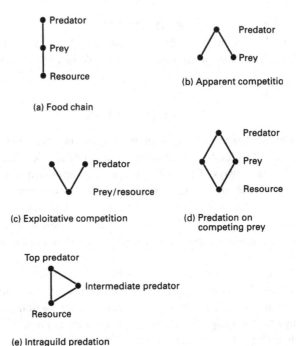

FIG. 17.1. Examples of community modules.

clearly raised our consciousness about the potential importance of indirect inter-actions, a major theme in community ecology over the past two decades (Schoener 1993; Wootton 1994; Menge 1995). Second, models of systems with as few as three species can exhibit cyclical or chaotic dynamics (e.g. Gilpin 1979), even if all constituent pair-wise interactions are stable. As a novel example, I show that unstable dynamics arise in simple models of intra-guild predation. Because most species live in species-rich communities, these theoretical studies have helped motivate the search for chaotic dynamics in natural populations (Hastings *et al.* 1993).

FOOD CHAINS

The food-chain module has received more attention than any other module, excepting exploitative competition. Simple food chain models (e.g. May 1973; Rosenzweig 1973; Oksanen *et al.* 1981; Hallam 1986; DeAngelis 1992) have helped clarify issues in the long-standing debate in ecology about the relative importance of natural enemies and resources in population regulation, and stimulated interest in the interplay of primary production and trophic interactions in determining community structure and ecosystem function (Carpenter & Kitchell 1993). In an influential paper, Oksanen *et al.* (1981) analysed a three-level model in which either herbivores or carnivores can potentially regulate their respective resources, and consumers only indirectly interact via such exploitation. This model led to some striking predictions: the number of trophic levels that could be sustained tended to increase with primary production, and the relation between plant standing crop and production varied qualitatively, in a step-like fashion, with changes in the number of trophic levels (Oksanen 1990).

The simplest qualitative prediction of food-chain models is that the addition or removal of a top enemy species indirectly produces dramatic changes in species abundances at lower trophic levels. In some cases, this 'trophic cascade' structures entire communities (Power 1992). Familiar examples include: (i) effects of piscivorous fishes on the composition and productivity of lake phytoplankton, via shifts in the abundance of intermediate planktivorous consumers (Carpenter & Kitchell 1993); (ii) the influence of sea otters on space-occupying macroalgae in the North Pacific, by altered abundances of invertebrate herbivores (Estes & Duggins 1995); (iii) the impact of West Indies lizards on leaf damage by herbivorous insects (Spiller & Schoener 1994; Dial & Roughgarden 1995). Specialist parasitism upon potentially abundant herbivores can also lead to spectacular trophic cascades (Dobson & Crawley 1994).

These qualitative predictions are consistent with a wide range of food-chain models. More specific predictions of simple food-chain models often seem to fail. In particular, simple chain models predict a relationship between food-chain length and primary productivity. Except possibly at very low productivities, this does not appear to be the case (Pimm 1991). Moreover, along gradients in productivity, all levels typically increase in abundance, once again seeming to contradict the predictions of the simplest models (Ginzburg & Akcakaya 1992). For instance, Balciunas and Lawler (1995) in a laboratory study of a three-link food chain (bacteria–bacterivore–top predator) showed that manipulating either nutrient or predator levels led to an increase in the abundance of the intermediate level, unlike predictions of simple Lotka–Volterra food-chain models.

Two broad categories of explanations have been proposed for these discrepancies between simple food-chain theory and data. First, adding realistic features to the trophic interaction and direct density-dependence alters the impact of primary production on higher trophic levels; for instance, incorporating direct interference implies all levels should increase along gradients in productivity (Ginzburg & Akcakaya 1992; Schmitz 1992). Saturating (type 2) functional responses dilutes the response of consumer populations to basal enrichment (Oksanen 1990) and can induce cyclic or chaotic dynamics (Hastings & Powell 1991), particularly at high K for the basal resource (Abrams and Roth 1994). Given unstable dynamics, the mean abundance of higher trophic levels may decline with increasing K (Abrams & Roth 1994; Lundberg & Fryxell 1995). Similar dynamic complexities occur in plant–herbivore–pathogen systems (Grenfell 1992).

Second, the 'stacked specialist' food-chain model may be a misleading caricature of whole communities (Strong 1992; Polis & Strong 1996). Even if distinct trophic levels are present, multiple species within each level can confound theoretical expectations: consumption by one level shifts the composition of lower levels toward more resistant prey (Leibold 1989; Abrams 1993; Grover 1995), therefore reducing the shunt of production to higher levels. Moreover, in many systems omnivory or intra-guild predation blurs the distinctness of trophic levels (Polis & Strong 1996). Elser *et al.* (1995) describe an example of manipulation of rainbow trout in a lake leading to unexpected increases in primary production, because of omnivory.

In spite of present disagreement about how to interpret discrepancies between simple food-chain theory and observed patterns, it seems to me that analyses of the simple food-chain module were crucial. Such work helped focus attention on a number of important issues, such as the importance of temporal heterogeneity, predator interference, and non-linear functional responses in community dynamics. As shown below, these same factors are important in governing the strength of apparent competition among prey.

It is useful to stand back for a moment from these details, and note that the simplest attribute of a food chain is that it describes sequential trophic dependencies among species. This fact alone has consequences, given that all local communities are assembled by colonization (MacArthur & Wilson 1967; Holt 1993).

Assume the food chain occurs in a landscape with numerous habitat patches, and that a consumer species is absent from any patch that is missing that consumer's required resource. The simplest descriptor of a patch is chain length, which can change by colonization (lengthening the chain) or extinction (shortening it). Describe the landscape by the fraction of patches, P_i, with chains of length i. May (1994) describes a two-level metapopulation model for a specialist predator attacking a prey species. The prey can use only a fraction k of habitat patches in the landscape. We can generalize this model to a three-link food chain as follows (see Holt 1993, 1995, 1996, for more details and alternative models):

$$\frac{dP_1}{dt} = (c_{01}P_1 + c_{01}P_3)(k - P_1 - P_2 - P_3)P_1 - c_{12}P_1P_2 - e_{10}P_1$$

$$\frac{dP_2}{dt} = c_{12}P_1P_2 - c_{23}P_2P_3 - e_{20}P_2$$

$$\frac{dP_3}{dt} = c_{23}P_2P_3 - e_{30}P_3$$

The c_{ij} and e_{ij} characterize rates of colonization and extinction taking patches from state i to j. Colonization is sequential; the basal prey always precedes the intermediate predator, followed by the top predator; the intermediate predator when alone suppresses prey colonization. All extinctions are assumed to involve the basal species; if it goes extinct, so do species directly or indirectly dependent upon it.

The basal species invades when rare, provided $k > e_{10}/c_{01}$. The intermediate predator in turn invades, given that the basal species is at equilibrium, when $k > e_{10}/c_{01} + e_{20}/c_{12}$. Finally, given that the intermediate species has equilibrated, for the top predator to invade, $k > e_{10}/c_{01} + e_{20}/c_{12} + e_{30}/c_{23}[(c_{12} + c_{01})/c_{01}]$. May (1994) notes that the requirement for a specialist predator to persist in a metapopulation is more stringent than the requirement faced by its prey. Comparing the above inequalities shows that as trophic levels are added, the condition for persistence of the top level becomes increasingly stringent.

The parameter k measures the sparseness (or ubiquity) of the habitat utilized by the basal species, and is in a way akin to K for the basal species in a standard Lotka–Volterra food-chain model. If there are constraints on species' colonization abilities, and local extinctions occur, specialist food chains are likely to be shorter in sparser habitats (Holt 1996). Because maximal k is unity, spatial dynamics may limit food-chain length even in widespread habitats. Moreover, high rates of extinction for the basal species (e.g. in successional habitats) are likely to restrict food-chain length.

As with the simple Lotka–Volterra model for local chain dynamics, however, the strength of this conclusion may, in the end, be tempered by a consideration of more realistic dispersal scenarios (e.g. lattice models; Hassell *et al.* 1994), or the effects of multiple species and complex trophic interactions in each trophic level. None the less, this model for a food-chain module clearly highlights the potential for spatial dynamics and habitable area to constrain food-chain length, in a fashion reminiscent of the role of productivity in more standard food-chain models.

APPARENT COMPETITION

The second module I consider is shared predation. Just as consumer species can reciprocally reduce each other's abundance via depleting a shared resource, prey species can indirectly depress each other by increasing the abundance of a shared natural enemy (Holt 1977; Holt & Lawton 1993). Apparent competition can, in principle, generate all the community patterns produced by standard competition for resources (e.g. habitat partitioning, Holt 1984) and act as a force limiting local species richness. If shared predation typically generates $(-, -)$ interactions between alternative prey, shared natural enemies could provide one mechanism by which local communities become saturated in their species composition, via the exclusion of species drawn from some larger species pool (Cornell & Lawton 1992).

Though I believe such patterns are important and still largely underappreciated, in broad comparisons among communities or along environmental gradients, predation may in the end prove to have inconsistent effects on community saturation. To set the stage for this argument, it is useful to reprise the basic logic of apparent competition. In Holt (1977), I showed that apparent competition was a generic phenomenon in spatially homogeneous, multiprey systems where: (i) the predator was strictly food limited; (ii) the predator had a positive numerical response to each prey; and (iii) the system settled into a point equilibrium. Prey with high values for r/a [= (intrinsic growth rate)/(attack rate)] can withstand high predator numbers and potentially exclude prey with lower values. The likelihood of exclusion depends upon the predator's capacity for limiting dominant prey numbers to well below K; increases in K indirectly increase predator numbers, magnifying the potential for exclusion by apparent competition. Similar effects emerge in multihost–parasitoid (Holt & Lawton 1993) and multihost–pathogen models (Holt & Pickering 1985; Begon & Bowers 1995), and models with mixed exploitative and apparent competition (Holt et al. 1994; Grover 1995).

These models depict reasonably well the outcome of shared predation in some systems. Qualitatively, there are now numerous well-documented examples of apparent competition arising from shared predation in laboratory microcosms, field experiments, and 'natural' experiments, in a wide variety of taxa and habitats (Holt & Lawton 1994); for instance, Nakajima and Kurihara (1994) studied a laboratory microcosm consisting of mixed clones of E. coli attacked by the protozoan Tetrahymena thermophila. Clones with either higher growth rates or lower predation rates (namely, higher r/a) dominated, because these prey traits increased predator equilibrium density, thereby leading indirectly to exclusion of the alternative prey clone (with lower r/a). The laboratory experiments of Lawler (1993) and Balciunas and Lawler (1995) similarly show that in two-prey cultures, the prey with lower r/a tends to be excluded. Shared predation can lead to local exclusion even in quite complex systems (e.g. Hochberg et al. 1994). Mobile predators can generate apparent competition even between prey living in distinct habitats, because of predator 'spillover' (Holt 1984), or coupled colonization–extinction dynamics (Holt 1996).

As with exploitative competition, these theoretical and empirical findings suggest that one sensible research programme is to search for mechanisms of coexistence between prey (Holt & Lawton 1993); for instance, in a patchy environment, if one prey has a greater r/a, but the inferior prey is a superior disperser, coexistence may occur at a landscape scale (Hassell et al. 1994).

However, the module of shared predation is intrinsically more complex than that of shared resources. In exploitative competition, species indirectly interact only through effects on resource levels. By contrast, predators have functional responses – as well as numerical responses – to prey, leading to the possibility of indirect mutualisms; for instance, time spent handling one prey species necessarily reduces the time available to capture a second species.

With sufficient constraints on the numerical response, alternative prey might either not interact (if the functional response is linear), or experience a *net* effect of indirect mutualism (Holt & Lawton 1994). This possibility was noted in Holt (1977), some circumstances leading to apparent mutualism have been addressed (e.g. Holt & Kotler 1987; Holt 1987), and empirical examples of indirect mutualism with shared predation are known (Holt & Lawton 1994). Abrams and Matsuda (1996) have recently examined in some detail the conditions leading to apparent competition vs. apparent mutualism in specific models including both non-linear functional responses (producing an indirect (+, +) interaction between prey), and direct density dependence in the predator (reducing the potential for (–, –) apparent competition interactions). They argue that one should expect a mixture of negative and positive net indirect interactions between prey in communities.

Here, I take a different but complementary approach to that of Abrams and Matsuda (in press), using a graphical model that encapsulates important qualitative features of many specific models. I argue that shared predation should have a variable effect on community saturation, and in particular that it may not lead to saturation at all in some circumstances.

Consider a focal prey species (species 2) invading a local community with resident predators, supported by prey species 1. Prey species 2 has a net growth rate when rare of $r_2 - a_2P$, where P is predator density, a_2 is the per predator attack rate imposed upon prey 2 (when rare), and r_2 is the intrinsic growth rate of this prey. The invader is excluded if $r_2 < a_2P$.

The attack rate on prey 2 can vary directly with resident prey density; for instance, assume predator feeding follows a two-species disc equation, $a_i(R_1,R_2) = a_i'/(1 + a_1'h_1R_1 + a_2'h_2R_2)$, where a_i' is the maximal attack rate on prey species i, and h_i the handling time (Murdoch & Oaten 1975). With this model, the per predator attack rate on the invading prey declines with resident prey density.

But predator density also varies with the density and productivity of the resident prey. Assume the predator has a growth equation ($dP/dt = PF(R_1,R_2,P)$, such that F increases with each R_i, and declines with P, and where prey densities are fixed (say experimentally). Let $R_2 \approx 0$. Equilibrial predator abundance can be determined by setting the predator equation to zero and solving for predator density. Below some prey abundance, the predator cannot persist. Greater prey numbers sustain more predators, but other limiting factors should become progressively more important at higher prey levels, leading to a concave-down relation between realized predator numbers and ambient prey levels. Combining the functional and numerical responses determines the total rate of predation (i.e. aP) upon an invader as a function of resident prey abundance. In general, this relation will be an asymmetrical hump, with the greatest resistance to invasion by a novel prey species at intermediate values of resident prey number (see Fig. 17.2).

Total removal of resident prey eliminates the predator, and so always facilitates invasion by additional prey species. In this broad sense, an invading prey always experiences apparent competition with the entirety of the resident predator's food supply. However, substantial reductions in resident prey abundance (without total removal) may sometimes not markedly alter the intensity of predation experienced by invading prey, and sometimes even lead to more intense predation on the invader. To illustrate this point, consider the following simple model for predator dynamics (the model assumes a type II functional response, a linear relation between prey consumption and predator reproduction, and linear predator density dependence):

$$\frac{dP}{dt} = P\left[\frac{a_1'b_1R_1 + a_2'b_2R_2}{1 + a_1'h_1R_1 + a_2'h_2R_2} - iP - m\right]$$

At low R_2, given that the predator is at equilibrium, the relation between resident prey abundance and predator abundance is $P(R_1) = ([b_1a_1'R_1/(1+a_1'h_1R_1)] - m)/i$. The saturating functional response weakens the predator's numerical response as prey numbers rise. The per prey attack rate on the invading prey is $a_2(R_1,0)P(R_1)$.

Manipulating the expression for a_2P reveals that, overall, exclusion of an invading prey species at any given R_1 is more likely if: (i) the resident prey is high quality (high b_1); (ii) the predator does not easily satiate (low h_1); (iii) direct density dependence in the predator is weak (small i); and (iv) the predator has a low inherent death rate (low m). These parameter combinations enhance the numerical response of the predator to the resident prey and reduce the magnitude of indirect mutualism via the functional response, and thereby ensure that predation upon an invading prey increases with R_1 over a wide range of resident prey densities. Converse parameter choices (e.g. high h_1) vitiate the numerical response, enhance apparent mutualism resulting from the saturating functional response, and at higher prey levels lead to inverse relations between resident prey abundance and predation pressure on invading prey. There is always some range of prey densities over which increasing R_1 heightens predation upon an invading prey, but for some parameter choices this may be observed only for a narrow range of resident prey densities.

There are two ways to interpret the R-axis in Fig. 17.2. Increasing the productivity of a single-prey species tends to increase both prey and predator densities, given predator interference. So the figure could describe how predator impact on an invading prey species varies along an environmental gradient in productivity. Alternatively, one could imagine that several roughly equivalent prey species are present, each with their own exclusive resource. Given direct density dependence in the predator, increasing the number of similar prey species should be reflected in increased total prey numbers (Holt, unpublished data). In this case, the *R*-axis is a reasonable proxy for total prey species richness in the resident community.

For the parameter values leading to Fig. 17.2a, the prey 'guild' (those prey sharing a given predator) could readily exhibit saturation; as prey abundance/species richness rises, additional species should find it increasingly difficult to invade. In Fig. 17.2b, there is little effect of prey abundance/richness, except at low densities; overall there could be considerable variation in the resident prey community, with little effect upon the chance of invasion by additional species. And finally, in Fig. 17.2c, increases in prey abundance/richness may actually facilitate invasion by other prey; a given prey species may be able to invade at very low, or quite high, R, but be excluded at intermediate values.

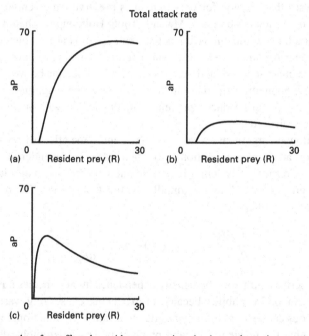

Fig. 17.2. Examples of net effects by resident prey on invasion by an alternative prey (aP), expressed as variation in total mortality along a gradient in resident prey abundance, R (implicitly, gradients in prey productivity or species richness). (a) The predator has a small handling time for the resident prey, and over most of the observed range in resident prey abundance, predation pressure on the invader increases with R ($i = 0.1$, $b_1 = 0.6$, $a_1 = 5$, $h_1 = 0.01$, $m = 5$, $a_2 = 1$). (b) The handling time is larger; this reduces both the attack rate on the invader, and the number of predators sustained by any given prey density (parameters as in Fig. 17.3a, but $h_1 = 0.03$. (c) The predator has weaker density dependence, but a larger handling time, than in (a) or (b); the greatest net effect of the resident prey on the invader is at rather low resident prey densities (parameters as above, but $i = 0.05$, $h_1 = 0.06$).

The above heuristic argument highlights the importance of jointly considering limiting factors other than prey availability, and of non-linear functional responses, in determining the potential for apparent competition, vs. apparent mutualism, in prey guilds. Predator density dependence alone weakens apparent competition, but on its own does not lead to apparent mutualism; without such direct density dependence in the predator, the short-term non-linear functional response, which tends towards indirect mutualism between prey, is overshadowed by the longer-term numerical response, leading to a net effect of apparent competition. However, putting these factors together weakens apparent competition and increases the importance of apparent mutualism via the functional response, particularly at higher resident prey densities.

It should be noted that this graphical argument is incomplete, because the realized abundance of the resident prey is a dependent variable of the system. However, fuller analyses that account for prey dynamics preserve the essential features of the above conclusions (Abrams & Matsuda 1996; Holt, unpublished data). More-over, the graphical argument assumes that the system is at a stable equilibrium. Recently, Peter Abrams, James Roth and I (unpublished data) have investigated the influence of non-equilibrial dynamics on the net interaction between alternative prey. Briefly, population fluctuations tend to reduce the magnitude of apparent competition, and can produce apparent mutualism even without direct density dependence in the predator.

The following argument reveals a key element in the effect of non-equilibrial dynamics on apparent competition to be direct density dependence in the prey themselves. Consider a system where a predator can persist with either of two alternative prey species. Prey i is initially the resident species. The resident prey dynamics are described by:

$$\frac{1}{R_i}\frac{dR_i}{dt} = r_i(t)g_i(R_i) - a_i P, \quad \frac{dg_i}{dR_i} < 0$$

where g_i describes intra-specific density dependence in prey species i. Populations may show temporal variability because of exogenous causes (e.g. variation in r), or endogenous causes (e.g. limit cycles due to a saturating functional response).

We assume the resident prey and predator persist indefinitely, despite such fluctuations. Levins (1979) argued that if a population of density X_i varies between an upper bound and a lower bound > 0, the long-term time average of the per capita growth rate must be zero:

$$E\left[\frac{1}{X_i}\frac{dX_i}{dt}\right] = \lim_{t \to \infty} \int_0^t \left(\frac{1}{X_i}\frac{dX_i}{dt}\right)dt = 0$$

The expectation operator E is linear, which simplifies consideration of the average consequences of temporal variability in the above model.

Using an overbar as shorthand for expected value, if the resident prey population (say species 1) persists, then

$$\overline{r_1 g_1} = \overline{a_1 P}$$

In other words, average prey productivity must equal average mortality due to predation.

Now consider the invader, prey species 2. Using time averaging again, the invader increases when rare if

$$\overline{r_2} > \overline{a_2 P}$$

Assume that the relative predation pressure imposed on the resident and the invader is constant, or $a_2 P / a_1 P = u$; for instance, with the two-species disc equation, $u = a_2'/a_1'$. After substitution, the condition for invasion now involves a comparison of the average intrinsic growth rate of the invader, and the average productivity of the resident prey, so that prey 2 invades only if

$$\overline{r_2} > u \overline{r_1 g_1}$$

Now assume that the predator is consistently effective at limiting prey numbers (for each species) well below the level where prey experience density dependence, so that g_i is approximately unity. (For persistence of the predator–prey system, this usually requires density-dependence in the predator, such as direct interference, or the induced density dependence provided by a small trickle of immigrant predators, Holt 1993.) This implies that

$$\overline{r_1 g_1} \approx \overline{r_1}.$$

The condition for invasion is now simply

$$\overline{r_2} > u \overline{r_1}$$

Were prey 2 initially present alone, sustaining an efficient predator, one can repeat the above line of argument to show that prey species 1 can invade when rare, provided

$$\overline{r_1} > u^{-1} \overline{r_2}$$

For most combinations of relative attack rates and intrinsic growth rates, it is impossible for both these inequalities to hold; the prey with the higher average intrinsic growth rate, or lower relative attack rate (or both), will exclude the alternative prey from the community, even if the predator experiences direct density dependence, the prey experience indirect mutualism via the predator's functional response and the environment fluctuates (without affecting relative attack rates). Holt and Lawton (1993) present a similar argument for a two-host, one-parasitoid model.

What this argument shows is that if environmental variability weakens apparent competition, it must be because variability enhances the effects of direct density dependence in the prey. The basic idea is that given a saturating functional response, predator population growth is more sharply and negatively affected by prey declines, than positively affected by prey increases. Thus, for a predator population to match its own density-independent mortality, higher average resource levels are needed in a fluctuating environment, than in a constant environment. With direct density dependence in the prey, higher numbers imply lower prey productivity, which (from above) directly translates into a lower average rate of predation upon the resident prey, and in turn upon the invading prey.

Hence, the magnitude of apparent competition is reduced by temporal fluctuations in prey abundance, which magnify the importance of limiting factors (e.g. resource competition) other than the shared predator.

As prey species are added to a community, their shared predator in effect experiences an enriched resource base. In single-prey species models, enrichment can lead to a decline in average predator abundance, given unstable dynamics (Abrams & Roth 1994). With multiple-prey species, the magnitude of population fluctuations increases with increasing prey species richness, diminishing the overall numerical response by the predator to its prey base and thus making the beneficial effect via the functional response potentially more important. However, if the unstable dynamics induced by high productivity include excursions to low densities by resident prey, the risk of prey extinction due to demographic stochasticity is also enhanced. If enrichment is destabilizing this facilitates initial colonization but can also increase local extinction rates.

Thus, interesting complications may arise if high productivity causes trophic dynamics to become destabilized. This weakens or reverses the expected relationship between productivity and predator abundance, and so tends to reduce the magnitude of apparent competition upon invading prey, but also may increase the extinction rate of resident prey species. The overall effect of these opposing forces on the likelihood of prey community saturation cannot be addressed outside the context of rather detailed models.

The above results permit us to identify major axes of variation among prey communities; different natural enemy–prey ensembles should be more (or less) likely, *inter alia*, to exhibit local saturation in defined prey 'guilds' (prey exploited by the same suite of predators) because of shared predation (namely competition for enemy-free space; Jeffries & Lawton 1984). For instance, prey communities in variable environments should experience apparent competition less intensely than do prey communities in constant environments, and may be less likely to exhibit saturation because of shared predation.

Along a gradient in productivity, given stable dynamics, saturation in a prey community (defined by difficulty of invasion for non-resident prey) should be most likely at intermediate points on the gradient. At high productivity, the shared predator is likely to be increasingly limited by factors other than prey availability, including higher-order predators and specialist pathogens, reducing the importance of apparent competition relative to indirect mutualism in determining the impact of resident prey on invasion by additional prey species.

However, different prey guilds are likely to reach this maximal level of resistance to invasion at different points along any given environmental gradient, due to idiosyncratic differences in predator and prey traits, and in the suite of limiting factors (particularly other species) impinging on both the prey and their shared predator. The net effect is that saturation due to shared predation may be difficult to discern at the level of entire communities.

INTRA-GUILD PREDATION

The final module I briefly discuss is 'intra-guild predation' (IGP), which arises whenever predators and their prey also compete for resources (Polis *et al.* 1989). This module combines the elements of exploitative and apparent competition, but in a different way than keystone predation: the intermediate predator competes with the top predator for the basal resource, and the basal resource, by sustaining the top predator, indirectly increases mortality on the intermediate predator. IGP has received much less theoretical attention than the other modules of Fig. 17.1.

Simple models of IGP highlight a substantial problem of species coexistence (Pimm & Lawton 1978; Pimm 1991; Polis & Holt 1992). A necessary condition for coexistence is that the intermediate species be a superior competitor for the basal resource (Pimm 1991; Polis & Holt 1992). Even if this holds, however, exclusion may occur if the top predator imposes too high a rate of mortality on the intermediate predator (Holt & Polis, in press).

Moreover, simple IGP models reveal that highly variable dynamical behaviour can arise, further hampering sustained coexistence. Consider the simplest IGP model, a Lotka–Volterra food chain with an added link between the top and bottom species:

$$\frac{dR}{dt} = R[r(1 - R/K) - aN - a'P]$$

$$\frac{dN}{dt} = N[abR - m - \alpha P]$$

$$\frac{dP}{dt} = P'[\alpha\beta N + a'b'R - m']$$

Numerical studies of conditions for joint invasibility, and stability of the resulting equilibrium, reveal that relatively small amounts of IGP can strongly destabilize the system (Holt & Polis, in press). Figure 17.3 shows an example. In the case shown, both the top and intermediate predator, when alone, persist stably with the resource, and each can invade the community containing the other. Such invasions, however, set up dramatically violent oscillations, which in practice would foster local extinctions.

Lotka–Volterra models for food chains and multiprey, single-predator systems, always reach stable point equilibria. Complex dynamics in three-species Lotka–

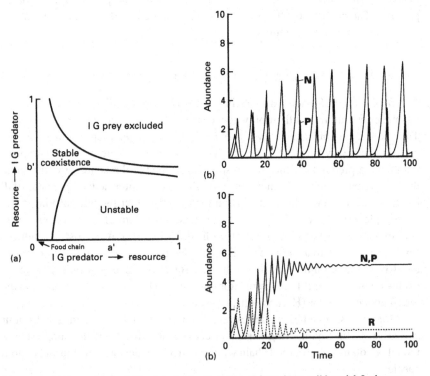

FIG. 17.3. (a) Example of stability domains for the Lotka–Volterra intra-guild model. In the case depicted, $r = 1$, $K = 10$, $a = 1$, $\alpha = 1$, $b = 1$, $m = 0.1$, $m' = 0.5$. (See Holt & Polis, in press for more details). For the stable and unstable parameter sets, each predator can increase when rare. The a'-axis is the rate of attack by the top predator on the basal resource ($a' = 0$ leads to a simple food chain); the b'-axis is the benefit the predator derives from such attacks. (b) Illustrative samples of dynamical behaviour. Parameter values are as in (a), with $b' = 0.3$ and $a' = 0.01$ (bottom), and $a' = 0.3$ (top).

Volterra systems have previously been reported for three directly competing species (May & Leonard 1975), and for a generalist predator feeding upon two directly competing prey (Gilpin 1979). To my knowledge, this is the first demonstration of unstable dynamics in a Lotka–Volterra model arising when all the inter-specific interactions are linear trophic interactions (though it should be noted that dynamical instability is known in comparable host–pathogen systems, see Hochberg and Holt (1990).

The dynamical instability produced by IGP illustrated in Fig. 17.3 makes it even less likely that simple IGP modules will persist. Yet, there are numerous empirical systems in which IGP is conspicuous (Diehl 1993, 1995; Polis *et al.* 1989). This raises the interesting possibility that something systematic may be missing from the simple models; for example, Diehl (1993) notes in his review that in many examples, the IGP prey has a refuge from predation, and Pimm (1991) notes other recurrent features of systems with IGP. If top predators consistently experience strong intra-specific density dependence, for instance, that would suffice for persistence of the intermediate predator (Holt & Polis, in press). Whether any of these suggestions provide a general, robust explanation for the paradox of systems persisting with strong IGP remains to be seen. What the simple model of the IGP module does is to alert one to an interesting question regarding species persistence in complex communities, and provide guidelines as to key parameters that should be measured, or additional processes that should be considered.

CONCLUSIONS

Community modules provide community ecologists with a research path that with any luck skirts both the Scylla of unrealistic simplicity, and the Charybdis of unmanageable complexity. Analyses of modules help crystallize our understanding of core processes, which can then be discerned (albeit at times obscurely) as a driving force in many disparate systems. They provide fresh hypotheses for empirical studies. It may be even more interesting when the predictions of a particular module model fail; characterizing the possible reasons for failure helps provide a conceptual framework for organizing the complexity of natural communities, and may provoke the search for new approaches (e.g. constraints on food-chain length may need be sought at the level of regional processes, e.g. colonization–extinction dynamics, rather than local, production-driven dynamics). The examples of community modules explored above all provide examples of this healthy intellectual dynamic.

REFERENCES

Abrams, P. A. (1993). Effects of increased productivity on the abundances of trophic levels. *American Naturalist*, **141**, 351–371.

Abrams, P. A. & Matsuda, H. (1996). Positive indirect effects between prey species that share predators. *Ecology*, **77**, 610–616.

Abrams, P. A. & Roth, J. D. (1994). The effects of enrichment of three-species food chains with nonlinear functional responses. *Ecology*, **75**, 1118–1130.

Balciunas, D. & Lawler, S. P. (1995). Effects of basal resources, predation, and alternative prey in microcosm food chains. *Ecology*, **76**, 1327–1336.

Begon, M. & Bowers, R. G. (1995). Beyond host–pathogen dynamics. *Ecology of Infectious Diseases in Natural Populations* (Ed. by B. T. Grenfell & A. P. Dobson), pp. 478–509. Cambridge University Press, Cambridge.

Carpenter, S. R. & Kitchell, J. F. (1993). *The Trophic Cascade in Lakes.* Cambridge University Press, Cambridge.

Cornell, H. V. & Lawton, J. H. (1992). Species interactions, local and regional processes, and limits to the richness of ecological communities: a theoretical perspective. *Journal of Animal Ecology*, **61**, 1–12.

DeAngelis, D. L. (1992). *Dynamics of Nutrient Cycling and Food Webs.* Chapman & Hall, London.

Dial, R. & Roughgarden, J. (1995). Experimental removal of insectivores from rain forest canopy: direct and indirect effects. *Ecology*, **76**, 1821–1834.

Diehl, S. (1993). Relative consumer sizes and the strengths of direct and indirect interactions in omnivorous feeding relationships. *Oikos*, **68**, 151–157.

Diehl, S. (1995). Direct and indirect effects of omnivory in a littoral lake community. *Ecology*, **76**, 1727–1740.

Dobson, A. & Crawley, M. (1994). Pathogens and the structure of plant communities. *Trends in Ecology and Evolution*, **9**, 393–397.

Elser, J. J., Luecke, C., Brett, M. T. & Goldman, C. R. (1995). Effects of food web compensation after manipulation of rainbow trout in an oligotrophic lake. *Ecology*, **76**, 52–69.

Estes, J.A., & Duggins, D. O. (1995). Sea otters and kelp forests in Alaska: generality and variation in a community ecological paradigm. *Ecological Monographs*, **65**, 75–100.

Ginzburg, L. R. & Akcakaya, H. R. (1992). Consequences of ratio-dependent predation for steady-state properties of ecosystems. *Ecology*, **73**, 1536–1543.

Gilpin, M. E. (1979). Spiral chaos in a predator–prey model. *American Naturalist*, **113**, 306–308.

Grenfell, B. T. (1992). Parasitism and the dynamics of ungulate grazing systems. *American Naturalist*, **139**, 907–929.

Grover, J. P. (1995). Competition, herbivory, and enrichment: nutrient-based models for edible and inedible plants. *American Naturalist*, **145**, 746–774.

Hallam, T. G. (1986). Community dynamics in a homogeneous environment. *Mathematical Ecology* (Ed. by T. G. Hallam & S. A. Levin), pp. 241–285. Springer, Berlin.

Hassell, M. P., Comins, H. N. & May, R. M. (1994). Species coexistence and self-organizing spatial dynamics. *Nature*, **370**, 290–292.

Hastings, A. & Powell, T. (1991). Chaos in a three-species food chain. *Ecology*, **72**, 896–903.

Hastings, A., Hom, C. L., Ellner, S., Turchin, P. & Godfray, H. C. J. (1993). Chaos in ecology: is Mother Nature a strange attractor? *Annual Review of Ecology and Systematics*, **24**, 1–34.

Hochberg, M. E. & Holt, R. D. (1990). The coexistence of competing parasites. I. The role of cross-species infection. *American Naturalist*, **136**, 517–541.

Hochberg, M. E., Clarke, R. T., Elmes, G. W. & Thomas, J. A. (1994). Population dynamic consequences of direct and indirect interactions involving a large blue butterfly and its plant and red ant hosts. *Journal of Animal Ecology*, **63**, 375–391.

Holt, R. D. (1977). Predation, apparent competition, and the structure of prey communities. *Theoretical Population Biology*, **12**, 197–229.

Holt, R. D. (1984). Spatial heterogeneity, indirect interactions, and the coexistence of prey species. *American Naturalist*, **124**, 377–406.

Holt, R. D. (1987). Prey communities in patchy environments. *Oikos*, **50**, 276–290.

Holt, R. D. (1993). Ecology at the mesoscale: the influence of regional processes on local communities. *Species Diversity in Ecological Communities* (Ed. by R. Ricklefs & D. Schluter), pp. 77–88. University of Chicago Press, Chicago.

Holt, R. D. (1995). Food webs in space: an island biogeographic perspective. *Food Webs: Contemporary Perspectives* (Ed. by G. Polis & K. Winemiller), pp. 313–323. Chapman & Hall, London.

Holt, R. D. (1996). From metapopulation dynamics to community structure: some consequences of spatial heterogeneity. *Metapopulation Dynamics: Ecology, Genetics, and Evolution* (Ed. by I. Hanski & M. Gilpin) pp. 149–164. Academic Press, New York.

Holt, R. D. & Kotler, B. P. (1987). Short-term apparent competition. *American Naturalist*, 130, 412–430.

Holt, R. D. & Lawton, J. H. (1993). Apparent competition and enemy-free space in insect host-parasitoid communities. *American Naturalist*, 142, 623–645.

Holt, R. D. & Lawton, J. H. (1994). The ecological consequences of shared natural enemies. *Annual Review of Ecology and Systematics*, 25, 495–520.

Holt, R. D. & Pickering, J. (1985). Infectious disease and species coexistence: a model of Lotka–Volterra form. *American Naturalist*, 126, 196–211.

Holt, R. D. & Polis, G. A. (in press). A theoretical framework for intraguild predation. *American Naturalist*.

Holt, R. D., Grover, J. & Tilman, D. (1994). Simple rules for interspecific dominance in systems with exploitative and apparent competition. *American Naturalist*, 144, 741–777.

Jeffries, M. J. & Lawton, J. H. (1984). Enemy free space and the structure of ecological communities. *Biological Journal of the Linnaean Society*, 23, 269–286.

Jones, T. H., Hassell, M. P. & May, R. M. (1994). Population dynamics of host-parasitoid interactions. *Parasitoid Communities* (Ed. by B. Hawkins), pp. 371–394. Oxford University Press, Oxford.

Lawler, S. P. (1993). Direct and indirect effects in microcosm communities of protists. *Oecologia*, 93, 184–190.

Lawton, J. H. (1995). Ecological experiments with model systems. *Science*, 269, 328–331.

Leibold, M. (1989). Resource edibility and the effects of predators and productivity on the outcome of trophic interactions. *American Naturalist*, 134, 922–949.

Leibold, M. (1996). A graphical model of keystone predators in food webs: trophic regulation of abundance, incidence and diversity patterns in communities. *American Naturalist*, 147, 784–812.

Levins, R. (1979). Coexistence in a variable environment. *American Naturalist*, 114, 765–783.

Lundberg, P. & Fryxell, J. M. (1995). Expected population density versus productivity in ratio-dependent and prey-dependent models. *American Naturalist*, 146, 153–161.

MacArthur, R. H. & Wilson, E. O. (1967). *The Theory of Island Biogeography*. Princeton University Press, Princeton.

May, R. M. (1973). Time-delay versus stability in population models with two and three trophic levels. *Ecology*, 54, 315–325.

May, R. M. (1994). The effects of spatial scale on ecological questions and answers. *Large-scale Ecology and Conservation Biology* (Ed. by P. J. Edwards, R. M. May & N. R. Webb), pp. 1–17. Oxford University Press, Oxford.

May, R. M. & Leonard, W. J. (1975). Nonlinear aspects of competition between three species. *SIAM Journal of Applied Mathematics*, 29, 243–253.

Menge, B. A. (1995). Indirect effects in marine rocky intertidal interaction webs: patterns and importance. *Ecological Monographs*, 65, 21–74.

Morin, P. J. (1995). Functional redundancy, non-additive interactions, and supply-side dynamics in experimental pond communities. *Ecology*, 76, 133–149.

Murdoch, W. W. & Oaten, A. (1975). Predation and population stability. *Advances in Ecological Research*, 9, 1–13.

Nakajima, T. & Kurihara, Y. (1994). Evolutionary changes of ecological traits of bacterial populations through predator-mediated competition. 1. Experimental analysis. *Oikos*, 71, 24–34.

Oksanen, L. (1990). Predation, herbivory and plant strategies along gradients of primary productivity. *Perspectives on Plant Competition* (Ed. by J. B. Grace & D. Tilman), pp. 445–575. Academic Press, New York.

Oksanen, L., Fretwell, S. D., Arruda, J. & Niemala, P. (1981). Exploitation ecosystems in gradients of primary productivity. *American Naturalist*, **118**, 240–261.

Paine, R. T. (1992). Food-web analysis through field measurement of per capita interaction strength. *Nature*, **355**, 73–75.

Pimm, S. L. (1991). *The Balance of Nature*? University of Chicago Press, Chicago.

Pimm, S. L. & Lawton, J. H. (1978). On feeding on more than one trophic level. *Nature*, **275**, 542–544.

Polis, G. A. & Holt, R. D. (1992). Intraguild predation: the dynamics of complex trophic interactions. *Trends in Ecology and Evolution*, **7**, 151–155.

Polis, G. A. & Strong, D. R. (1996). Food web complexity and community dynamics. *American Naturalist*, **147**, 813–846.

Polis, G. A., Myers, C. A. & Holt, R. D. (1989). The ecology and evolution of intraguild predation. *Annual Review of Ecology and Systematics*, **20**, 297–330.

Power, M. E. (1992). Top-down and bottom-up forces in food webs: do plants have primacy? *Ecology*, **73**, 733–746.

Rosenzweig, M. L. (1973). Exploitation in three trophic levels. *American Naturalist*, **107**, 275–294.

Schmitz, O. J. (1992). Exploitation in model food chains with mechanistic consumer-resource dynamics. *Theoretical Population Biology*, **41**, 161–183.

Schoener, T. W. (1993). On the relative importance of direct versus indirect effects in ecological communities. *Mutualism and Community Organization* (Ed. by H. Kawanabe, J. E. Cohen & K. Iwasaki), pp. 365–411. Oxford University Press, Oxford.

Spiller, D. & Schoener, T. S. (1994). Effects of top and intermediate predators in a terrestrial food web. *Ecology*, **75**, 182–196.

Strong, D. R. (1992). Are trophic cascades all wet? Differentiation and donor-control in speciose ecosystems. *Ecology*, **73**, 747–755.

Tilman, D. (1982). *Resource Competition and Community Structure*. Princeton University Press, Princeton.

Wootton, J. T. (1994). The nature and consequences of indirect effects in ecological communities. *Annual Review of Ecology and Systematics*, **25**, 443–466.

18. OUTLINES OF FOOD WEBS IN A LOW ARCTIC TUNDRA LANDSCAPE IN RELATION TO THREE THEORIES ON TROPHIC DYNAMICS

L. OKSANEN*†, M. AUNAPUU‡, T. OKSANEN¶,
M. SCHNEIDER¶, P. EKERHOLM¶, P. A. LUNDBERG*,
T. ARMULIK‡, V. ARUOJA‡ AND L. BONDESTAD#

*Department of Ecological Botany, Umeå University, S-901 87 Umeå, Sweden;
†Department of Biology, University of Oulu, FIN-905 70 Oulu, Finland;
‡Department of Zoology and Hydrobiology, University of Tartu, EE-2400,
Estonia; ¶Department of Animal Ecology, Umeå University, S-901 87 Umeå,
Sweden; #Unit for Environmental Protection, Provincial Government of
Västerbotten, S-901 86 Umeå, Sweden

INTRODUCTION

Generalizations derived from published food webs have been criticized recently by several ecologists (Paine 1988; Hall & Raffaelli 1991, this volume; Martinez 1991; Polis 1994). Food-web charts are abstractions, normally produced to illuminate central properties of different ecosystems. They tend to be neither too complex (since complex charts are difficult to interpret and thus impractical), nor too simple (because simple systems can be described verbally). Computing statistical properties of published food webs is thus likely to be a study of heraldics of graphical communication (Raffaelli & Hall 1995). Paine (1988, 1992) not only questioned the validity of established conclusions but even the utility of constructing connectivity food webs, since such webs need have little to do with the dynamics of the system they represent. Energy flow webs, where quantitative information has been added to recorded trophic connections, can be equally uninformative, because minor energy transfers can have major regulatory impacts and vice versa (Power 1990, 1992; Polis 1994). This problem is especially severe in terrestrial ecosystems, where many organisms obtain most of their energy from plants, but where no consensus exists about their potential impact on standing crops and population processes of plants. Many terrestrial ecologists argue that trophic interactions in general and herbivore–plant interactions in particular are donor controlled and the impact of top-down forces is modest or non-existent (Haukioja & Hakala 1975; White 1978; Price et al. 1980; Rhoades 1985; Hunter & Price 1992; Strong 1992; Seldal et al. 1994; Polis & Strong 1996).

The hypothesis of donor controlled trophic interactions in terrestrial ecosystems is an interesting and testable proposition. It predicts that interactions between herbivores and plants should not change their characteristics when the density of predators is reduced. This prediction has been tested in a wide array of systems, including: greenhouse tables, where one vole amounts to an extraordinarily high population density (Moen *et al.* 1993a); small, isolated islands, where voles or hares have spread spontaneously (Bergman 1970; Soikkeli & Virtanen 1975; Pokki 1981; Angerbjörn & Hjernquist 1984) or where they have been introduced (Oksanen *et al.* 1987; Lundberg, Ekerholm & Oksanen, unpublished data); islands, where important predators of folivorous insects are absent (Spiller & Schoener 1990); various kinds of fenced herbivore populations (Desy & Batzli 1989; Atlegrim 1989; Marquis and Whelan 1994; Ostfeld 1994); relatively large, predator-free islands where rabbits (Werth 1928) or reindeer (Klein 1987; Leader-Williams 1988) have been introduced; areas invaded by feral reindeer (Höglund & Eriksson 1973); an island where wolves are suffering from inbreeding depression and are thus unable to respond to an increase in moose density (McLaren & Peterson 1994); and an island, where a novel herbivore has evolved in the absence of predators (Hnatiuk *et al.* 1976; Merton *et al.* 1976).

Except for cases where herbivores quickly reach a size that is too large to be included in the predator diet (Schmitz 1993), the story is always the same. Areas with exceptionally low predator numbers are characterized by exceptionally high herbivore density and by exceptionally intense herbivory pressure. Cascading impacts to plants have been weaker in the case of folivorous insects than for grazing vertebrates, but this pattern may have a technical explanation. Experiments on herbivorous invertebrates have normally only considered insectivorous birds and lizards, yet there is persuasive circumstantial evidence for the importance of pupal predators (Hanski 1987) and parasitoids as regulators of folivores (Caughley & Lawton 1981). Gall midges appear to be involved in a strong interaction with parasitoids (Strong & Larsson 1994), and predatory nematodes appear to regulate root herbivores (Strong *et al.* 1995a; Strong *et al.* 1995b). Moreover, most invertebrate work has been done with univoltine species during a single growing season, which precludes the study of folivore population growth between generations. We suspect that stronger impacts could be observed in greenhouse experiments, where folivores could be effectively protected against all natural enemies.

The case of Aldabra, with its strictly resource-limited grazing tortoises and its tightly clipped grazing lawns, is especially instructive. Contrary to the argument of Strong (1992), the tortoises are not aberrantly efficient. They are rather bizarre grazers for this habitat, needing shade and large amounts of fresh water, which makes it impossible for them to eliminate woody plants from the island (Hnatiuk *et al.* 1976; Merton *et al.* 1976). They are currently under threat of being outcompeted by an expanding population of feral goats, which are more efficient herbivores than the tortoises (Hambler 1984; Coblentz *et al.* 1990). Moreover, Aldabra has received 'much of the rich mainland flora', including relatively unpalatable plants, and plants that are morphologically quite different from the prostrate constituents of grazing lawns. Many species of woody plants are present, the atoll being largely covered by woody vegetation at the start of this century, when the tortoise population had been overexploited by man (Stoddart 1971a, b). The ongoing decimation of woody vegetation and the re-expansion of grazing lawns is a direct response to increasing grazing pressure (Hnatiuk *et al.* 1976; Merton *et al.* 1976) and can be reversed by excluding grazers (Gibson *et al.* 1983). This natural experiment even has an evolutionary component. The dominant constituents of grazing lawns belong to genera which are represented by totally different types of plants in adjacent mainland areas (e.g. tall graminoids in the case of *Scirpus* spp., *Panicum aldabarense* and *Sporobolus* spp., exceptionally unpalatable dicots in the cases of *Euphorbia* spp.), indicating that the persistent, intense herbivory has selected for prostrateness and against investments in unpalatable, woody shoots (Braithwaite *et al.* 1973; Merton *et al.* 1976).

Another line of empirical evidence is provided by the traditional, extensive grazing systems of temperate Eurasia, where a few, initially native grazers (cattle and horses) and browsers (sheep, goats, and in arid regions, even camels) were protected against predators. While specialized cattle ranching frequently leads to the invasion of woody plants, due to the high degree of grass specialization by bovids, traditional grazing with mixed livestock has invariably led to the replacement of forests by secondary grasslands, heath or cushion plant communities (Walter 1964, 1968; Sjörs 1971; Gimingham 1972; Ellenberg 1978). In areas where grazing has been long lasting and intense, it has led to the evolution of endemic, prostrate plants and to the development of peculiar plant communities, which are now threatened by the relaxation of grazing pressure (Pettersson 1959). The results of this continent-wide experiment clearly show that there is no such thing as an inedible plant. Even the least palatable trees are vulnerable as seedlings (Ostfeld & Canham 1993). Elimination of woody plants by browsing mammals may take some time, but in the end it will always succeed. To witness this, one does not have to travel to Aldabra; corresponding grazing lawns, created by sheep or rabbits, are commonplace all over England, and forest regeneration in many areas has been restricted to the period when rabbits were still vulnerable to myxomatosis (Crawley 1983).

The hypothesis of donor control in terrestrial grazing chains has been tested and, in our opinion, thoroughly falsified. Donor-controlled trophic interactions between animals and plants do indeed occur, involving thousands of species of granivores, frugivores, nectarivores, sapsuckers and other consumers of high-quality plant organs or products. These consumers might live in White's (1978) 'passively harsh world', where the main regulating factor is the production of adequate food. However, there are also true folivores, which have evolved a symbiotic relationship with cellulase-producing microbes, and can thus subsist and reproduce by exploiting vegetative plant organs. Some of these folivores have for millennia been used for the production of meat, milk and wool, while others occasionally defoliate forests and cropfields, causing heavy economic loss. Consequently, top-down forces in terrestrial food webs are not only an interesting academic issue, but vital for resource management as well.

THREE HYPOTHESES AND THEIR PREDICTIONS

From the current debate on trophic dynamics in terrestrial food webs, the *hypothesis of ratio-dependent predation* (Arditi & Ginzburg 1989; Arditi *et al.* 1991; Akçakaya *et al.* 1995) stands out as the one with least-surprising and seemingly most plausible predictions. The core of the hypothesis is that the functional response of predators in the population dynamic time scale depends on the amount of prey per predator. Consequently, the structure and dynamics of food webs is predicted to remain unchanged across productivity gradients, as equilibria at all trophic levels are similarly influenced by primary productivity. If predators are assumed to survive at relatively low prey/predator ratio, the hypothesis of ratio-dependent predation can be regarded as a formalization of the green world hypothesis of Hairston *et al.* (1960) and Hairston and Hairston (1993), according to which plants and carnivores and resource limited and herbivores are predation controlled in all terrestrial ecosystems. With other parameter values, the hypothesis of ratio-dependent predation can be made to approach the donor-controlled ideas of Hunter and Price (1992).

Entirely different predictions are generated by *the hypothesis of exploitation ecosystems* (Oksanen *et al.* 1981; see also Fretwell 1977, 1987), which was deduced by modelling trophic interactions across productivity gradients by means of Rosenzweig's (1971, 1973, 1977) elaborations of Lotka–Volterra type exploitation models. According to this hypothesis, relatively productive ecosystems (e.g. forests and their successional stages) are characterized by three-link trophic dynamics – that is, by resource-limited predators, predation-controlled herbivores and resource-limited plants, as proposed by Hairston *et al.* (1960). Predator and plant communities are structured by resource-based competition, whereas apparent competition (Holt 1977) is the structuring force of herbivore communities. In relatively unproductive ecosystems, such as tundras, steppes and semideserts, depletion of the vegetation arises from a lower density of herbivores than would be required to support a population of predators. Consequently, two-link trophic dynamics prevail. Carnivores can only be present during grazer outbreaks, and are thus relegated to the role of donor-controlled consumers. Herbivores are resource limited and their communities are structured by resource-based competition, which is predicted to create clear niche segregations and to favour those species which can break even at low plant biomass and utilize low-quality forage. Plants are subjected to intense natural grazing pressure, and the vegetation is structured by apparent competition. In the most extreme environments, such as polar deserts, grazers are predicted to be absent. The only trophic interaction in these one-link ecosystems is between plants and physical resources. Consequently, the scanty vegetation is structured by pre-emptive competition for the few sites where growth is possible.

The patterns described above should be found in large-scale productivity gradients. Trophic dynamics within relatively small patches of unproductive habitat can be profoundly influenced by 'spillover predation' (Holt 1984), due to despotic habitat selection in predators (Fretwell 1972; T. Oksanen 1990) and opportunistic behaviour of transient predators (Oksanen *et al.* 1992a). The most simple version of the hypothesis (Oksanen *et al.* 1981) predicts that equilibrium biomass at different trophic levels increases in a step-wise manner along gradients of increasing primary productivity. In two-link ecosystems, the equilibrium herbivore biomass increases whereas standing crop of plants is constant. In three-link ecosystems, equilibrium plant biomass increases, equilibrium herbivore biomass stays constant and equilibrium carnivore biomass may or may not increase (depending on whether herbivores are saturated). Especially in the case of three-link ecosystems, these biomass patterns can become modified by plausible evolutionary and behavioural responses to intense exploitation pressure (Abrams 1984, 1993; Oksanen 1992). This may also happen in nutrient-poor two-link ecosystems, where the evolutionarily stable strategy of plants includes relatively high investment in defence and relatively low palatability (Bryant *et al.* 1983; L. Oksanen 1990a; Herms & Mattson 1992).

In their recent contributions, Polis *et al.* (1995) and Polis and Strong (1996) have vigorously argued for a third alternative, to be referred to as the *hypothesis of energy shunts*. The basic premises are that omnivory is rampant in all food webs, making the concept of trophic level meaningless (Cousins 1980, 1987) and that most energy fixed in terrestrial ecosystems goes inevitably directly to the detritus chain (Odum 1971). The causal background of this phenomenon was not discussed in detail; possible, uncontroversial causes include the high turnover rate of fine roots, the copious amounts energy directed to mycorrhizas, and the problems of grazers tracking the seasonal pulse of plant growth. Moreover, Polis and Strong argue that detritus-based food chains and food chains based on green plants are tightly intertwined, because consumers ignore the feeding history of their resources (Cousins 1987). The impact of these internal energy shunts can be enhanced by energy transfers from other ecosystems, especially where relatively barren ecosystems are juxtaposed to substantially more productive ones (Polis & Hurd 1995).

Unfortunately, the interesting ideas of Polis and his coworkers have been only vaguely outlined, and aspects of the contributions are logically inconsistent. On one hand, the authors argue for the invulnerability of plants and for the unimportance of herbivory, on the other hand they argue for the hypothesis of Menge and Sutherland (1976), according to which basal organisms are intensely exploited by omnivorous top consumers and exploitation predation pressure increases monotonically from the top of the food web to its base. For us, a reasonable compromise is to assume that there are true folivores, capable of exploiting vegetative plant tissues and of decimating plant biomass (see above) and omnivores, which function as predators of folivores but, in addition, obtain energy from detrital food webs and plants in a donor-controlled way. The variant of the energy shunt hypothesis thus developed is partially our own and does not necessarily conform with the views of Polis and Strong.

The dynamics of such energy shunt systems have been formally studied by L. Oksanen (unpublished data); here we restrict ourselves to intuitively obvious predictions. In the case of productive ecosystems, the dynamics of energy shunt systems by and large converge with the behaviour of exploitation ecosystems and with the green world hypothesis of Hairston *et al.* (1960). Omnivory creates intra-guild predation (Polis *et al.* 1989), which on the guild level dynamically corresponds to direct density dependence and thus tends to reduce the predation pressure experienced by folivores (Wollkind 1976; T. Oksanen 1990). On the other hand, availability of detritus-based prey and direct energy transfers from plants to carnivores increase the equilibrium standing crop of carnivores and the predation pressure on folivores. Thus, the impacts of the additional assumptions incorporated in the energy shunt hypothesis tend to cancel in the case of relatively productive ecosystems. The behaviour of energy shunt systems along gradients of decreasing primary productivity is radically different from the predictions of the two other hypotheses. In unproductive habitats, where foliage lies close to the ground, carnivores can simultaneously search for herbivores and for detritus-based predators of similar size. This creates intense apparent competition between herbivores and detritus-based predators, and in apparent competition, folivores are handicapped by the load of their internal 'fermentation factory' (see Oksanen 1992). Consequently, decreasing primary productivity leads to elimination of folivores from energy shunt systems even in the absence of direct energy transfers from plants to carnivores.

Standing crops of plants, herbivores and carnivores along gradients of primary productivity have been frequently used in the debate on trophic dynamics (Oksanen *et al.* 1981; Oksanen 1983; McNaughton *et al.* 1989; Arditi & Ginzburg 1989; Arditi *et al.* 1991; Moen & Oksanen 1991; Oksanen *et al.* 1992; Persson *et al.* 1992; Diehl *et al.* 1993; T. Oksanen *et al.* 1995; Akçakaya *et al.* 1995; Crête & Manseau 1996). While many interesting points have been raised by such data, we doubt that they will ever settle the issue. Heavy human impacts on top predators are commonplace, and many populations of terrestrial herbivores are hunted and managed, making the relevance of most data points dubious. Standing crops of

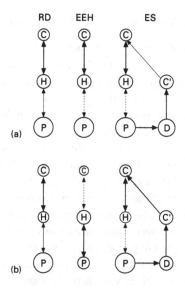

Fig. 18.1. Outlines of food chain dynamics in (a) productive and (b) unproductive terrestrial ecosystems according to the hypotheses of ratio-dependent predation (RD), exploitation ecosystems (EEH) and energy shunts (ES).

plants are normally measured at the end of the growing season, whereas the prediction of constant plant biomass in two-link ecosystems refers to the end of the winter or the dry season. Moreover, the hypothesis of exploitation ecosystems predicts that many two-link ecosystems are characterized by non-equilibrium dynamics (Oksanen *et al.* 1981; L. Oksanen 1990b). In such systems, mean standing crop of plants should correlate positively with primary productivity, even if the theoretical equilibrium standing crop remained constant (Abrams & Roth 1994; Lundberg & Fryxell in press).

Predictions concerning the structure and dynamics of food webs, summarized in Fig. 18.1, are more robust. Statistical analyses of published food webs (Pimm 1982; Briand & Cohen 1987; Cohen *et al.* 1990) have been supportive to the hypothesis of ratio-dependent predation. However, according to a re-analysis of these food webs by L. Oksanen *et al.* (1995), predaceous connections between herbivorous and carnivorous vertebrates become less frequent along the gradient from boreal to high arctic areas. The same trend appears to exist in a mesoscale gradient in northern Fennoscandia. The shortcomings of the chapter, Oksanen *et al.* (1995), include a clear vertebrate bias and heavy reliance on unpublished data from northernmost Fennoscandia. Below, we broaden the taxonomical perspective and provide the reader with access to those primary data that have not yet been published and which are critical to the food webs outlined by L. Oksanen *et al.* (1995).

STUDY AREA AND METHODS

Since 1977, we have studied interactions between plants, microtine rodents and small predators on the tundra plateau of Finnmarksvidda, northernmost Norway. The area of 16 km², where our studies have been conducted since 1986, has been described by Oksanen *et al.* (1992b), L. Oksanen *et al.* (1995) and Oksanen & Schneider (1995). Its southern part consists of lowland tundra, underlain by nutrient-poor Precambrian rocks and dominated by barren lichen heaths and palsa bogs (Oksanen & Virtanen 1995). We have subdivided this lowland area in two parts, the divide (representing typical, barren low arctic tundra landscape), and the valley, lying in the vicinity of the escarpment of the Scandinavian mountain formation and harbouring relatively large patches of more productive habitats (mesic scrublands and mesic birch woodlands). The slope immediately below the escarpment is south facing, moist, nutrient rich and occupied by moderately productive (e.g. mesic heaths) and luxuriant habitats (e.g. herb-rich willow thickets and birch woodlands). Above the escarpment, there is a highland plateau, underlain by nutrient-poor silicaceous nappes and occupied by various types of unproductive alpine vegetation (lichen-rich dwarf birch–crowberry heaths, sedge–cottongrass bogs and moss–dwarf willow snowbeds, see Oksanen & Virtanen 1995).

Plants and microtine rodents have been studied both experimentally and by means of descriptive methods. In the context of these studies, we have also recorded observations of other herbivorous vertebrates (reindeer, mountain hare, willow grouse, rock ptarmigan). The composition of the guild of grazing vertebrates in different subareas has been presented by L. Oksanen *et al.* (1995), the numerical dynamics of microtine rodents during 1977–90 have been summarized by Oksanen and Oksanen (1992). Mammalian predators have been studied by means of snow-tracking and live-trapping (Oksanen & Oksanen 1992; Oksanen *et al.* 1992b; Oksanen & Schneider 1995; Aunapuu, unpublished data). The study area has been annually surveyed for breeding avian predators. In the peak year of 1988, the raptor study area was extended to cover an area of 70 km², and the survey was initiated in the spring, when snow allowed rapid moving and made the arriving predators visible. Since 1993, we have used the same method, but extended the raptor study area to 100 km², and recorded inhabited fox dens. The impact of mammalian predators on lemmings in the highland was studied in spring 1989, immediately after a lemming crash, by recording indications of winter nest predation (fur-lined winter nests, see MacLean *et al.* 1974) and by looking at signs of physical injury (e.g. broken spinal cords) in dead lemmings. When live-trapping microtine rodents, we have also recorded shrews.

We have studied invertebrates by means of pitfall traps in 12 localities, on a transect from the divide through the valley and slope to the highland. Pitfall samples have been identified to species, counted, converted to biomass equivalents using the third power of the body weight as a scaling factor and pooled into four trophic categories (carnivores, herbivores, detritivores and others). L.O. conducted line transect studies of breeding passerine birds and waders in the same general area in 1976, with the Finnish line transect method (transect length 50 km, main belt width 50 m, transects followed co-ordinate lines to yield an unbiased sample of the landscape; for details, see Järvinen and Väisänen 1976). The areas covered by this survey can be regarded as representative of lowland and highland habitats in Fennoscandia, whereas the slope subarea is essentially more productive than the areas covered by the transect lines. We thus conducted similar line transect survey (transect length 2 km) in the most luxuriant part of the slope subarea in June 1995.

HERBIVOROUS VERTEBRATES AND THEIR PREDATORS

The dominant components of the guild of grazing vertebrates are the man-managed reindeer and the microtine rodents. In the luxuriant slope area and in the most productive lowland habitats, microtines have had regular, cyclic density fluctuations, where declines have been accompanied by relatively high numbers of small mustelids (Oksanen & Oksanen 1992). Tracking in these habitats has indicated high mustelid activity (Oksanen *et al.* 1992b). In the 1990s weasels have remained rare, and the period of the vole cycle has increased from 4 to 5–6 years, the typical peak syndrome with high densities in spring has not been observed, and minimum vole densities have been exceptionally high. Except for the years of lowest vole numbers (1990 and 1995), the density of stoats within the slope subarea has consistently exceeded one individual km^{-2}. The number peaked in late summer 1994, when 10 stoats were captured (corresponding to 3.4 stoats km^{-2}). In the valley part of the lowland subarea, stoat density has ranged from 0.2 to 0.8 km^{-2} during our tracking and trapping periods in the 1990s. Both in the lowland and in the valley, the activity of small mustelids has been concentrated in the most productive habitats, but not restricted to them (Table 18.1; Oksanen *et al.* 1992b). The dynamics of the microtine–mustelid part of the grazing chain in the slope and productive valley habitats thus seems to conform to the model of L. Oksanen (1990b), where productive habitats are predicted to be characterized by a stable mustelid–microtine limit cycle (for a slightly different approach to the same issue, see Hanski *et al.* 1991 and Hanski *et al.* 1993). In any case, predators show a persistent presence and the impact of microtines on the vegetation is modest (Moen 1993).

In the divide, mustelid activity has been sporadic, mainly occurring during the final phases of microtine crashes in 1989 and 1995 when densities were low in all parts of the study area, and concentrated in the most productive habitats (Table 18.1; Oksanen *et al.* 1992b). The highest density was recorded in summer 1995, when four stoats were trapped (a density of about one individual km^{-2}).

TABLE 18.1. Activity indices for stoats and weasels for the three sample areas during 1991–95 and for the main habitat categories within them. The indices represent percentages of 14 × 14-m grids visited by small mustelids during a 5-day tracking period. Index values smaller than 1 are denoted by + . Unproductive habitats include lichen heaths, heath snowbeds and open bogs. Moderate habitats include scrublands, heathlands, woodlands with a closed field layer, wetlands with herbs and willows and herb-rich snowbeds. Productive habitats are tall herb meadows plus woodlands and scrublands with tall herbs. The percentage contribution of the habitat categories to each subarea are noted in parentheses.

	Stoats	Weasels
Divide		
Over-all	0	1
Unproductive (96%)	0	+
Moderate (4%)	0	16
Slope-valley		
Over-all	8	+
Unproductive (67%)	4	+
Moderate (28%)	14	0
Productive (5%)	39	10
Highland		
Over-all	0	0
Unproductive (99%)	0	0
Moderate (1%)	0	0

In the highland, the occurrence of small mustelids has been even more sporadic. Occasionally, individuals have been tracked along the margins of the highland, but observations from its interior have been confined to December 1988 (one stoat tracked once), and to August 1995 (two juvenile stoats trapped). In spring 1989, after the crash of the lemming population (Oksanen & Oksanen 1992), we located 193 winter nests in the highland; of these only 13 (7%) showed signs of winter predation. Of 51 lemming bodies recovered, only eight (16%) showed any kind of external injury (wounds or broken spinal cord). Except for two preyed-upon winter nests (found in the same area where the stoat had been tracked), these indications of predation were confined to the marginal parts of the highland (Table 18.2). Within our permanent highland plots, the moss cover was totally destroyed and graminoid cover was reduced in 1989 (Moen *et al.* 1993b).

During the 1990s, medium-sized mammalian predators have been relatively common in our study area. American mink (*Mustela vison*) was first trapped in 1990 (three individuals), and again in 1991 (seven individuals). All mink captures were made along creeks in the lowest part of the slope. Mink tracks have been observed outside the vicinity of creeks in the lower parts of the slope in late autumn and early winter. Red foxes (*Vulpes vulpes*), which had been been largely absent during the early and mid-1980s (probably due to the sarcoptic mange) appeared in

TABLE 18.2. Indications of winter nest predation (a) and signs of predation or scavenging recorded in observed dead bodies of lemmings (b) in spring 1989 in the highland proper and in marginal areas (representing transition towards the slope). The border between the two areas was based on limits between drainage systems. Differences between the two subareas were statistically significant in both cases. For winter nest predation, $\chi^2 = 12.62$, $P < 0.001$; for lemming bodies, $\chi^2 = 15.98$, $P < 0.001$.

(a) Winter nests	Predated	Intact
Interior highland	2	117
Transition	11	63

(b) Lemming bodies	Damaged	Intact
Interior highland	0	32
Transition	8	11

1991. The hunting activity of foxes and mink could be assessed by recording areas where microtine live traps had been overturned systematically even in poorly visible sites. In 1993, five red fox dens were located within the raptor study area or close to its borders (Fig. 18.2). One of these was also within the study area proper (in the highland subarea, in an abandoned arctic fox den). The activity of this litter was concentrated around the piece of luxuriant slope immediately south of the den, where we had to pile stones over small mammal traps in order to continue live trapping. The activity of medium-sized mammalian predators thus appeared to be even more confined to the most productive slope habitats than the activity of small mustelids.

Avian predators had been relatively numerous in 1983 (Oksanen & Oksanen 1992), but the only subsequent year with comparable densities was 1988. Highest densities of avian predators were then observed in the slope, where the dominating species was the rough-legged buzzard (*Buteo lagopus*) (Fig. 18.2; notice that the rough-legged buzzards breeding on the escarpment cliff hunted almost exclusively on the slope and in the valley). However, the most numerous avian predator, the long-tailed jaeger (*Stercorarius longicaudus*), preferred wetlands of the divide. Owls, which had been relatively common during previous peaks (Oksanen & Oksanen 1992) were only represented by a single breeding pair. During the 1990s, the highest density of avian predators was recorded in 1993. Except for merlins (*Falco columbarius*), densities were considerably lower than in 1988 (Fig. 18.2). The diet samples analysed so far indicate that rough-legged buzzards are pronounced microtine specialists (96% of samples included microtine remains, birds and shrews were found in 12% and 8% of samples respectively). For long-tailed jaegers and

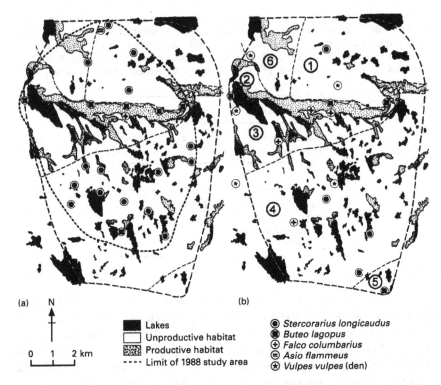

FIG. 18.2. Breeding sites of avian predators in the extended study area in (a) 1988 and (b) 1993, the two summers with highest densities of avian predators between 1986 and 95. Numbers in (b) refer to subdivisions of the extended study area (1, extended highland; 2, extended slope; 3, extended valley; 4, extended divide; 5 and 6 are relatively productive marshy areas, which cannot be directly related to the subdivisions of the main study area.) The merlins attempting to breed on the divide in 1993 failed; other symbols refer to successful breeding.

merlins, microtines and birds are about equally important, occurring in 40–60% of diet samples. Jaegers frequently utilized invertebrates (24%) and plant material (44%) which other predaceous birds did not use at all (see also Andersson 1971, 1976). Dietary data thus amplify the difference in intensity of avian predation of microtines between the slope and other subareas.

INVERTEBRATES, SHREWS AND PASSERINE BIRDS

The dominant trophic groups in all pitfall samples were carnivores and 'others' (Fig. 18.3), the latter group consisting of animals involved in obviously donor-controlled trophic interactions (e.g. feeding on sap or mould). Detritivores were numerous, but their biomass was low due to the predominance of tiny Collembola.

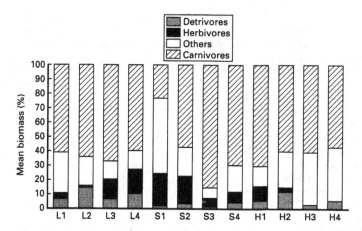

FIG. 18.3. Abundance relationships between main trophic guilds of invertebrates in different parts of the study area according to pitfall trap samples, converted to biomass equivalents. Locations of trapping sites are as follows: L1, southern part of the divide, lichen tundra; L2, valley, lichen tundra; L3, valley, mesic heath; L4, valley, willow scrubland; S1, slope, herb-rich woodland; S2, slope, mesic woodland; S3, slope, timberline; S4, slope, mesic heath; H1, slope-highland transition, mesic health; H2, highland, creekside with low willows; H3, highland, alpine tundra; H4, highland, top plateau.

Herbivores were most abundant in the wooded parts of the slope (sites S1 and S2), moderately abundant in shrubby lowland, slope and highland sites (L3, L4, S3, S4, H1 and H2), rare in typical lowland tundra with scattered dwarf birch shrubs (L1 and L2) and absent from alpine sites with only prostrate vegetation (H3 and H4). Notice that herbivores are more trappable by pitfall traps in tundra habitats, where the dominant taxa (homopterans, hemipterans and chrysomelid beetles; see Solhöy *et al.* 1975) move along the ground, than in the forest, where the dominating taxa (sawfly and moth larvae; Haukioja & Koponen 1975) live in the canopy. The concentration of herbivores in forested slope habitat was thus probably even more pronounced than shown in our data.

Shrews (mainly *Sorex araneus*) were largely confined to the slope; in 1992, the abundance of shrews was at its highest, and we trapped four in the highland and three in the valley, while slope captures were in double figures. Densities of passerine birds declined from luxuriant birch forests to barren tundra habitats, but not dramatically, and this was balanced by an opposite trend in the abundance of waders (Table 18.3), especially in palsa bogs of the lowland. The exceptionally high density and species diversity of waders in the tundra in general, and palsa bogs in particular, is a regional phenomenon (Järvinen & Väisänen 1976, 1978) and obviously detritus based. In terms of primary productivity, palsa bogs are more barren than lichen heaths (Sonesson *et al.* 1975).

TABLE 18.3. Estimated densities of breeding birds (pairs km^{-2}) in the main habitat categories of Finnmarksvidda on the basis of line transect censuses (main belt observations only, main belt width 50 m). In the slope forests, birds observed outside the main belt but no further than 200 m from the transect line are denoted by +.

	SLOPE Forest	Scrub	LOWLAND Bog	Heath	HIGHLAND Bog	Heath
Passerines						
Anthus pratensis	–	8	19	4	17	6
Calcarius lapponicus	–	42	38	43	22	25
Carduelis flammea	+	8	–	–	–	–
Emberiza schoeniculus	+					
Fringilla montifringilla	50	17	–	–	–	–
Luscinia svecica	40	17	19	2	22	–
Oenanthe oenanthe	–	–	–	–	–	3
Phylloscopus trochilus	130	33	8	1	9	–
Plectrophenax nivalis	–	–	–	–	–	2
Turdus iliacus	–	–	3	3	–	–
Total	**220**	**125**	**87**	**53**	**70**	**36**
Waders						
Calidris alpina	–	–	8	–	4	2
C. temmincki	–	–	5	–	–	–
Eudromias morinellus	–	–	–	–	–	2
Philomachus pugnax	–	–	8	1	9	–
Phalaropus lobatus	–	–	5	–	–	–
Pluvialis apricaria	–	–	3	13	4	13
Tringa glareola	–	–	8	–	–	–
T. erythropus	–	–	3	–	–	–
Total	**–**	**0**	**40**	**14**	**17**	**17**
Larids						
Stercorarius longicaudus	–	–	–	1	–	–
Sterna paradisaea	–	–	5	–	–	–
Gallinaceous birds						
Lagopus lagopus	–	17	–	1	–	–
Corvids						
Corvus corone	+	–	–	–	–	–
Raptors						
Falco columbarius	+	–	–	–	–	–
km of transect line	2	2	7	21	5	12

OUTLINES OF FOOD WEBS

The data presented above can be summarized in the form of two food webs (Fig. 18.4). The one for slope, productive valley and divide habitats (Fig. 18.4a) is characterized by three-link trophic dynamics in all parts of the grazing web. The vegetation is little influenced by herbivory and dominated by obviously competition-adapted, broad-leaved plants. Herbivores are exploited by numerous predators, especially in the most luxuriant slope habitats. The dynamics of microtines appear to be changing from weasel–vole limit cycle, which used to be typical for boreal Fennoscandia (L. Oksanen 1990b; Hanski *et al.* 1991, 1993), to seasonal density fluctuations where generalist (fox, mink) and intermediate predators (stoats) play a central role. The absence of typical peak years with high spring densities has

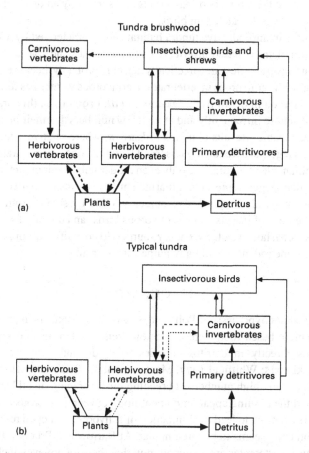

FIG. 18.4. Food-web outlines for the most productive parts of (a) tundra–brushwood landscape at Joatkanjávri and (b) typical low arctic and alpine tundra within the study area. Arrows pointing from resources to consumers represent energy transfers; their thickness indicates the relative magnitude of the flow. Arrows pointing from consumers to resources indicate top-down impacts, their thickness indicates the strength of the top-down control. Dashed lines from consumers to resources indicate potential top-down impacts which are not realized due to top-down controls higher up in the food web.

been reflected by low numbers of avian predators. In the invertebrate branch, detritus-based predators abound, and shrews are likely to feed on both folivores and detritus-based predators. For passerines, the existence of a substantial detritus shunt is less certain, because of three common species, two are foliage gleaners and only one (the brambling *Fringilla montifringilla*) is a partial ground feeder. In somewhat less productive habitats, the detrital shunt is stronger, owing to the higher abundance of ground-feeding birds. The detrital shunt continues to carnivorous vertebrates but can hardly play a major part in their resource supply, as vole numbers normally exceed the numbers of passerines and shrews by an order of magnitude, and voles are easier to catch than birds.

Food webs in unproductive tundra habitats are characterized by a strong energy shunt from detritus-based invertebrates to passerines and waders (Fig. 18.4b). Herbivorous invertebrates are practically absent, probably because of apparent competition with carnivorous invertebrates. Herbivorous vertebrates form a separate branch, presented in detail by L. Oksanen *et al.* (1995a). In this branch, predators are present so infrequently and at such low numbers that their inclusion in the web is questionable. Results obtained at Barrow, where relatively high densities of owls and jaegers are frequently observed (Pitelka 1973) are not transferrable to Fennoscandian inland tundra. The only even moderately common predaceous bird, the long-tailed jaeger, represents a marine shunt in terms of its over-all biology and obtains much of its summer resources via energy shunts from the detritus chain (in the form of insects and insectivorous birds) and directly from the plants. In our study area, however, these energy sources do not suffice to support breeding, unless microtine rodents are already numerous in spring.

DISCUSSION

The energy shunt hypothesis of Polis and Strong (1996) appears to be relevant for the invertebrate-based food web. Its central premise – that most energy fixed by plants goes directly to detritus – applies to arctic and subarctic ecosystems (Wielgolaski 1975; Whitfield 1977; MacLean 1980). This energy flow suffices to maintain relatively high numbers of breeding birds foraging on all kinds of invertebrates, and the ensuing apparent competition between folivorous and predatory invertebrates appears to lead to elimination of folivores from open tundra habitats. The phenomenon seems to be circumpolar. At Barrow and Prudhoe Bay, Alaska, herbivorous invertebrates are so uncommon that they are not even included in the herbivore community of Batzli *et al.* (1980). On Truelove Lowland, Devon Island, Canada, the dominant folivorous insects consumed 0.0001–0.01% of the annual vascular plant production (Ryan & Hergert 1977). At least in our study area, the virtual absence of folivorous invertebrates did not seem to depend on low temperatures *per se*, as the phenomenon was associated with the absence of shrubby vegetation, and was observed both in the relatively warm lowland and on the cooler highland.

On Devon Island, energy shunts support jaegers (though their breeding success was low, see Pattie 1977) and even a few stoats (Riewe 1977), in a system where the densities of lemmings are often so low that predators could not possibly support themselves on herbivores only (Fuller *et al.* 1977). The same seems to apply to the Barrow area, where predation on lemmings can occasionally be intense (Pitelka 1973; MacLean *et al.* 1974; Batzli *et al.* 1980), in spite of the fact that lemming numbers can be very low for long periods, such that predators could not persist without alternative resources. In the inland areas of northern Fennoscandia, detrital energy shunts are insufficient to sustain predators of grazing vertebrates. The same seems to apply to Precambrian Shield areas in the inland of arctic Canada (Krebs 1964). We suspect that the difference reflects distance from the shoreline and the consequent difference in the impact of marine energy shunts (see Polis & Hurd 1995), and that the insufficiency of detrital shunts to support predaceous vertebrates is a generic feature of the tundra. The basic problem is that all arctic insectivores are homeothermic and thus characterized by low ecoenergetic efficiency. Moreover, they are absent in the winter when food would be most needed. In warm deserts, where detritus-based insects are consumed by snakes and lizards, the detrital shunt might be more important.

We agree with Polis and Strong (1996) that the hypothesis of exploitation ecosystems only applies to a small fraction of animal species. However, plants are included in all branches of the grazing chain. As herbivorous vertebrates are resource limited, the plant cover of the tundra is subjected to intense, natural grazing. There is thus no way to understand the ecology of arctic plants without due consideration of grazer–plant interactions, as emphasized already by Tihomirov (1959). Moreover, in spite of their relatively low species diversity, herbivorous mammals are fascinating research objects. The shift from resource limitation to predation control at the transition from arctic to boreal habitats seems to apply to ungulates too (Crête & Manseau 1996), and thus even has implications for game management. Moreover, the feeding ecology and guild structure of arctic mammals can be understood on the basis of classical resource competition, whereas apparent competition seems to be the organizing principle of boreal and temperate communities of grazing mammals (Oksanen 1992). The change in organizing principle is even reflected in the central role of predation at the southern distributional limits of primarily arctic mammals, such as the tundra hare (Holt 1977), the caribou (Bergerud 1988) and the Norwegian lemming (Oksanen 1993).

In research on trophic dynamics, it is useful to see different ideas as abstractions, which are neither perfectly right nor entirely wrong (unless derived in a logically flawed way), but have varying ranges of applicability with shifts in geography, taxonomy, etc. The green world hypothesis of Hairston *et al.* (1960) was one of the great generalizations in ecology, with a wide range of applicability in all respects. It is unlikely that equally big ideas remain to be discovered, and it is counter-productive not to give credit where credit is due. However, even the green world hypothesis has its limitations, which should be vigorously pursued in order to improve our understanding of nature and our ability to prudently manage and protect it. These limitations appear to be especially pronounced in relatively barren areas, where several processes, not included in the green world hypothesis, become critical for trophic dynamics.

ACKNOWLEDGEMENTS

Sincere thanks to Gary Polis and Don Strong, who have freely shared their interesting ideas on trophic dynamics. The data presented here could not have been gathered without the input of a large number of field assistants and practitioners from Umeå, Würzburg and Tartu, and without Helge and Britta Romsdal and Oskar Eriksen, always ready to solve our practical problems. The study has been supported by Naturvetenskapliga Forskningsrådet (Swedish Council for Natural Sciences) and Kungliga Vetenskapsakademin (Royal Academy of Science).

REFERENCES

Abrams, P. (1984). Foraging time optimization and interactions in food webs. *American Naturalist*, 124, 80–96.

Abrams, P. (1993). Effects of increased productivity of the abundance of trophic levels. *American Naturalist*, 141, 351–371.

Abrams, P. & Roth, J. D. (1994). The effects of enrichment of three-species food chains with nonlinear functional responses. *Ecology*, 75, 1118–1130.

Akçakaya, H. R., Arditi, R. & Ginzburg, L. R. (1995). Ratio-dependent predation: an abstraction that works. *Ecology*, 76, 995–1004.

Angerbjörn, A. & Hjernquist, B. (1984). A rapid summer decline in a mountain hare population on an island. *Acta Theriologica*, 29(6), 63–75.

Andersson, M. (1971). Breeding behaviour of the long-tailed skua *Stercorarius longicaudus* (Vieillot). *Ornis Scandinavica*, 2, 35–54.

Andersson, M. (1976). Population ecology of the long-tailed skua *Stercorarius longicaudus* (Vieill.). *Journal of Animal Ecology*, 45, 537–559.

Arditi, R. & Ginzburg L. R. (1989). Coupling in predator–prey dynamics: ratio-dependence. *Journal of Theoretical Biology*, 139, 311–326.

Arditi, R., Ginzburg, L. R. & Akçakaya, H. R. (1991). Variation in plankton densities among lakes: a case for ratio-dependent predation models. *American Naturalist*, 138, 1287–1289.

Atlegrim, O. (1989). Exclusion of birds from bilberry stands: impact on larval density and damage on bilberry. *Oecologia*, 79, 136–139.

Batzli, G. O., White, R. G., MacLean, S. F., Jr, Pitelka, F. A. & Collier, B. D. (1980). The herbivore-based trophic system. *An Arctic Ecosystem: The Coastal Tundra at Barrow, Alaska* (Ed. by J. Brown, P. C. Miller, L. Tieszen & F. Bunnell), pp. 335–410. Dowden, Hutchinson & Ross, Stroudsburg, PA.

Bergerud, A. T. (1988). Caribou, wolves and man. *Trends in Ecology and Evolution*, **3**, 68–72.

Bergman, G. (1970). Skogsharens (*Lepus timidus*) inverkan på björk (*Betula pubescens*) och blåbärsris (*Vaccinium myrtillus*) på en grupp ytterskär. *Memoranda Societatis pro Fauna et Flora Fennica*, **30**, 59–62.

Braithwaite, C. J. R., Taylor, J. D. & Kennedy, W. J. (1973). The evolution of an atoll: the depositional and erosional history of Aldabra. *Philosophical Transactions of the Royal Society of London, B.*, **266**, 307–340.

Briand, F. J. & Cohen, J. E. (1987). Environmental correlates of food chain lengths. *Science*, **250**, 956–960.

Bryant, J. P., Chapin, F. S. & Klein, D. J. (1983). Carbon/nutrient balance in boreal plants in relation to vertebrate herbivory. *Oikos*, **40**, 357–368.

Caughley, G. & Lawton, J. H. (1981). Plant herbivore systems. *Theoretical Ecology: Principles and Applications* (Ed. by R. M. May), pp. 132–166. Blackwell Scientific Publications, Oxford.

Coblentz, B. E., Van Vuren, D. & Main, M. B. (1990). Control of feral goats on Aldabra Atoll. *Atoll Research*, **337**, 1–14.

Cohen, J. E., Briand, F. & Newman, S. (1990). *Community Food Webs: Data and Theory.* Springer, Berlin.

Cousins, S. (1980). A trophic continuum derived from plant structure, animal size and detritus cascade. *Journal of Theoretical Biology*, **82**, 607–618.

Cousins, S. (1987). The decline of the trophic level concept. *Trends in Ecology and Evolution*, **2**, 312–316.

Crawley, M. J. (1983). *Herbivory: The Dynamics of Animal–Plant Interactions.* Blackwell, Oxford.

Crête, M. & Manseau, M. (1996). Top-down and bottom-up forces in food webs. *Evolutionary Ecology*, **10**, 51–62.

Desy, E. A. & Batzli, G. O. (1989). Effects of food availability and predation on prairie vole demography: a field experiment. *Ecology*, **70**, 411–421.

Diehl, S., Lundberg, P. A., Gardfjell, H., Oksanen, L. & Persson, L. (1993). *Daphnia*–phytoplankton interactions in lakes: is there a need for ratio-dependent consumer–resource models? *American Naturalist*, **142**, 1052–1061.

Ellenberg, H. (1978). *Vegetation Mitteleuropas mit den Alpen in ökologischer Sicht. 2. Aufl.* Ulmer, Stuttgart.

Fretwell, S. D. (1972). *Populations in a Seasonal Environment.* Princeton University Press, Princeton.

Fretwell, S. D. (1977). The regulation of plant communities by food chains exploiting them. *Perspectives in Biology and Medicine*, **20**, 169–185.

Fretwell, S. D. (1987). Food chain dynamics: the central theory of ecology? *Oikos*, **50**, 291–301.

Fuller, W. A., Martell, A. M., Smith, R. F. C. & Speller, S. W. (1977). Biology and secondary production of *Dicrostonyx groenlandicus* on Truelove Lowland. *Truelove Lowland, Devon Island, Canada: A High Arctic Ecosystem.* (Ed. by L. C. Bliss), pp. 437–466. University of Alberta Press, Edmonton.

Gibson, C. W. D., Guiford, T. C., Hamber, C. & Sterling, P. H. (1983). Transition matrix models and succession after release from grazing on Aldabra Atoll. *Vegetatio*, **52**, 151–159.

Gimingham, C. H. (1972). *Ecology of Heathlands.* Chapman & Hall, London.

Hairston, N. G., Jr & Hairston, N. G., Sr (1993). Cause–effect relationships in energy flow, trophic structure, and interspecific interactions. *American Naturalist*, **142**, 379–411.

Hairston, N. G., Smith, F. E. & Slobodkin, L. B. (1960). Community structure, population control, and competition. *American Naturalist*, **94**, 421–425.

Hall, S. D. & Raffaelli, D. (1991). Food web patterns: lessons from a species-rich web. *Journal of Animal Ecology*, **60**, 823–842.

Hambler, C. (1984). Goat threat. *New Scientist*, **104**(1432), 46.

Hanski, I. (1987). Pine sawfly populations: patterns, processes, problems. *Oikos*, **50**, 327–335.

Hanski, I., Hansson, L. & Henttonen, H. (1991). Specialist predators, generalist predators, and the microtine rodent cycle. *Journal of Animal Ecology*, **60**, 353–367.

Hanski, I., Turchin, P., Korpimäki, E. & Henttonen, H. (1993). Population oscillations of boreal rodents: regulation by mustelid predators leads to chaos. *Nature*, **346**, 232–235.

Haukioja, E. & Hakala, T. (1975). Herbivore cycles and periodic outbreaks: formulation of a general hypothesis. *Reports from Kevo Subarctic Research Station*, **12**, 1–9.

Haukioja, E. & Koponen, S. (1975). Birch herbivores and herbivory at Kevo. *Fennoscandian Tundra Ecosystems. 2. Animals and Systems Analysis (Ecological Studies 17)* (Ed. by F. E. Wielgolaski), pp. 318–388. Springer, Berlin.

Herms, D. A. & Mattson, W. J. (1992). The dilemma of plants: to grow or defend. *Quarterly Review of Biology*, **67**, 283–335.

Hnatiuk, R. J., Woodell, S. R. J. & Bourn, D. M. (1976). Giant tortoise and vegetation interactions on Aldabra Atoll. Part 2. Coastal. *Biological Conservation*, **9**, 305–316.

Höglund, N. & Eriksson, B. (1973). Förvildade tamrenarnas inverkan på vegetationen inom Lövhögsområdet. *SNV (Stockholm), Research Bulletin 7-61/72.*

Holt, R. D. (1977). Predation, apparent competition, and the structure of prey communities. *Theoretical Population Biology*, **12**, 197–229.

Holt, R. D. (1984). Spatial heterogeneity, indirect interactions, and the coexistence of prey species. *American Naturalist*, **124**, 377–406.

Hunter, M. D. & Price, P. W. (1992). Playing chutes and ladders: heterogeneity and the relative roles of bottom-up and top-down forces in natural communities. *Ecology*, **73**, 724–732.

Järvinen, O. & Väisänen, R. (1976). Species diversity of Finnish birds. II. Biotopes at the transition between taiga and tundra. *Acta Zoologica Fennica*, **145**, 1–35.

Järvinen, O. & Väisänen, R. (1978). Ecological zoogeography of North European waders, or Why do so many waders breed in the north? *Oikos*, **30**, 496–507.

Klein, D. R. (1987). Vegetation recovery patterns following overgrazing by reindeer on St. Mathew Island. *Journal of Range Management*, **40**, 336–338.

Krebs, C. J. (1964). The lemming cycle at Baker Lake, Northwest Territories, during 1959–62. *Arctic Institute of North America, Technical Paper 15.*

Leader-Williams, N. (1988). *Reindeer on South Georgia: The Ecology of An Introduced Population.* Cambridge University Press, Cambridge.

Lundberg, P. & Fryxell, J. M. (1995). Expected population density vs. productivity in ratio-dependent and prey-dependent models. *American Naturalist*, **146**, 153–161.

MacLean, S. F. (1980). The detritus-based trophic system. *An Arctic Ecosystem: The Coastal Tundra at Barrow, Alaska* (Ed. by J. Brown, P. C. Miller, L. Tieszen & F. Bunnell), pp. 411–457. Dowden, Hutchinson & Ross, Stroudsburg, PA.

MacLean, S. F., Fitzgerald, B. M. & Pitelka, F. A. (1974). Population cycles in arctic lemmings: winter reproduction and predation by weasels. *Arctic and Alpine Research*, **6**, 1–12.

McLaren, B. E. & Peterson, R. O. (1994). Wolves, moose, and tree rings on Isle Royale. *Science*, **266**, 1555–1558.

McNaughton, S. J., Oesterheld, M., Frank, D. A. & Williams, K. J. (1989). Ecosystem-level patterns of primary productivity and herbivory in terrestrial ecosystems. *Nature*, **341**, 142–144.

Marquis, R. J. & Whelan, C. J. (1994). Insectivorous birds increase growth of white oak through consumption of leaf-chewing insects. *Ecology*, **75**, 2007–2014.

Martinez, N. (1991). Artifact or attribute: effects of resolution on the Little Rock Lake food web. *Ecological Monographs*, **61**, 367–392.

Menge, B. & Sutherland, J. (1976). Species diversity gradients: synthesis of the roles of predation, competition and temporal heterogeneity. *American Naturalist*, **110**, 351–369.

Merton, L. F. H., Bourn, D. M. & Hnatiuk, R. J. (1976). Giant tortoise and vegetation interactions on Aldabra Atoll. I. Inland. *Biological Conservation*, **9**, 293–304.

Moen, J. (1993). *Herbivory and plant community structure in a subarctic altitudinal gradient.* PhD Thesis, University of Umeå, Sweden.

Moen, J. & Oksanen, L. (1991). Ecosystem trends. *Nature*, **353**, 510.

Moen, J., Gardfjell, H., Oksanen, L., Ericson, L. & Ekerholm, P. (1993a). Grazing by food-limited rodents on a productive experimental plant community: does the 'green desert' exist? *Oikos*, **68**, 401–403.

Moen, J., Lundberg, P. A. & Oksanen, L. (1993b). Lemming grazing on snowbed vegetation during a population peak, northern Norway. *Arctic and Alpine Research*, **25**, 130–135.

Odum, H. T. (1971). *Fundamentals of Ecology*. W. B. Saunders, Philadelphia.

Oksanen, L. (1983). Trophic exploitation and arctic phytomass patterns. *American Naturalist*, **122**, 45–52.

Oksanen, L. (1990a). Predation, herbivory, and plant strategies along gradients of primary productivity. *Perspectives on Plant Competition* (Ed. by D. Tilman & J. B. Grace), pp. 445–474. Academic Press, San Diego.

Oksanen, L. (1990b). Exploitation ecosystems in seasonal environments. *Oikos*, **57**, 14–24.

Oksanen, L. (1992). Evolution of exploitation ecosystems. I. Predation, foraging ecology and population dynamics of herbivores. *Evolutionary Ecology*, **6**, 15–33.

Oksanen, L. & Oksanen, T. (1992). Long-term microtine dynamics in north Fennoscandian tundra: the vole cycle and the lemming chaos. *Ecography*, **15**, 226–236.

Oksanen, L. & Virtanen, R. (1995). Topographic, altitudinal, and regional patterns in continental and suboceanic heath vegetation of northernmost Fennoscandia. *Acta Botanica Fennica*, **15**, 1–80.

Oksanen, L., Fretwell, S. D., Arruda, J. & Niemelä, P. (1981). Exploitation ecosystems in gradients of primary productivity. *American Naturalist*, **118**, 240–261.

Oksanen, L., Oksanen, T., Lukkari, A. & Sirén, S. (1987). The role of phenol-based inducible defense in the interaction between tundra populations of the vole *Clethrionomys rufocanus* and the dwarf shrub *Vaccinium myrtillus*. *Oikos*, **50**, 371–380.

Oksanen, L., Moen, J. & Lundberg, P. A. (1992). The time scale problem in exploiter–victim models: does the solution lie in ratio-dependent exploitation? *American Naturalist*, **140**, 938–960.

Oksanen, L., Oksanen, T., Ekerholm, P., Moen, J., Lundberg, P., Schneider, M. & Aunapuu, M. (1995). Structure and dynamics of arctic-subarctic grazing webs in relation to primary productivity. *Food Webs: Integrating Structure and Dynamics* (Ed. by K. Winemiller & G. Polis), pp. 231–242. Chapman & Hall, London.

Oksanen, T. (1990). Exploitation ecosystems in heterogeneous habitat complexes. *Evolutionary Ecology*, **4**, 220–234.

Oksanen, T. (1993). Does predation prevent Norwegian lemmings from establishing permanent populations in lowland forests? *The Biology of Lemmings* (Ed. by N. C. Stenseth & R. A. Ims), pp. 425–437. Linnean Society, London.

Oksanen, T. & Schneider, M. (1995). Predator–prey dynamics as influenced by habitat heterogeneity. *Landscape Approaches in Mammalian Ecology and Conservation* (Ed. by W. Z. Lidicker), pp. 122–150. University of Minnesota Press, Minneapolis, MN.

Oksanen, T., Oksanen, L. & Gyllenberg, M. (1992a). Exploitation ecosystems in heterogeneous habitat complexes. II. The impact of small-scale spatial heterogeneity on predator-prey dynamics. *Evolutionary Ecology*, **6**, 383–398.

Oksanen, T., Oksanen, L. & Norberg, M. (1992b). Habitat use of small mustelids in north Fennoscandian tundra: a test of the hypothesis of patchy exploitation ecosystems. *Ecography*, **15**, 237–244.

Oksanen, T., Power, M. E. & Oksanen, L. (1995). Ideal free habitat selection and consumer-resource dynamics. *American Naturalist*, **146**, 565–585.

Ostfeld, R. S. (1994). The fence effect reconsidered. *Oikos*, **70**, 340–348.

Ostfeld, R. S. & Canham, C. D. (1993). Effects of meadow vole population density on tree seedling survival in old fields. *Ecology*, **74**, 179–801.

Paine, R. T. (1988). Food webs: road maps of interactions or grist for theoretical development? *Ecology*, **69**, 1648–1654.

Paine, R. T. (1992). Food web analysis: field measurements of *per capita* interaction strength. *Nature*, 355, 73–75.

Pattie, D. L. (1977). Population levels and bioenergetics of arctic birds on Truelove Lowland. *Truelove Lowland, Devon Island, Canada: A High Arctic Ecosystem* (Ed. by L. C. Bliss), pp. 413–436. University of Alberta Press, Edmonton.

Persson, L., Diehl, S., Johansson, L., Andersson, G. & Hamrin, S. (1992). Trophic interactions in temperate lake ecosystems. *American Naturalist*, 140, 59–84.

Pettersson, B. (1959). Dynamik och konstans i Gotlands flora och vegetation. *Acta Phytogeographica Suecica*, 40, 1–288.

Pimm, S. L. (1982). *Food webs*. Chapman & Hall, London.

Pitelka, F. A. (1973). Cyclic pattern in lemming populations near Barrow, Alaska. *Arctic Institute of North America, Technical Paper*, 25, 199–215.

Pokki, J. (1981). Distribution, demography and dispersal of the field vole (*Microtus agrestis*) in the Tvärminne archipelago, Finland. *Acta Zoologica Fennica*, 164, 1–48.

Polis, G. A. (1994). Food webs, trophic cascades and community structure. *Australian Journal of Ecology*, 19, 121–136.

Polis, G. A. & Hurd, S. D. (1995). Allochtonous inputs across habitats, subsidized consumers, and apparent trophic cascades: examples from the ocean–land interface. *Food Webs: Integrating Structure and Dynamics* (Ed. by K. Winemiller & G. Polis), pp. 275–285. Chapman & Hall, London.

Polis, G. A. & Strong, D. R. (1996). Food web complexity and community dynamics. *American Naturalist*, 147, 813–846.

Polis, G. A., Myers, C. A. & Holt, R. D. (1989). The ecology and evolution of intraguild predation: potential competitors that eat up each other. *Annual Review of Ecology and Systematics*, 20, 297–330.

Polis, G. A., Holt, R. D., Menge, B. A. & Winemiller, K. O. (1995). Time, space and life history: influences on food webs. *Food Webs: Integrating Structure and Dynamics* (Ed. by K. Winemiller and G. Polis), pp. 435–460. Chapman & Hall, London.

Power, M. E. (1990). Effect of fish in river food webs. *Science*, 250, 411–415.

Power, M. E. (1992). Top-down and bottom-up forces in food webs: do plants have primacy? *Ecology*, 73, 733–746.

Price, P. W., Bouton, C. E., Gross, P., McPheron, B. A., Thompson, J. N. & Weis, A. E. (1980). Interactions among three trophic levels: influence of plants on interactions between insect herbivores and natural enemies. *Annual Review of Ecology and Systematics*, 11, 41–65.

Raffaelli, D. & Hall, S. (1995). Recurrent properties across terrestrial and aquatic webs. *The Bulletin* (British Ecological Society), 26(3), 181–184.

Rhoades, D. F. (1985). Offensive–defensive interactions between plants: their relevance in herbivore population dynamics and ecological theory. *American Naturalist*, 125, 205–238.

Riewe, R. R. (1977). Mammalian carnivores utilizing Truelove Lowland. *Truelove Lowland, Devon Island, Canada: A High Arctic Ecosystem* (Ed. by L. C. Bliss), pp. 413–436. University of Alberta Press, Edmonton.

Rosenzweig, M. L. (1971). Paradox of enrichment: destabilization of exploitation ecosystems in ecological time. *Science*, 171, 385–387.

Rosenzweig, M. L. (1973). Exploitation in three trophic levels. *American Naturalist*, 107, 275–294.

Rosenzweig, M. L. (1977). Aspects of biological exploitation. *Quarterly Review of Biology*, 52, 371–380.

Ryan, J. K. & Hergert, C. R. (1977). Energy budget for *Gynaephora groenlandica* (Homeyr) and *G. rossii* (Curtis) (Lepidoptera, Lymantriidae) on Truelove Lowland. *Truelove Lowland, Devon Island, Canada: A High Arctic Ecosystem* (Ed. by L. C. Bliss), pp. 395–409. University of Alberta Press, Edmonton.

Schmitz, O. J. (1993). Trophic exploitation in grassland food chains: simple models and a field experiment. *Oecologia*, 93, 327–335.

Seldal, T., Andersen, K.-J. & Högstedt, G. (1994). Grazing-induced proteinase inhibitors: a possible cause for lemming population cycles. *Oikos*, **70**, 3–11.

Sjörs, H. (1971). *Ekologisk botanik*, Almquist & Wiksell, Stockholm.

Soikkeli, M. & Virtanen, J. (1975). Ulkosaariston huipputiheät jäniskannat. (The extremely dense hare populations of the outer archipelago.) *Metsästys ja Kalastus*, **64**(3), 16–19.

Solhöy, T., Östbye, E., Kauri, H., Hagen, A., Lien, L. & Skar, H.-J. (1975). *Fennoscandian Tundra Ecosystems. 2. Animals and Systems Analysis (Ecological Studies 17)* (Ed. by F. E. Wielgolaski), pp. 28–45. Springer, Berlin.

Sonesson, M., Wielgolaski, F. E. & Kallio, P. (1975). Description of Fennoscandian tundra ecosystems. *Fennoscandian Tundra Ecosystems. 1. Plants and Microorganisms (Ecological Studies 16)* (Ed. by F. E. Wielgolaski), pp. 3–28. Springer, Berlin.

Spiller, D. A. & Schoener, T. W. (1990). A terrestrial field experiment showing the impact of eliminating top predators on foliage damage. *Nature*, **347**, 469–472.

Stoddart, D. R. (1971a). Land vegetation of Diego-Gardia. *Atoll Research Bulletin*, **149**, 127–142.

Stoddart, D. R. (1971b). Terrestrial fauna of Diego-Garcia and other Chagos atolls. *Atoll Research Bulletin*, **149**, 163–170.

Strong, D. R. (1992). Are trophic cascades all wet? Differentiation and donor-control in speciose ecosystems. *Ecology*, **73**, 747–754.

Strong, D. R. & Larsson, S. (1994). Is the evolution of herbivore resistance influenced by parasitoids? *Parasitoid Community Ecology* (Ed. by B. Hawkins & W. Sheenan), pp. 261–276. Oxford University Press, Oxford.

Strong, D. R., Maron, J. L. & Connors, P. G. (1995a). Top down from underground? The unappreciated influence of subterranean food webs on above-ground ecology. *Food Webs: Integrating Structure and Dynamics* (Ed. by K. Winemiller & G. Polis), pp. 170–175. Chapman & Hall, London.

Strong, D. R., Maron, J. L., Connors, P. G., Whipple, A., Harrison, S. & Jeffries, R. L. (1995b). High mortality, fluctuating numbers, and heavy subterranean insect herbivory in bush lupine, *Lupinus arboreus*. *Oecologia*, **104**, 85–92.

Tihomirov, B. A. (1959). Vzajmosvjazi životnogy mira i rastitel'nogo pokrova tundry. Trudy Botaničeskij institut V.L. Komarova, Akademija Nauk SSSR, Moscow.

Walter, H. (1964). *Die Vegetation der Erde in öko-physiologischer Betrachtung. I. Die tropischen und subtropischen Zonen*. Gustav Fischer, Jena.

Walter, H. (1968). *Die Vegetation der Erde in öko-physiologischer Betrachtung. II. Die gemässigten und arktischen Zonen*. Gustav Fischer, Jena.

Werth, E. (1928). Überblick über die Vegetationsgliederung von Kerguelen sowie von Possession–Eiland (Crozet–Gruppe) und Heard–Eiland. *Deutsche Südpolar-Expedition 1901–1903*. Teil 8. (Ed. by E. Von Drygalski), pp. 127–176. De Gryuter, Berlin.

White, T. R. C. (1978). The importance of relative shortage of food in animal ecology. *Oecologia*, **33**, 71–86.

Whitfield, D. W. A. (1977). Energy budgets and ecological efficiencies on Truelove Lowland. *Truelove Lowland, Devon Island, Canada: A High Arctic Ecosystem* (Ed. by L. C. Bliss), pp. 607–630. University of Alberta Press, Edmonton.

Wielgolaski, F. E. (1975). Functioning of Fennoscandian tundra ecosystems. *Fennoscandian Tundra Ecosystems. 2. Animals and Systems Analysis (Ecological Studies 17)* (Ed. by F. E. Wielgolaski), pp. 301–326. Springer, Berlin.

Wollkind, D. J. (1976). Exploitation in three trophic levels: an extension allowing intraspecies carnivore

19. COMPARTMENTALIZATION OF RESOURCE UTILIZATION WITHIN SOIL ECOSYSTEMS

J. C. MOORE* AND P. C. DE RUITER†

Department of Biological Sciences, University of North Colorado, Greeley, CO 80639 USA, and Natural Resource Ecology Laboratory, Colorado State University, Fort Collins, CO 80523, USA and †Research Institute for Agrobiology and Soil Fertility (AB-DLO), PO Box 129, 9750 AC Haren, The Netherlands

INTRODUCTION

The interplay of diversity, complexity and stability has been a recurrent theme in modern ecology. In his now classic paper, Hutchinson (1959) asked 'why are there so many kinds of animals?' and concluded that in part the answer was 'because a complex trophic organization of a community is more stable than a simple one'. As an afterthought, Hutchinson noted that, within communities, species tended to interact as aggregates in niche space. As these aggregations of species formed during the development of a community, they became more complex and more stable. While the notion that complexity favours stability was tenuous, the afterthought represented an early recognition that complex systems possess emergent properties (Simon 1962; Koestler 1967; Allen & Starr 1982; O'Neill et al. 1986) and that these properties were related to stability.

The conclusion that complex systems are more stable than simple ones was challenged by Gardner and Ashby (1970). Their studies of model systems demonstrated that, as the diversity and complexity of the models increased, the models were less likely to be stable. May (1972) applied the analyses to ecological communities and obtained similar results. However, if the trophic interactions among species were aggregated into 'blocks', with strong interactions within the blocks and weak interactions among the blocks, the models were more likely to be stable than models of a random assemblage of species of similar diversity and complexity. One interpretation of these results is that certain types of complexity favour stability. As a community increases in diversity and complexity, fewer of the possible combinations of interactions among species would produce a stable community. The subset of all possible combinations of interactions among species that produces stable communities is the compartmentalized type. The aggregation of the interactions amongst species into compartments is a type of complexity that favours stability.

The 'blocks' that May (1972) referred to were based on the arrangement of terms within the Jacobian matrix (community matrix) derived from the system of differential equations used to describe the dynamics of the species within the community. These elements of the matrix represent the interaction strengths that quantify the per capita effects of each species on the dynamics of others. A compartment within a community is defined by the patterning of the interaction strengths within the community matrix. Compartments are most likely to be formed along the principal niche axes of habitat, food and time (*sensu* Schoener 1974) – that is, compartments are a product of overlap in the utilization of resources; for example, soil invertebrates that feed on similar species of fungi, that are active during similar time periods and share habitat, are more likely to interact amongst themselves than species that feed on different prey, are active at different time periods and utilize different habitats. Hence, the prey, the invertebrates and the predators of the invertebrates would form a compartment within the community.

Our objectives are to present evidence that resource utilization by species is organized within communities in a compartmentalized fashion that favours stability. We begin by presenting an annotated history of the concept of resource compartments within communities. Next, we pose three questions. First, do the interactions among species within communities form compartments? We present evidence that soil communities form interactive assemblages of species that share common niche space. We draw from a rich data set that includes studies of soils from natural ecosystems and agricultural ecosystems across the globe. Second, are the assemblages of species observed in soils unique to soil communities or do they share features that are common to communities within other habitats? We argue that the assemblages of species observed in soils share features that are common to other ecosystems. We compare characteristics of the descriptions of our soil food webs to the descriptions of food webs in the literature. Third, why are communities compartmentalized as assemblages of species? We submit that communities that are compartmentalized in this manner are more likely to be stable than ones that are not. Finally, we propose a general hypothesis that integrates ecosystem structure, ecosystem energetics and stability; namely, that the energetic organization of communities forms the basis of ecosystem stability.

COMPARTMENTS WITHIN COMMUNITIES

The work of Hutchinson (1959) and May (1972) led to several studies that strongly indicated that communities were organized into assemblages of species. Pimm and Lawton (1980) defined a community that possessed assemblages of species as being *compartmented* and one that did not possess compartments as being *reticulate* (see Fig. 19.1). Within a compartmented community, species would form interactive assemblages based on food type, habitat usage and time that they are active, while

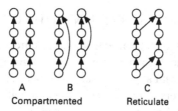

FIG. 19.1. Simple binary representations of compartmented (A and B) and reticulate (C) food webs.

a reticulate community would not possess this type of organization. In terms of classic niche theory, if the resource utilization of each species were projected into their respective niche hypervolumes and the compartmented hypothesis were true, we could envision clusters of species within niche space. Alternatively, under the reticulate hypothesis, resource utilization would be represented as a uniform cloud within niche space.

Most ecologists agree that communities exhibit some degree of compartment-alization. The debate seems to be centred on the reasons that communities are compartmented. Hutchinson (1959) and May (1972) proposed that compartmented systems were more stable than reticulate systems based on empirical observations, deduction, and on the results of Monte Carlo simulations using abstract models. Pimm and Lawton (1980) analysed food-web descriptions taken from the literature and argued that compartments within communities were rare, and that if they were present, they occurred for biological reasons rather than through population dynamics (stability). For example, marine shore birds forage within the littoral zone, but reproduce and are preyed upon within the terrestrial environment. A description of the food web that includes the shore birds might well include the intertidal habitat and the terrestrial habitat (habitat-based compartments). For terrestrial systems, the above-soil portion of the community that begins with the primary production of plants is often viewed as being separate from the soil portion of the community that is based largely on detritus, for example leaf litter, dead roots, faeces, and carcasses (food-based compartments). The biological reasons that could generate a compartmented system correspond to the dominant niche axes of food, habitat, and time, discussed above. From this point of view, resource utilization (biological reasons) may form the basis of compartments within communities, but it is seen as a constraint that is separate from that imposed by population dynamics (stability).

Is resource utilization related to stability? To answer this question, we will make use of field data from soils and simulation models of nutrient flow within these ecosystems to estimate population densities, the rates of material flow associated with each trophic interaction, and the per capita effects (interaction strengths) of each species on the dynamics of the other species.

COMPARTMENTS WITHIN SOIL COMMUNITIES

Moore and Hunt (1988) hypothesized that food resources at the base of food webs (basal resources) initiated trophic compartments termed 'energy channels'. An energy channel begins with a basal resource, includes all species that feed on the basal resource, consumers of these species, and ultimately terminates with a top predator (Cohen 1978). The food webs depicted in Fig. 19.2 possess roots and detritus as basal resources. An energy channel may take the form of a simple food

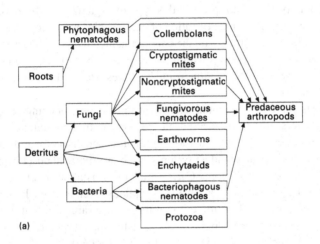

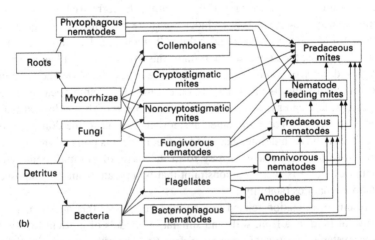

FIG. 19.2. *Above and opposite.* Soil food webs from four sites across the globe. (a) Horseshoe Bend Experimental Farm, GA, USA (Hendrix *et al.* 1986); (b) Central Plains Experimental Range (CPER), Nunn, CO, USA (Hunt *et al.* 1987).

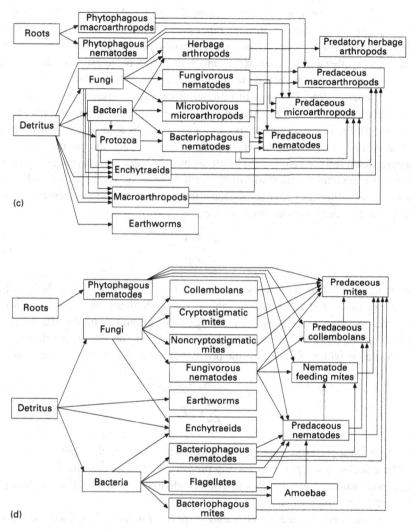

FIG. 19.2. *Continued.* (c) Kjettslinge Experimental Farm, (Sweden (Andrén *et al.* 1990); (d) Lovinkhoeve Experimental Farm, Markinesse, The Netherlands (Brussaard *et al.* 1988). These represent the binary descriptions referred to in the text. The webs are described in terms of functional groups of species based on feed sources, feeding mode, life history and habitat use (Moore *et al.* 1988).

chain as seen in the root pathway of the Horseshoe Bend food web (root to phyto-phagous nematodes to predaceous arthropods, see Fig. 19.2a), or a more highly interactive subweb as illustrated by the detritus energy channel within the Kjettslinge food web (Fig. 19.2c). The root and detritus energy channels are forms of the grazer and detritus pathways (Odum 1971; Marples 1966).

The fraction of energy a species obtains from a basal resource either directly by feeding on it or indirectly through other prey is a measure of the species membership in that energy channel (Moore & Hunt 1988; Moore *et al.* 1988). To estimate a species membership in an energy channel, we first have to estimate the flow of energy among species within the whole food web. We will use nitrogen flow as an index of energy flow. The flow of nitrogen from a basal resource to and among consumers within an energy channel is estimated with simulation models (Hunt *et al.* 1987). Feeding rates among species are estimated by assuming the populations are at steady state. Under the assumption of steady state, the population death (natural death and predation) must be off-set by population growth for each species. Hence, the amount of material lost to death must be overcome by enough consumption to compensate for the loss, the efficiency at which prey is consumed, and the costs associated with metabolism. Let:

$$F_j = \frac{d_j B_j + P_j}{a_j p_j}$$ (1)

where for species j, F_j represents the feeding rate (gN m^{-2} year^{-1}), d_j is the specific death rate (year^{-1}), B_j is the density (gN m^{-2}), P_j is the death rate due to predation, a_j is the assimilation efficiency (percentage of killed prey assimilated), and p_j is the production efficiency (percentage of assimilated prey converted to new biomass). For predators with multiple prey, the preference for each prey type is considered:

$$F_{ij} = \frac{w_{ij} B_j}{\sum_{k=1}^{n} w_{kj} B_K} F_j$$ (2)

where w_{ij} is the preference weighting factor for prey i among the n prey of predator j. The calculations begin with the top predators, since we assume that no other species consumes them, i.e. $P_j = 0$. The calculations proceed backwards through the web to the basal resources.

We estimated the feeding rates among species for each of the food webs presented in Fig. 19.2 and the membership of each functional group in the root, detritus, bacterial and fungal energy channels (Table 19.1). We will use the predatory arthropods that are top predators in each of the food webs to illustrate the concept. From the diagrams, pathways that end at the top predators can be traced back to either roots or detritus. Hence, the biomass of the predaceous arthropods within each web can be thought of as either root or detritus in origin. For predaceous mites at the Central Plains Experimental Range (CPER), of the 1.4 mg N m^{-2} year^{-1} consumed, 21.9% originated from plant roots and 78.1% from detritus (see Table 19.1). The detritus energy channel can be further separated into a bacterial and

TABLE 19.1. The estimated proportion of energy that each taxon derives from bacteria, fungi and roots for the Central Plains Experimental Range, CO, USA (Hunt *et al.* 1987) and the Lovinkhoeve Experimental Farm, The Netherlands (De Ruiter *et al.* 1995), (see Fig. 19.2). The estimates are based on nitrogen and carbon flux rates as calculated by equations 1 and 2. Energy that originates from bacteria, fungi and roots passes through distinct assemblages of species referred to as energy channels (Moore & Hunt 1988).

| | Energy channels | | | | | |
| | Central Plains Experimental Range | | | Lovinkhoeve Experimental Farm | | |
Functional group	Bacteria	Fungi	Roots	Bacteria	Fungi	Roots
Protozoa						
Flagellates	100	0	0	100	0	0
Amoebae	100	0	0	100	0	0
Nematodes						
Phytophages	0	0	100	0	0	100
Fungivores	0	90	10	0	100	0
Bacteriavores	100	0	0	100	0	0
Omnivores	100	0	0	— With bacteriavores —		
Predators	68.7	3.5	27.8	89.1	10.6	0.4
Microarthropods						
Fungivorous Collembola	0	90	10	0	100	0
Fungivorous oribatid mites	0	90	10	0	100	0
Fungivorous prostigmatid mites	0	90	10	0	100	0
Nematophagous mites	66.7	3.8	29.5	53.9	13.9	32.3
Predaceous Collembola	— Not in description —			16.2	24.7	59.2
Predaceous mites	39.5	38.6	21.9	22.0	64.9	13.1
Annelids				Detritus	Bacteria	Fungi
Enchytraeids	— Not in description —			52.1	47.4	0.4

fungal channel when the habitat usage and temporal dynamics of the primary consumers of detritus and detritivores are considered. Organisms within the bacterial energy channel tend to be small, aquatic or dependent on water films, and have short generation times and rapid turnover rates when compared to organisms within the fungal energy channel. The predaceous mites at the CPER obtain 39.5% from bacteria and 38.6% from fungi (the near equal percentages were coincidental).

Species that are absent from the diagram are as informative about the structure of communities as those that are present. The root and detritus energy channels also provide insight into linkages with the ecosystem above ground. There is a direct linkage from plant roots to plant shoots to herbivores in each system. Furthermore, the earthworms present in the soils at Horseshoe Bend (Fig. 19.2a), Kjettslinge (Fig. 19.2c) and The Lovinkhoeve (Fig. 19.2d) are preyed upon by birds and small mammals. Additional linkages occur via predaceous arthropods to their consumers above ground (e.g. centipedes, ground beetles, etc.).

The compartmentalization of trophic interactions is not unique to soil communities (Root 1967; Cummins 1973, 1974). There is a clear parallel between the microbial loop described in the aquatic ecology literature, the bacterial energy channel within soils and the complex interaction between plants, microbes and microbivores that occurs within the rhizosphere along plant roots (Pomeroy 1970; Elliott et al. 1980; Clarholm 1985; Coleman 1994). The similarity goes beyond the compartmentalization of energy inputs, but includes the utilization of the inputs as well. For both soil and lentic aquatic systems in temperate regions, the ratio of production to respiration that results from the activities of the producer-based energy channels and the detritus-based energy channels parallel one another on a seasonal basis. Moreover, many of the published food-web descriptions (including those in Fig. 19.2) are arguably descriptions of habitat compartments. Food-web descriptions often exclude top predators or consumers that do not reside in the habitat, but that may exploit the habitat for prey items (Briand & Cohen 1989; Moore et al. 1989; Polis et al. 1989). For example, many descriptions of the rocky intertidal focus on the marine invertebrates and algae within the habitat. These descriptions do not include marine mammals and fish, or birds and land mammals that forage within the intertidal during the high and low tide cycles.

RESOURCE COMPARTMENTS AND SYSTEM STABILITY

To this point, the compartments we have identified could be defined on the basis of biological reasons (*sensu* Pimm 1982) rather than for reasons of dynamic stability (*sensu* May 1972). For example, the separation of energy pathways into detritus-based and root-based energy channels may represent nothing more than the development of pathways that are based on the two fundamental means by which organisms utilized energy – aerobic respiration and photosynthesis. The division of the detritus energy channel into the bacterial and fungal energy channels could be explained by the physiological and morphological attributes of their respective taxa. Organisms within the bacterial energy channel lack or possess poorly developed integuments making them susceptible to desiccation, while fungi and their consumers possess more highly developed integuments or external walls to thwart desiccation. The predators that unite the energy channels are adapted to attack morphologically similar prey regardless of where the prey might obtain energy, making the webs appear highly reticulate at intermediate trophic levels (Fig. 19.2).

Many of the biological reasons used to explain the compartmentalized structure of communities have also been shown to affect population dynamics and stability; for example, the dynamics of the detritus energy channel are donor controlled (detritus has no specific birth rate *per se*), while the dynamics of the root energy channel fit the more traditional Lotka–Volterra form (May 1972; Pimm 1982; Moore *et al.* 1993). Additionally, theoretical models of the population dynamics of species that possess differences in enrichment rates (primary productivity or detritus inputs), generation times and turnover times that parallel the bacterial and fungal energy channels differ in their local stability, resilience to minor disturbances and the steady-state densities of the species (Rosenzweig 1971; May 1973; DeAngelis *et al.* 1989; Moore *et al.* 1993).

Does the compartmentalized structure described above favour stability? De Ruiter *et al.* (1995) compared the stability of the food webs presented in Fig. 19.2 (life-like webs) to webs of similar structure, but with random selections of plausible interaction strengths (random webs). This analysis did not affect the positioning of zero and non-zero elements of the community matrices for each food web. The life-like webs were more stable than the random webs, suggesting that life-like webs possess structures that favour stability. The question that remains open is – *what* attribute of community organization favours stability?

The results of De Ruiter *et al.* (1995) suggest that there is an added dimension to the nature of compartmentalization of ecosystems. Recall that compartments are based on the pairing of species and the strengths of those pairings in terms of the effects that each species has on each other's dynamics. The food-web diagrams describe the pairings amongst species, but do not provide estimates of the strengths of interaction. The estimates of the flow of nitrogen indicate that communities are compartmentalized in terms of energy utilization, although the connection between energy utilization, interaction strength and stability has yet to be established. When the estimates of interaction strengths and the fluxes of nitrogen are arranged by trophic position an interesting pattern emerges (Fig. 19.3). The bulk of material flux occurs at the lower trophic positions and decreases with increased trophic position ($P \leq 0.05$). As for the interaction strengths, the negative per capita effects of predators on prey were disproportionately large when compared to the net positive

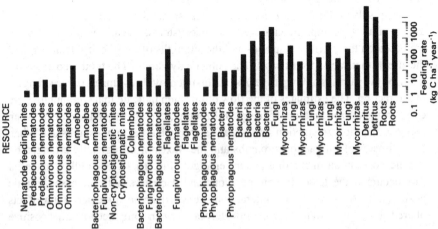

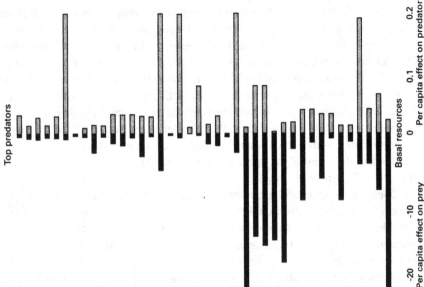

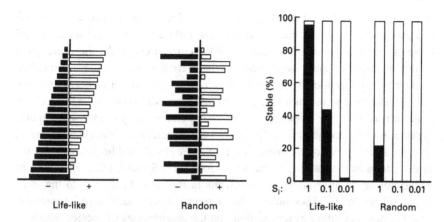

FIG. 19.4. Stability of food webs that possess patterning of interaction strength with trophic position (life-like webs) vs. webs with the same interactions but a random permutation of interaction strength with trophic position (random webs). The random and theoretical webs were evaluated under three different levels of self-limitation (diagonal terms of the community matrix).

per capita effects of prey on predators ($P \leq 0.05$) for the lower trophic positions with the converse being true at higher positions.

The results presented above and those of the analysis of the stability of the life-like and random webs provide the linkage between material flow rates, interaction strength, and stability (Fig. 19.4). The energetic organization of communities forms the basis of ecosystem stability, and an important aspect of community organization that favours stability is the compartmentalization of the utilization of resources.

RESOURCE COMPARTMENTALIZATION AND HIERARCHICAL ORGANIZATION

Several authors have postulated that the compartmentalized structure described above represents a hierarchical type of organization (Simon 1962; Koestler 1967; Allen & Starr 1982; O'Neill *et al.* 1986). A feature of the soil webs that supports this conclusion is the emergent property that stability is enhanced by the energetic organization of the system outlined above. Each of the pair-wise interactions used

FIG. 19.3. *Opposite.* Relationship between feeding rates and interaction strength (per capita effects) and the trophic position of the predator–prey interaction for all interactions within the CPER food web (Fig. 19.2b). Resources are given on the left of the figure and their consumers on the right. The stacked bar graph to the left depicts the feeding rates for each predator–prey interaction, while that on the right depicts the strengths associated with each interaction. These interaction strengths are the off-diagonal elements of the community matrix and were estimated using the procedures outlined in De Ruiter *et al.* (1995), and Moore *et al.* (1996).

in the analysis for both the real and random food webs were feasible (all species possessed positive steady states) and plausible (all parameters were selected from those observed in nature). The stability of the community could not be predicted by piecing together pair-wise interactions that were known to be stable, but rather, could only be understood by observing all pair-wise interaction together as a system.

The soil food webs possess two additional properties common to hierarchical organization. First, we proposed that the energy channels represent compartments or subunits of the ecosystems. If the system were organized in a hierarchical manner, then the stability of compartments within food webs should be governed by the same criteria as the stability of the whole food web. Second, minor disturbances would not necessarily affect compartments in the same fashion or to the same degree (Pimm 1982). This should not be interpreted as a repudiation of the argument raised above, but rather as recognition that the assemblages of species within an energy channel could respond to a minor disturbance as a unit. We restrict our comments to minor disturbances, since a large disturbance has the potential to overwhelm the entire community.

Are energy channels structured in the same manner as whole webs, and are the stability of energy channels and whole webs governed in the same way? An analysis of the 40 food webs collected from the literature by Briand (1983) reveals that energy channels are structured in the same fashion as the webs that they originated from (Moore & Hunt 1988). These 40 food webs possessed 138 separate energy channels. The majority of the energy channels originated from primary producers (63% algae, bryophytes or vascular plants). Detritus accounted for 20% of the basal resources, while consumers accounted for 17%. Moore and Hunt (1988) calculated the connectance (C), species richness (S), and maximum interaction strength (i) for each food web and energy channel. Connectance was measured as the proportion of off-diagonal non-zero elements of the community matrix. Species richness was estimated as the number of species or functional groups used in the description. Maximum interaction strength was obtained from the inequality established by May (1972) for food-web stability $\{i(SC)^{1/2} < 1\}$. Consistent with the theory advanced by Gardner and Ashby (1970) and May (1972) and subsequent empirical work (Rejmanek & Stary 1979), both C and i declined with increased S for whole food webs. The same relationships were established for the primary producer and detritus-based energy channels that were derived from these webs, and the energy channels did not differ from one another. The relationship between the negative and positive interaction strengths for each trophic interaction with increased trophic position that was established by De Ruiter et al. (1995) for the whole webs applies to the energy channels as well (Fig. 19.5). The asymmetry between the positive and negative per capita effects with increased trophic position that was observed for the whole web was also observed within the root and detritus energy channels (Fig. 19.5). To illustrate this, the negative and positive interaction strengths were standardized by dividing each by the mean of the negative and positive interaction strengths respectively. The ratio of the standardized means (absolute value of negative/positive) decreased with increased trophic positions within the energy channels as they did for the whole web. Regardless of the basal resource, predators have a disproportionately large negative effect on the dynamics of their prey at lower trophic positions, while prey have a disproportionately large positive effect on the dynamics of the predators at higher trophic positions.

Additional evidence which suggests that soil food webs are compartmentalized is found in studies of the effects of disturbances on soil communities. Energy channels process matter at different rates and exhibit different responses to disturbance. Coleman *et al.* (1983) observed that the utilization of plant by-products along plant roots (mucigels, sloughed cells, exudates and dead roots) could be separated into a 'fast' cycle dominated by bacteria and their consumers (bacterial energy channel), and a 'slow' cycle dominated by fungi and their consumers (fungal energy channel). Moore (1986) observed that dynamics of soil biota under two different management practices of winter wheat grown in soils of the Shortgrass Steppe of Colorado differed markedly. The densities of organisms within the fungal energy channel were lower in soils under the more intensive management, and these organisms occurred later in the growing season. The different management practices showed little or no effect on organisms within the bacterial energy channel. The same phenomenon was observed in a comparison of two winter wheat rotations in The Netherlands (Moore & De Ruiter 1991).

A fundamental tenet of hierarchy theory is that rates and/or frequencies of processes represent boundaries between compartments within systems. The organisms

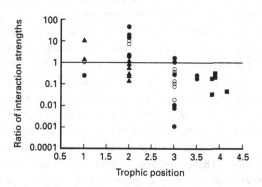

FIG. 19.5. Asymmetry in interaction strengths with increased trophic position for the CPER food web that was observed in Fig. 19.3 can also be found within each energy channel (bacteria channel ●, fungal channel ○, and root channel ▲). Each symbol represents a predator–prey interaction and the energy channel in which the interaction resided. Since predators obtain energy from more than one energy channel, they were considered separately (predators ■).

within the bacterial and fungal energy channels are different not only in terms of the habitats and substrates that they utilize, but also in terms of their growth rates, turnover times and ecological efficiencies (Table 19.2). The observation that disturbances have a more lingering effect on the fungal energy channel than the bacterial channel can be explained by these differences. The resilience of simple food chains is an increasing function of the rate of material input (primary productivity or detritus input rates) and on the turnover times (residence time) of material (DeAngelis *et al.* 1989; DeAngelis 1992; Moore *et al.* 1993). Hence, the temporal shifts between compartments can be explained by the high intrinsic rates of increase and turnover times that organisms within the bacterial energy channel possess relative to organisms at similar trophic positions within the fungal energy channel. Studies of freeze–thaw cycles using microcosms in the laboratory, and wetting–drying cycles using whole cylinders of intact soils in a greenhouse (mesocosms) have demonstrated that the bacterial energy channel has the capacity to recover from disturbances more rapidly than the fungal energy channel (Allen-Morley & Coleman 1989; Hunt *et al.* 1989). A result of minor perturbation would be to cause temporal displacements in the activity of one channel relative to another. This appears to have occurred within winter wheat fields in Colorado and The Netherlands, and in the microcosm and mesocosm studies involving soil biota of the bacterial and fungal energy channels (Moore 1986; Hunt *et al.* 1989; Allen-Morley & Coleman 1989; Moore & De Ruiter 1991).

RESOURCE COMPARTMENTALIZATION AND COMMUNITY DEVELOPMENT

O'Neill *et al.* (1986) argued that the formation of assemblages of species is a natural outcome of the evolution of complex systems. As an ecosystem develops, stable subunits must develop before further stages can develop (Simon 1962). Although stability routinely has been used as a necessary condition of ecosystem structure and an endpoint in development, little is understood about the mechanisms behind community assembly that results in a stable ecosystem. The biogeography literature offers some important insights into this process. In their review of ecosystem structure and development, O'Neill *et al.* (1986) noted that many different forms of stable subsystems have been described. Wilson (1969) identified these assem-blages of species as 'assortative equilibria', while Diamond (1975) called them 'permissible species combinations'. These conclusions were based on the obser-vations that only a small subset of the possible combinations of species that could occur on islands were actually observed. Why do some assemblages persist and others not?

The studies by Simberloff and Wilson (1969a, b) and the re-analysis of their work by Heatwole and Levins (1972) provide the beginnings of an answer. The experimental manipulations of small islands by Simberloff and Wilson (1969a, b)

TABLE 19.2. Habitat use, life-history characteristics, and energetic properties of the taxa that are common to soil food webs (Hunt et al. 1987; Coleman 1994).

Taxon Habitat	Bacteria Water/ surfaces	Fungi Free/ surfaces	Protozoa Water/ surfaces	Microbivorous nematodes Water films/ surfaces	Collembola Free	Mites Free
Minimum generation time (h)	0.5	4–8	2–4	120	720	720
Turnover times (season^{-1})	2–3	0.75	10	2–4	2–3	2–3
Assimilation efficiency	1.00	1.00	0.95	0.38–0.60	0.5	0.3–0.9
Production efficiency	0.4–0.5	0.4–0.5	0.4	0.37	0.35	0.35–0.40

provided results that supported the Theory of Island Biogeography as proposed by MacArthur and Wilson (1967). Species richness increased on each of the fumigated mangrove islands in the manner predicted by the theory based on the size of the islands and their distance from the mainland. The re-analysis of their data by Heatwole and Levins (1972) provided an added dimension to the process of colonization and extinction. They assigned the species that were present on the islands prior to fumigation and those that recolonized the islands after one year into crude assemblages based on function (e.g. decomposers, predators, ants, etc.). Fewer species were present on the islands 1 year after fumigation than before. However, the proportion of species in each functional group on the islands before and after fumigation was similar. These results indicated that functional diversity preceded species diversity. While the chance that a species colonizes the site could be viewed as a random phenomenon, the chance of extinction depended on the functional development of the community at the time of colonization and how the colonizing species utilized resources.

Our results suggest that the process of community development goes far beyond the simple example of a predator with a narrow choice of prey being less likely to survive in a newly colonized habitat than one with a wider diet. We propose that the energetic organization of the community plays an important role in governing the colonization and extinction process. As new species enter a community, the interactions that they engage in alter the pattern of material flow and the patterning of interaction strengths within the community. If the patterns of interaction strength were altered to an unstable configuration, one consequence would be the local extinction of species. Given this, would a reticulate or compartmented community be more likely to develop? The answer clearly depends on how many basal resources can be established within the habitat. In the extreme case where the habitat possessed only a single basal resource, the diversity and structure of the system would be limited the productivity of the basal resource and by dynamic constraints as successive trophic positions are added (Rosenzweig 1971; Moore et al. 1993). If a second basal resource were to become established within the habitat, would the biota that subsequently colonize the habitat have a greater chance of utilizing the original resource or the new basal resource? With all things being equal, save the conditions allowing the new resource being able to establish itself, there are fewer restrictions to colonizing the new basal resource than the original resource. If this process were to continue two things would occur: (i) the habitat would be comprised of many parallel unconnected food chains; and (ii) the species diversity of consumers within a community would be positively correlated with the number of basal resources within the community. The first condition occurs to a point, as predators eventually link the chains. The empirical evidence supports the second condition, as the diversity of consumers is positively correlated to the number of basal resources (Moore & Hunt 1988).

ACKNOWLEDGEMENTS

We thank two anonymous reviewers for their comments and encouragement. Special thanks to V. Brown, A. Gange and R. May for organizing the conference. This work was supported by the British Ecological Society and the US National Science Foundation.

REFERENCES

Allen, T. F. H. & Starr, T. B. (1982). *Hierarchy: Perspectives for Ecological Complexity*. University of Chicago Press, Chicago.

Allen-Morley, C. R. & Coleman, D. C. (1989). Resilience of soil biota in various food webs to freezing perturbations. *Ecology*, 70, 127–141.

Andrén, O., Linberg, T., Bostrom, U., Clarholm, M., Hanson, A. C., Johnansson, G., Lagerloff, J., Paustain, K., Persson, J., Petterson, R., Schnurer, J., Sohlenius, B. & Wivstad, W. (1990). Organic carbon nitrogen flows. *Ecological Bulletins* 40, 85–125.

Briand, F. (1983). Environmental control of food web structure. *Ecology*, 64, 253–263.

Briand, F. & Cohen, J. E. (1989). *Response to*: Moore, J. C., Walter, D. E. & Hunt, H. W. (1989). Habitat compartmentation and environmental correlates to food chain length. *Science*, 243, 238–240.

Brussaard, L., Van Veen, J. A., Kooistra, M. J. & Lebbink, J. (1988). The Dutch programme of soil ecology and arable farming systems. I. Objectives, approach and some preliminary results. *Ecological Bulletins*, 39, 35–40.

Clarholm, M. (1985). Possible roles of roots, bacteria, protozoa, and fungi in supplying nitrogen to plants. *Ecological Interactions in Soil: Plants, Microbes, and Animals* (Ed. by A. H. Fitter, D. Atkinson, D. J. Read & M. B. Usher), pp. 355–365. Blackwell Scientific Publications, Oxford.

Cohen, J. E. (1978). *Food Webs in Niche Space*. Princeton University Press, Princeton, NJ.

Coleman, D. C. (1994). The microbial loop concept as used in terrestrial soil ecology studies. *Microbial Ecology*, 28, 245–250.

Coleman, D. C., Reid, C. P. P. & Cole, C. V. (1983). Biological strategies of nutrient cycling in soil systems. *Advances in Ecological Research*, Vol. 13. (Ed. by A. Macfayden & E. D. Ford), pp. 1–55. Academic Press, New York.

Cummins, K. W. (1973). Trophic relations of aquatic insects. *Annual Review of Entomology*, 18, 183–206.

Cummins, K. W. (1974). Structure and function of stream ecosystems. *Bioscience*, 24, 631–641.

DeAngelis, D. L. (1992). *Dynamics of Nutrient Cycling and Food Webs*. Chapman & Hall, London.

DeAngelis, D. L., Bartell, S. M. & Brenkert, A. L. (1989). Effects of nutrient cycling and food chain length on resilience. *Nature*, 134, 778–805.

De Ruiter, P. C., Neutel, A. & Moore, J. C. (1995). Energetics, patterns of interaction strengths, and stability in real ecosystems. *Science*, 269, 1257–1260.

Diamond, J. M. (1975). Assembly of communities in ecocystems. *Ecology and Evolution of Communities* (Ed. by M. L. Cody & J. M. Diamond), pp. 342–444. Harvard University Press, Cambridge, MA.

Elliott, E. T., Anderson, R. V., Coleman, D. C. & Cole, C. V. (1980). Habitable pore space and microbial trophic interactions. *Oikos*, 35, 327–335.

Gardner, M. R. & Ashby, W. R. (1970). Connectance of large, dynamical (cybernetic) systems: critical values for stability. *Nature*, 228, 784.

Heatwole, H. & Levins, R. (1972). Trophic structure stability and faunal change during colonization. *Ecology*, 53, 531–534.

Hendrix, P. F., Parmelee, R. W., Jr, Crossley, D. A., Coleman, D. C., Odum, E. P. & Groffman, P.

M. (1986). Detritus food webs in conventional and no-tillage agroecosystems. *Bioscience*, 36, 374–380.

Hunt, H. W., Coleman, D. C., Ingham, E. R., Ingham, R. E., Elliott, E. T., Moore, J. C., Reid, C. P. P. & Morley, C. R. (1987). The detrital food web in a shortgrass prairie. *Biology and Fertility of Soils*, 3, 57–68.

Hunt, H. W., Elliott, E. T. & Walter, D. E. (1989). Inferring trophic transfers from pulse-dynamics in detrital food webs. *Plant and Soil*, 115, 247–259.

Hutchinson, G. E. (1959). Homage to Santa Rosalia, or why are there so many kinds of animals? *American Naturalist*, 93, 145–159.

Koestler, A. (1967). *Ghost in the Machine*. MacMillan, New York.

MacArthur, R. H. & Wilson, E. O. (1967). *The Theory of Island Biogeography*. Princeton University Press, Princeton, NJ.

Marples, T. G. (1966). A radionuclide tracer study of arthropod food chains in a *Spartina* salt-marsh ecosystem. *Ecology*, 47, 270–277.

May, R. M. (1972). Will large and complex systems be stable? *Nature*, 238, 413–414.

May, R. M. (1973). *Stability and Complexity in Model Ecosystems. Monographs in Population Biology*, Vol. 6. Princeton University Press, Princeton, NJ.

Moore, J. C. (1986). *Micro-mesofauna dynamics and functions in dryland wheat-fallow agroecosystems*. Phd Thesis, Colorado State University.

Moore, J. C. & De Ruiter, P. C. (1991). Temporal and spatial heterogeneity of trophic interactions within belowground food webs. *Agriculture, Ecosystems, and Environment*, 34, 371–397.

Moore, J. C. & Hunt, H. W. (1988). Resource compartmentation and the stability of real ecosystems. *Nature*, 333, 261–263.

Moore, J. C., Walter, D. E. & Hunt, H. W. (1988). Arthropod regulation of micro- and mesobiota in below-ground detrital food webs. *Annual Review of Entomology*, 33, 419–439.

Moore, J. C., Walter, D. E. & Hunt, H. W. (1989). Habitat compartmentation and environmental correlates to food chain length. *Science*, 243, 238–239.

Moore, J. C., De Ruiter, P. C. & Hunt, H. W. (1993). The influence of productivity on the stability of real and model ecosystems. *Science*, 261, 906–908.

Moore, J. C., De Ruiter, P. C., Hunt, H. W., Coleman, D. C. & Freckman, D. W. (1996). Microcosm in soil ecology: critical linkages between field and modelling research. *Ecology*, 77, 694–705.

Odum, E. P. (1971). *Fundamentals of Ecology*, 3rd edn. W. B. Saunders, Philadelphia.

O'Neill, R. V., DeAngelis, D. L., Waide, J. B. & Allen, T. F. H. (1986). *A Hierarchical Concept of the Ecosystem*. Princeton University Press, Princeton, NJ.

Pimm, S. L. (1982). *Food webs*. Chapman & Hall, London.

Pimm, S. L. & Lawton, J. H. (1980). Are food webs divided into compartments? *Journal of Animal Ecology*, 49, 879–898.

Polis, G. A., Myers, C. A. & Holt, R. D. (1989). The ecology and evolution of intraguild predation – potential predators that eat each other. *Annual Review of Ecology and Systematics*, 20, 297–330.

Pomeroy, L. R. (1970). The strategy of mineral cycling. *Annual Review of Ecology and Systematics*, 1, 171–190.

Rejmanek, M. & Stary, P. (1979). Connectance in real biotic communities and critical values for stability of model ecosystems. *Nature*, 280, 311–313.

Root, R. (1967). The niche exploitation pattern of the blue-gray gnatcatcher. *Ecological Monographs*, 37, 317–350.

Rosenzweig, M. L. (1971). Paradox of enrichment: destabilization of exploitative ecosystems in ecological time. *Science*, 171, 385–387.

Schoener, T. W. (1974). Resource partitioning in ecological communities. *Science*, 185, 27–39.

Simberloff, D. S. & Wilson, E. O. (1969a). Experimental zoogeography of islands: the colonization of empty islands. *Ecology*, 50, 267–278.

Simberloff, D. S. & Wilson, E. O. (1969b). Experimental zoogeography of islands: defaunation and monitoring techniques. *Ecology*, **50**, 278–296.

Simon, H. (1962). The architecture of complexity. *Proceedings of the American Philosophical Society*, **106**, 467–482.

Wilson, E. O. (1969). The species equilibrium. *Diversity and Stability in Ecological Systems* (Ed. by G. M. Woodwell & H. H. Smith), pp. 38–47. Brookhaven National Laboratory, New York.

20. FOOD-WEB PATTERNS:
WHAT DO WE REALLY KNOW?

S. J. HALL* AND D. G. RAFFAELLI†

School of Biological Sciences, Flinders University of South Australia, GPO Box 2100, Adelaide, 5001, SA, Australia and †Culterty Field Station, University of Aberdeen, Newburgh, Ellon, Aberdeenshire AB4 0AA, UK

INTRODUCTION

Describing and understanding the patterns and functional relationships between communities of predators and their prey remains a fundamental endeavour for ecologists. One prominent element of this endeavour has been the assembly of food-web data sets which attempt to capture some of the broad features of these relationships. At their simplest, food-web data sets are matrices in which rows represent predators and columns represent prey. A value of one in row i at column j denotes that predator i eats prey j, a zero denotes that it does not. Alternatively, data may be presented in the familiar pictorial form with nodes in a web representing species and lines connecting those nodes for which there are trophic links.

The underlying stimulus for efforts to analyse food-web data is that such community representations may offer insights for understanding the natural world. However, in view of their palpable simplicity, it is not surprising that the utility of food-web data sets for this purpose has been questioned. Questions can be asked at two levels. First, how good are the existing data and are the various features that are apparent in web data sets genuine properties of ecological systems? A second, more fundamental, issue is whether even the most fully described food-web data sets can offer valuable insight and understanding. For the most part, we focus on the easier (former) issue, but we return briefly to this more serious question towards the end of our contribution.

We are acutely aware of the excellent treatments of food-web issues that have gone before (e.g. Lawton 1989; Pimm *et al.* 1991) and we have previously addressed some of these issues (Hall & Raffaelli 1993). Nevertheless, we believe that an assessment of the options for future development of the discipline is timely, particularly in view of the enormous efforts that will be required of field ecologists if they are to respond to calls for new and better data. Our goal is not to re-open old debates or to reiterate earlier criticisms (although this is hard to avoid) and, for clarity, some review will be necessary. What we do wish to do, however, is try and build on achievements, explore possibilities and offer our opinions on the way forward.

When discussing food-web research, it is important to define the subject area. For our purposes, we mean analyses that are concerned with, or are directly motivated by, binary food-web matrices. This includes the analysis of static patterns in food-web data, modelling to explore potential explanations for the patterns, and experimental work to test specific hypotheses generated from pattern analysis. The literature which falls into this category forms a distinct and coherent body of work. More general community properties such as species abundance patterns, or general processes such as trophic cascades are not considered here.

In the first part of this chapter, we focus on the patterns that have been observed in food-web data sets and their current status in the light of recent analyses. We then go on to consider the variable criteria that are used to construct food webs and some of the consequences of these choices for emergent patterns. Some of the other types of data that are beginning to emerge to supplement basic food-web descriptions are then reviewed and the effect that the inclusion of such data has on earlier findings is examined. Finally, we review the utility of experimental manipulations for testing hypotheses that arise from food-web analyses by examining what has been and what could be done.

STRUCTURAL REGULARITIES?
THE SEARCH FOR PATTERNS

Formal analysis of patterns in food-web data dates from the late 1970s, and was brought to prominence largely by the work of Cohen, Briand and coworkers (e.g. Cohen 1977, 1978; Briand 1983; Cohen & Briand 1984; Cohen *et al.* 1990). Using a collection of food webs gleaned from the literature and edited to conform to certain criteria (see below), these and other authors derived a set of easily calculated statistics for the webs and a number of static features, or patterns, were identified. This pattern-seeking approach received considerable criticism, particularly with regard to the quality of the data obtained from published food-web diagrams. It was argued that the original authors of these diagrams almost certainly simplified the depiction of the system on which they were working – an objection best captured by Paines' (1988) phrase 'artistic convenience'. Another criticism concerned the differing degrees of taxonomic resolution to which the web entities were resolved in different studies, or even within the same web, and the effect that this might have on the perception of patterns (see, e.g., Hall & Raffaelli 1991, 1993).

To be fair, most of the criticisms that have been raised about the source data are openly accepted by those who have used them, but they are argued to be irrelevant to the general properties that emerge. Although we take the view that many of the criticisms of the original data sets are valid, it would be difficult to argue that there are no structural regularities, even though the patterns have been modified in the light of new data (Hall & Raffaelli 1993). For example, one of the most prominent claims of the early 1980s was that certain properties were scale invariant – that is, constant over a range of web sizes, but these claims were qualified at the time by the caveat that some scaling may emerge as the quality and amount of data increased. More recent analyses convincingly demonstrate that these properties are not insensitive to web size (Cohen *et al.* 1990; Pimm *et al.* 1991; Martinez 1993; Bersier & Sugihara, unpublished data). In Table 20.1 we summarize the original patterns that were identified and their current status in the light of more recent data sets that were not available when the original analyses were performed.

OBTAINING BETTER DATA

There can be few ecologists who do not value additional data and for students of food webs the most usual call is not just for more, but also better data. Discussion of the form that such improvements should take is provided by Lawton (1989) and Cohen *et al.* (1993a) and one can but hope that all those who gather food-web data will give due consideration to the issues discussed in these papers. Most importantly, in the absence of agreed standards for data collection and analysis, clear explanation of the methodological details about the compilations of any given food web is vital if the robustness of patterns is to be determined. This is not a trivial point and it is important to realize just how variable the criteria for constructing food webs actually are, even with data collected specifically for the purpose. From our own reading of the literature, we have become increasingly aware that a food-web data set represents different things to different people. Thus, we have identified three types of webs in the literature, here termed *empirical webs, likely webs* and *imaginable webs*. In empirical webs, all links are documented from gut contents and feeding trials; likely webs have most links documented, but some are inferred from local knowledge; imaginable webs are compiled from a species list of the area or habitat and discussions with experts or referral to the literature on the probable feeding links. All three types of web are present in the web data base (e.g. the Ythan (empirical), Coachela Valley (likely) and Little Rock Lake (imaginable) (Hall & Raffaelli 1991; Polis 1991; Martinez 1991, respectively)). Each type of web reflects a rather different perspective, ranging from a requirement for absolute empirical proof for each link in the web to an approach which produces webs describing what is possible, but not necessarily real, for that location and assemblage of organisms.

To illustrate the differences that can result from the use of these alternative criteria we have compiled imaginable versions of the Ythan web (where links are inferred) and compare these with the empirical web that we have previously described (Hall & Raffaelli 1991). In constructing an imaginable web, the first step is to obtain a complete species list for that location. Our imaginable web is

TABLE 20.1. Patterns identified in food-web data.

Pattern	Description	Current status
Scale-invariant proportions of top, intermediate and basal species	Top species have no predators, basal species have no prey and intermediate species have both predators and prey. The proportions of these species were considered to be independent of the size (number of species) in the web (Cohen & Briand 1984; Briand & Cohen 1984)	Recent work suggests that the proportion of intermediate species increases with web size, whereas the proportion of top and basal species declines (Martinez 1993)
Scale-invariant proportions of feeding links between top-intermediate, top-basal, intermediate-intermediate and intermediate-basal elements	Links between the categories defined above were also thought to be independent of web size	Recent work suggests that the proportion of intermediate-intermediate links increases with web size, whereas the proportion of top-basal links declines (Martinez 1993)
Constant linkage density (hyperbolic decline in connectance)	Linkage density, the number of links per species, was originally examined in terms of connectance (the number of links as a proportion of the total possible number of links), which was judged to decline hyperbolically with increasing web size. This decline is equivalent to a constant linkage density	Although an increase in linkage density with web size was noted over a decade ago (Cohen & Briand 1984), the idea that this property is scale invariant has persisted (Cohen et al. 1990). More recent data, however, re-affirm the idea that linkage density increases with web size (Martinez 1993; Warren 1994)
Rarity of feeding loops	A feeding loop is where feeding links can be traced in a circuit back to the starting point, i.e. species A eats B, B eats C and C eats A. These were reported to be rarer than would be expected by chance (Pimm & Lawton 1978; Pimm 1982)	No new tests, but several data sets show that loops are more common than previously thought

TABLE 20.1. *Continued.*

Pattern	Description	Current status
Short food chains	The length of a food chain is the number of links running from a top predator to a basal species. Chain lengths tend to be short, typically with only three or four links and chains involving more than six species are rare (Hutchinson 1959; Pimm 1982; Cohen *et al.* 1986)	Average and maximum food-chain lengths appear to increase for food-web data sets which have greater taxonomic resolution. Independent of this effect, however, there is evidence that these statistics increase with web size (Hall & Raffaelli 1993). The recent inclusion of larger webs and improved resolution have probably both contributed to an overall increase in the average and maximum values for the catalogue as a whole
Rarity of omnivory	Omnivores are organisms that feed on more than one trophic level, a behaviour that appears to be less common in some kinds of real webs than in randomly generated webs (Pimm 1982)	Omnivory now appears to be rather more prevalent than previously thought (Sprules & Bowerman 1988; Hall & Raffaelli 1993; Polis 1994)
Most webs are interval, rigid-circuit and do not have topological holes	Intervality refers to the food niche overlaps of predators in a web. For most webs these overlaps can be ordered along a single dimension and such webs are said to be interval. A web is rigid circuit if, for every path in a niche overlap graph with more than three vertices a shorter path can be traced. Topological holes are rather technical properties of the prey overlap or resource graphs, which are rarely observed in real webs. (For more complete descriptions of intervality, rigid-circuitry and topological holes see Hall & Raffaelli (1993))	Probability of a web being interval decreases if parasites are included (Huxham *et al.* 1996)

therefore based on the estuary's recorded species list which is more extensive than that used by Hall and Raffaelli (1991). In an earlier analysis, only those 95 entities for which actual trophic data were available were included in the web. Here, we extend the web to include a further 64 species not in the original empirical web (Hall & Raffaelli 1991) because they are either very uncommon (e.g. fulmar, teal, tufted duck, snipe and common seal), or were never detected in predator gut contents, as is the case for the additional invertebrates. Inclusion of these additional species brings the total to 159, comprising nine basal, 101 intermediate and 49 top predator species.

Whilst using the complete species list reflects the practice adopted by compilers of imaginary webs, our comparisons between the real and extended imaginary versions may be confounded by their differences in size (95 vs 159 species respectively). We therefore repeated the analyses on a constrained imaginary web of the original 95 species, but where the linkages were inferred as for the larger version.

Linkages were inferred from discussions with experts on probable feeding links and reported trophic interactions in the literature. Trophic data on avian predators were obtained largely from Cramp and Simmons (1977–83). Likely fish diets were provided by Dr Mike Elliott, based largely on his extensive studies in the Forth and Humber estuaries. Invertebrate diets were determined from a variety of literature sources and discussions with experts on marine invertebrates. It was our intention to compile a web based solely on the likely feeding links between species as opposed to those actually documented. For this reason, it was necessary to exclude experts familiar with, and literature based on, the Ythan.

A comparison of several web properties in the real and imaginary versions of the Ythan web is shown in Table 20.2. The number of trophic links in the extended imaginary web is three times that in the real web, but this is not simply due to the increased number of species in the imaginary web; even in the constrained version the number of links per species (linkage density) is almost twice that of the real web. The real web contains about 5500 food chains running from basal species to top predators, whereas there are almost half a million separate food chains in the extended imaginary web. The maximum and modal lengths of these chains are considerably longer in the extended imaginary web (Table 20.2, Fig. 20.1). Interestingly, the so-called scale invariant properties of webs are not appreciably dissimilar in the different versions. The difference in web size between the real and extended imaginary versions of the Ythan web does not account for their different web properties. The constrained imaginary web had properties similar to those of the extended version (Table 20.2, Fig. 20.1).

Many of the differences noted above probably simply reflect the broader diets of consumers in the imaginary web, each of which took, on average, about twice as many prey species as those in the real web (Table 20.3, Fig. 20.2). One could argue that this result reflects inadequate sampling and description of the real Ythan web

TABLE 20.2. Food web statistics for real and imaginable Ythan food webs.

Statistic	Real web (Hall & Raffaelli 1991)	Constrained imaginable web (true species)	Extended imaginable web (true species)	Extended imaginable web (trophic species)
No. of species (S)	95	95	159	147
No. of links (L)	409	899	1248	1133
Linkage density (L/S)	4.31	9.46	7.85	7.71
No. of food chains	5518	41 532	494 542	–
Food chain length:				
Minimum	1	1	1	–
Maximum	9	10	13	–
Mode	5	7	7	–
Species proportions:				
% Top	0.28	–	0.31	0.31
% Intermediate	0.68	–	0.64	0.63
% Basal	0.04	–	0.06	0.06
Prey : predator ratio*	66 : 89 (0.75)	–	110 : 150 (0.69)	101 : 138 (0.73)

– Denotes values not calculated.
* Prey : predator ratios are defined as the number of species which are preyed upon divided by the number of species which are consumed.

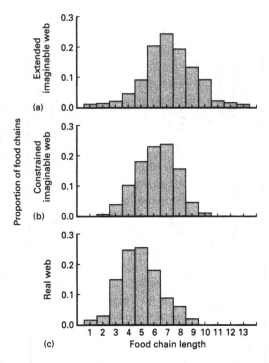

FIG. 20.1. The distribution of food-chain lengths in real and imaginable versions of the Ythan food web.

TABLE 20.3. Diet breadth of consumers in real and imaginable versions of the Ythan web.

Diet breadth	Real web	Constrained imaginable web	Extended imaginable web
All consumers			
Number of consumers	88	88	150
Mean (SD)	4.68 (6.04)	9.56 (9.64)	8.29 (8.71)
Mode	1	1	1
Maximum	40	40	42
All consumers except			
those feeding on basal resources			
Number of consumers	51	–	87
Mean (SD)	7.2 (6.68)	–	12.7 (8.75)
Mode	4	–	16
Maximum	40	–	42

– Denotes values not calculated.

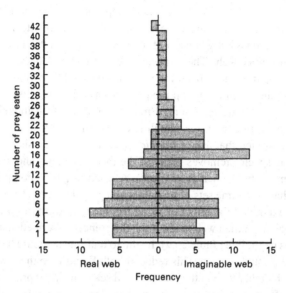

Fig. 20.2. Frequency distribution of diet breadths for real and imaginary versions of the Ythan food web.

and that we have simply failed to find some of the prey items which are rarely eaten. Conversely, one might view the imaginable web as a gross overestimation of the true number of links present. Clearly, the truth is likely to lie somewhere between the two extremes, but we believe that sampling problems are not likely to be great – that is, we are close to the top of the sampling curve. Data for the construction of the Ythan web have accumulated over many years and cover the vast majority of the metazoan taxa. Dietary analyses have also been very intense and detailed. We feel it unlikely, therefore, that we have grossly underestimated the true number of links in the real web. If anything, one could argue that the opposite were true and that our real web suffers from the same overestimation criticism as other webs accumulated over long time periods (Warren 1989). It seems more likely that, although predators could eat potential prey which occur on the Ythan and are known to eat them elsewhere, they do not in fact do so. In other words, the imaginable web represents a regional-scale construction that is not completely realized at this location. Fox and Morrow (1981) have similarly pointed out that herbivorous insects may be specialists at a local level, but feed on a much wider range of species at regional scales. The message is clear to us; webs constructed from dietary data from the literature may grossly overestimate the number of links in a local web.

It is difficult to predict a priori the impact of such an overestimation on different web patterns. The present analysis indicates significant effects on food-chain lengths

(Fig. 20.1), although not on scale-invariant properties (Table 20.2). The effect on food-chain lengths is largely due to the increased frequency of interactions between fish species in the imaginary version of the web. (Note that this effect also occurs even in the constrained web.) These species occupy an intermediate trophic position so that additional links have been inserted into the middle of food chains rather than stretching the chain at the top predator or basal levels. This is an important observation, since we never observed many of the fish–fish interactions in the imaginary web, despite intensive sampling over many years.

It has been argued that use of *trophic species* in the analysis of web patterns rather than true species removes many of the inconsistencies and biases inherent in the literature webs (Cohen *et al.* 1990). A trophic species comprises any real species that share the same sets of predators and prey. For the sake of consistency, we therefore examined the effect of using trophic species in our analyses of the extended imaginary Ythan web to see if the differences described above persisted. The effect was negligible (Table 20.2). In all, 20 real species could be grouped into eight separate trophic species. This reduced both S and L so that linkage density remained much higher than in the real web and scale-invariant properties were not significantly affected (Table 20.2). Food-chain lengths are expected to decrease with the decreasing taxonomic resolution generated by the trophic species approach (Hall & Raffaelli 1991; Martinez 1991) but, in view of the small number of species involved, the effect would probably be minimal. We return to the utility of trophic species later.

The above analysis clearly illustrates the need for clear explanations of the protocols adopted for gathering information. Better still would be an agreed set of standards for data acquisition so that webs can be compared and common features identified with greater confidence. But could a consensus on standards ever be arrived at? We suspect not. Food-web data will continue to be, for the most part, by-products of individual studies with widely differing objectives. What is likely, however, is that in view of the simplicity of traditional food-web representations, ecologists will continue to gather new kinds of information about food webs that might help address some of the problems in this area. Below we consider some of the information that is starting to emerge.

OTHER INFORMATION

Almost all of the food webs in the literature contain only limited information on interactions between web species. Most webs portray only the trophic interactions between predators and their prey and full interaction webs (*sensu* Menge 1995) are rare. Mutualistic interactions are difficult to quantify or even observe, yet may be crucial to the dynamics of communities (Cohen 1993; Paine 1994). Similarly, parasite/pathogen–host interactions are acknowledged as important in population and community dynamics, yet we know of only one case where an attempt has been made to include parasite–host links in a food web (Huxham *et al.* 1995b). Parasites are particularly interesting web elements in that many species can be classed as life-history omnivores and omnivory may therefore be much more prevalent than revealed in analyses of webs lacking information of parasites. Also, parasites are common and can be highly interactive with their host prey (Laukner 1988). For instance, in the Ythan food web, digenean parasites probably have as much, if not more, impact on their snail host populations than do avian or fish predators (Huxham *et al.* 1994, 1995a; Raffaelli & Hall 1996). Parasites are consumers of species traditionally recognized as 'top predators' and as such must assume top predator status themselves. Recognizing parasites as top predators inevitably lengthens food chains by one, but also reverses the predator–prey size hierarchy so obvious in most webs (Cohen *et al.* 1993b). If inclusion of parasites, and perhaps even pathogens, in webs is thought important because of their strong interactions with their prey (see below), then this will increase the difficulties of obtaining good web data by several orders of magnitude. Finally, if parasites are to be included in webs, then the trophic species approach (see above) may largely become redundant. Trophic species share the same predators and prey, but the high specificity of parasite/pathogen–host interactions means that few species are likely to have exactly the same consumers.

Like parasites and pathogens, decomposers have largely been neglected in descriptions of food webs. Perhaps because of the perceived difficulties in un-ravelling interactions towards the base of webs, decomposer organisms have often been lumped into very coarse taxonomic categories or ignored altogether. In this respect, soil web ecologists have made major contributions to our understanding of the nature and importance of this part of the web (De Ruiter *et al.* 1995). The effect of including well-described decomposer systems in the analyses of web patterns is likely to be dramatic but, to our knowledge, it has yet to be attempted. At the very least, many more species would be added to the web, food chains would be elongated and scale invariant properties would need to be re-examined if large numbers of microbial basal species are taken into account. In view of the dynamic importance of decomposers in probably all natural systems, it would be difficult to defend analyses of patterns based on data sets lacking resolution of the decomposer black-box.

Another kind of data that may be expected to be particularly revealing would be the routine inclusion of body size information for each of the species in a web. Interest in the potential of body size to explain food-web patterns stems largely from the development of the cascade model by Cohen and coworkers in the early 1980s (see Cohen *et al.* 1990). The model is so called because its central assumption is that species in the web can be ordered a priori into a feeding hierarchy. The model assumes that a species can never feed on those above it and that feeding links with those below it are chosen at random with a probability c/S, where c is a positive constant less than S (the number of species in the web). The number of links per species must be specified from available data before the model can be used. As Warren and Lawton (1987) point out, one of the most obvious ways in which such a hierarchy might be formed is on the basis of body size. Warren and Lawton's analysis of body size relationships in invertebrate predator–prey systems suggested that the upper triangularity[1] generated by a body size hierarchy could explain the success of the cascade model for predicting food-web patterns. Lawton (1989) further argued that the body size hierarchy that would lead to a lower triangular matrix for a parasite web should also result in the successful predictions of food-chain lengths by the cascade model. These ideas are intuitively appealing, but the recent inclusion of parasites, a hitherto ignored group, in food webs casts doubt on their legitimacy because neither upper or lower triangularity occur in this web, yet the cascade model still predicts food-chain lengths well (Huxham et al. 1995b). This seems to us to be an important result since it implies that species are indeed ordered in a trophic hierarchy, but that body size is insufficient to explain it.

A further approach to sanitizing the existing web database to minimize problems arising from its rather variable quality and completeness is to focus on those trophic links which are deemed to be dynamically important (Pimm et al. 1991). Essentially, this means removing all weak links from web data, so that the web remaining contains the really important interactions (and species). Whilst the idea might seem attractive, there are major difficulties in assessing the importance of links, not least because the amounts of material or energy which flow through a link from prey to predator offers no guide to its importance (Paine 1980; Raffaelli & Hall 1996). Measuring interaction strength as the per capita effects of a predator on a prey is a more sensible way forward (Paine 1992; Lawton 1992) although there are practical difficulties here as well (Raffaelli & Hall 1996), particularly in view of the now widely recognized likelihood of indirect effects (Schoener 1993; Menge 1995). Realistically, the effort required to measure interaction strength is probably too demanding and for the majority of webs in the catalogue such information can never be obtained. We also see no prospect of obtaining better data sets by editing existing literature webs without heroic guesswork, particularly since a species' importance may be independent of its body size, taxonomic position or biomass. The effects of parasites, for example, has been shown to affect the interpretation of the results of the introduction of white-tailed deer (*Odocoileus hemionus*) into Nova Scotia and the consequent extinction of caribou (*Rangifer tarandus*). Although

---

[1] If the predator species in a food-web matrix are arranged in increasing order of body size along rows and the prey species in increasing order down columns and if most predators are larger than their prey, most of the non-zero elements in the matrix will occur above the leading diagonal and it is termed upper triangular.

this extinction was originally attributed to the competitive exclusion of caribou by deer, it later appeared that differential susceptibility to a meningeal parasite may have been responsible (Anderson 1965; see also Huxham *et al.* 1995b).

IF THERE ARE PATTERNS IN FOOD WEBS WHAT DO (MIGHT) THEY TELL US?

If we accept that structural regularities are present in food-web data sets, it is reasonable to ask why. For most, but not all patterns, a range of explanations has been advanced, few of which are mutually exclusive. Lawton (1989) provides a table summarizing patterns and explanatory theories and we have found little to add to this. However, for completeness we have paraphrased Lawton's summary here (Table 20.4). Approaches to deciding on the relative importance of competing explanations include careful exploration of the available data sets, the development of theoretical models and the design and prosecution of experiments.

In 1989, Lawton highlighted the need for more experimental manipulations to test the hypotheses generated by food-web research. Below, we consider the progress that has been made with the experimental approach since his review. Although we focus here on experiments, this is not to say that the alternative approaches are any less valuable; in our view, however, the alternatives have received far more attention, probably because theoreticians and those who mine available data have been more active in this field than experimentalists, but also because developing the theory has, perhaps, proved easier than devising experimental tests.

EXPERIMENTAL APPROACHES?

Experimental tests of hypotheses about mechanisms underlying food-web patterns are few and far between. In fact, most of the experimental studies on food webs tackle the related issue of the relative importance of top-down and bottom-up processes (see reviews by Power 1992; Polis 1994; Balciunas & Lawler in press). The absence of direct tests of theories to explain patterns is at first sight surprising given the popularity of the manipulative approach to community ecology over the last 20 years. On reflection, however, we believe that few of the hypotheses proposed are directly testable using the conventional experimental approach. Many of the regularities listed in Table 20.2 are appealing to ecologists because they promise insights into processes conferring dynamic stability. It is argued that the arrangements of species and linkages are constrained to be as they appear because other arrangements are not tenable. Altering the patterns (e.g. by changing the ratios of top, intermediate and basal species, or adding or deleting omnivores, or altering linkage densities) and analysing the stability properties of the resultant web would provide a direct test of these ideas. Whilst this approach has been outstandingly

TABLE 20.4. Summary table of explanatory theories that have been proposed to explain food-web patterns (from Lawton 1989).

Pattern	Explanatory theories	Current status
Scale-invariant proportions of top, intermediate and basal species	Generated by the cascade model, which requires that species be ordered in a trophic hierarchy	Model confirms predictions
Scale-invariant proportions of feeding links between top–intermediate, top–basal, intermediate–intermediate and intermediate–basal elements	As above	As above
Constant linkage density (hyperbolic decline in connectance)	1 Dynamical constraints 2 Natural history, optimal foraging and other evolutionary constraints	1 Lotka–Volterra models confirm predictions (see Pimm et al. 1991 for review) 2 Arguments support predictions
Rarity of feeding loops	1 Dynamical constraints 2 Generated by the cascade model, which requires that species be ordered in a trophic hierarchy	1 Lotka–Volterra models confirm predictions (see Pimm et al. 1991 for review) 2 Model confirms predictions
Short food chains	1 Generated by the cascade model, which requires that species be ordered in a trophic hierarchy 2 Energetic constraints 3 Dynamic constraints 4 Donor control Dynamics 5 Natural historical/evolutionary constraints	1 Model confirms predictions. It was originally thought that the hierarchical ordering of species was on the basis of body size, but the inclusion of parasites in a web still leads the cascade model to generate correct predictions (see text) 2 Lotka–Volterra models and observation confirm predictions (e.g. Pimm 1982), but experimental data are equivocal (see text) 3 Lotka–Volterra models confirm predictions, but experimental data are equivocal (see text) 4 Models confirm predictions. Donor control not generally held to be important except in decomposer systems, but experimental data may overestimate the prevalence of strong interactions 5 Difficult to test

TABLE 20.4. *Continued.*

Pattern	Explanatory theories	Current status
Rarity of omnivory	1 Dynamical constraints 2 Evolutionary constraints	1 Lotka–Volterra models confirm predictions (e.g. Pimm 1982) and some experimental evidence (Schoener & Spiller 1987; Morin & Lawler 1993, 1996) 2 Arguments and data support predictions (e.g. Stenseth 1983; Yodzis 1984)
Most webs are interval, rigid-circuit and do not have topological holes	Generated by the cascade model, which requires that species be ordered in a trophic hierarchy. Also predicted by the assembly models of Yodzis and Sugihara.	Model confirms predictions

successful using mathematical models of ecological systems (e.g. May 1973; Pimm 1982; Yodzis 1989), it is impossible for most real webs. Notwithstanding the debate over whether the patterns really exist at all, there are several operational hurdles that must be overcome when manipulating real webs. First, the large size of many real webs means it is impossible to find similar replicate webs, including controls, to meet the rigorous demands of subsequent analyses. Second, there are no rigorous operational definitions of stability which can be easily applied in the field or even agreed on by ecologists (Connell & Sousa 1983; Pimm 1984; Law & Blackford 1992). Third, many of the patterns have such a high degree of variance associated with them that it is difficult to imagine what one would have to do to produce a web which differed significantly in its configuration from the 'norm'; the so-called scale invariant patterns are a prime example of this (see Fig. 6 in Hall & Raffaelli 1993).

Comparisons of the relative stabilities of large, real webs whose patterns have been experimentally altered will therefore be very difficult and the results of such experiments are likely to be ambiguous (see also below). It is no surprise then, that the few experiments that have been performed on food web patterns do not directly manipulate the patterns themselves, but the processes suspected of underlying the patterns. For obvious practical reasons, these experiments have been conducted in quite small webs for which replication is possible. Experimental webs include phytotelm webs in Australia (Pimm & Kitching 1987; Jenkins *et al.* 1992) and small freshwater ponds in the UK (Warren & Spencer in press), where the focus has been on the effects of energy supply on food-chain length and the 'stability' (sensitivity to a perturbation (disturbance)) of webs with different chain lengths. Other examples include freshwater plankton assemblages captured in mesocosms, where the effects of stress on food-chain length and linkage density have been examined (Havens 1994), and protist webs constructed in laboratory microcosms in order to explore the effects of trophic structure on population dynamics (Morin & Lawler 1993, 1996).

Phytotelms are small water bodies that occur in or on plants: water-filled tree holes, pitcher plants, bromeliads, bamboo internodes and axial waters are good examples. The food webs within these water bodies are simple, small scale and have clear boundaries. Most importantly, they can be easily replicated in artificial containers. In an early experiment, Pimm and Kitching (1987) manipulated energy inputs to such webs by a factor of 0.5–4 times the natural rates of litter input, but the effects on food-chain length were small and not convincing (Lawton 1989). The experiments were repeated using 10–100 times reduction in energy input (Jenkins *et al.* 1992) whereupon chain length decreased by one link, compared to webs receiving natural levels of litter. This decrease occurred due to the loss of top predators. A natural perturbation (low rainfall) had the greatest effect on webs with longer food chains implying that these systems were less likely to persist.

Warren and Spencer's (in press) approach using artificial ponds is similar in many ways to that used in the tree hole web studies described above. However, they found no statistically significant effects of energy supply on food-chain length. If anything, the trend was towards shorter chains in the high input treatments. Whilst this is at odds with the results of the tree-hole studies, the authors point out that the relationship between energy inputs and food-web patterns, like chain length, which are ultimately dependent on species richness (see below), may not be monotonic so that depending on where a food web is normally located on that relationship, changes in energy supply could shorten or lengthen food chains. Interestingly, the percentage of top and intermediate species did differ between treatments and linkage density was greater in low-energy systems. However, Warren and Spencer point out that the greatest effect in their experiments was on species richness, low-energy treatments having more species. Given the non-independence of species richness and most food-web properties (Martinez 1993; Bengtson 1994), they found it difficult to interpret experimental effects of the treatments on patterns *per se*. Indeed, if the correlation between species richness and web statistics is accounted for and removed from their analysis, few food-web responses remained significant. Disturbance (draining of ponds) had no differential effects in the two treatments implying that 'stability' was similar in the differently configured webs.

In Haven's (1994) study, plankton webs were exposed to a single dose of biocide (copper sulphate) in replicate field-based mesocosms. This stress reduced species richness and community diversity, and food-chain length and linkage density were lower in stressed assemblages. However, the dependence of these latter properties on species richness and the likely non-monotonic response of food webs to stress (see Warren and Spencer, above) were not examined.

A complementary, and to us, an extremely promising approach to the relation between structure and dynamics in webs has been taken by Morin and Lawler (1993, 1996). They assembled simple protist food chains in replicate laboratory microcosms and examined the effects of trophic structure on species population dynamics. Dynamics were more variable (and presumed less stable) in those systems with longer food chains. Also, omnivores attained higher densities and sometimes (but not always) less variable population dynamics than strict predators, presumably because they could cope better with fluctuations in prey species. Both these observations are in agreement with predictions of food-web theory.

DISCUSSION

There is one conclusion that we feel dominates all others – there is still considerable uncertainty as to whether many of the patterns described for food webs actually exist. There seem to be at least two reasons for this distrust: the quality of some of the data and the analytical approaches used. There is justifiable concern over the quality of many of the descriptions and pictures of food webs which appear in much of the earlier literature, and which constitute the bulk of the existing web catalogue. Many of these are less than adequate representations of nature, being little more than pencil-sketch simplifications drawn to provide a back-drop or to make a specific point. They were never intended for the kind of analyses (torture?) to which they have been subjected by food-web theorists. Given this inadequacy, attempts to sanitize these data and hence remove them one step further still from what the real web may look like, might be seen as a complete waste of time. This also applies to imaginary webs which tend to overestimate linkage density, food-chain lengths and perhaps other properties not analysed here. Patterns which emerge from the analyses of such artificial data sets may be compelling, but it is not surprising that they are viewed with suspicion.

On the other hand, several authors (e.g. Martinez 1993) are encouraged by the consistent properties which emerge from such analyses. The most frequent approach to coping with the acknowledged variable quality of much of the food-web data is to abandon the notion of the biological species and to work with functional groupings such as trophic species. A trophic species is defined as 'a collection of organisms that have the same diets and the same predators' and may be 'a biological species of plant or animal, or several species, or a stage in the life cycle of one biological species' (Cohen et al. 1990). The concept was originally designed as a laudable attempt to cope with artefacts in web data sets created by the tendency for ecologists to resolve furry and feathery predator species rather better than their prey species (Cohen et al. 1990). Now, it has become a powerful tool for smoothing off the rough corners in even the best web data sets and hence standardizing the quality of webs. The reduction of a complex web by collapsing biological species to trophic species does allow the relationships between functional groupings to be explored. This is a perfectly justifiable approach – one which paid major dividends in Elton and Lindman's classic analyses of the trophic structure of ecological systems. However, lumping quite unrelated species into functional groupings and perhaps splitting others into functional life-history classes produces a food web which is an abstraction of the original. The analysis of such abstraction might yield some interesting features, but it is not always clear how these relate to the original biological species web. Furthermore, inclusion of parasites and pathogens will in itself prohibit the use of the trophic species concept, since it is highly unlikely that species will have identical sets of these consumers.

Our uncertainty regarding the validity of the patterns will increase considerably when one considers what the topological structure of webs might look like with the inclusion of hitherto ignored groups such as decomposers, pathogens or parasites. Does studying the patterns in webs that ignore such groups make sense? We feel it does not if our goal is to explore the relationships between structural properties and dynamic behaviour – the utility of analysing patterns derived from webs which exclude such important functional groups must surely be limited. In our view, the solution to this deficiency probably does *not* lie in the assembly of ever more complete food-web data sets. Quite apart from the logistic problems involved in such an enterprise, we doubt that more detail would shed more light. In particular, we have no faith in webs constructed from censuses and regional information on predator diets.

We return now to the question posed in the introduction as to whether pattern analysis of even the best possible web data sets can offer valuable insights. This seems unlikely. We believe that it is by examining the dynamic behaviour of real and model (including simple assembled) food webs that most progress will be made. Ecologists justify much of their effort by arguing that their science can provide an understanding of how the real world will *behave* under changing circumstances. Such understanding ought to allow the formulation of practical solutions for the management, protection or restoration of natural systems. It is easier to see how an understanding of dynamics developed through experimental manipulations, or through carefully directed theoretical modelling, can be of help in this regard than it is to see the value of knowing the ratio of predators to prey or the proportion of intermediate tropho-species there should be in a web of a given size. It is reasonable to ask whether studying food-web patterns has had any impact, beyond the population of scientists who read and publish in the ecological journals. It is by no means clear to us whether history will see the analysis of food-web patterns as an academic backwater or a pursuit of wider significance and utility.

We have found that there have been relatively few experimental studies which test hypotheses derived from food-web pattern analysis and it is easy to see why those which are available involve very small webs with clear spatial boundaries (e.g. phytotelms, ponds and laboratory microcosms). This small-scale approach with simple, replicable enclosed systems looks especially helpful, particularly when it can be combined with theoretical modelling efforts designed to capture the key determinants of observed dynamics. Aquatic systems lend themselves well to this kind of approach, but, with the exception of the closed 'ecotron' systems now being developed (Lawton *et al.* 1993) there seem to be few tractable terrestrial systems which offer opportunities. We doubt that manipulating larger real systems to examine the food-web properties discussed here will ever be realistic. However, experimental tests of other aspects of food webs are possible for larger systems and are providing exciting results. Notable in this regard is the experimental effort that has been put into testing Hairston *et al.*'s early ideas (Hairston *et al.* 1960) and the 'trophic cascade' studies which have sprung from it (Power 1992; Polis 1994).

In conclusion, we feel that the pattern-seeking approach to food webs has been, and will probably continue to be, unsuccessful and unconvincing. Indeed, we would argue that a much better understanding of how ecological systems function has been obtained from focusing on underlying dynamics in the absence of any hypotheses derivable from food-web patterns. This is *not* to say that we feel the pattern-seeking approach has no place in ecology or that efforts to explore food-web data sets have not been worthwhile. There can be little doubt that regularities in nature can provide exciting insights into underlying mechanisms and guide further research. Unfortunately, however, the problems inherent in convincingly documenting the trophic interactions for most real webs are enormous and patterns built on demonstrably shaky foundations are unlikely to convince many that it is worthwhile devoting diminishing resources to subject them to experimental test. The scarcity of data on competitive interactions is also worthy of comment here since most of the predator manipulations which have demonstrated strong controlling effects on community dynamics have highlighted the importance of suppressing competitively superior prey. Moreover, in terrestrial systems such effects most often occur with herbivores controlling plant species – a taxonomic group which is conspicuously poorly resolved in food-web data sets.

In fairness, we must acknowledge that there are those who would disagree with the above assessment. Some see many of the existing food-web patterns as robust constraints on biological organization that it is unwise to ignore, even if explanations are not immediately apparent, and believe that characterization of food-web data sets provides a vehicle to 'unify the insights of many specialists into taxonomically comprehensive trophic portraits of ecological organisation' (Martinez 1993). In response, we can only say that we wish that this were so, but fear it is not.

ACKNOWLEDGEMENTS

We thank Mike Elliot and Dave Basford for their help in constructing the imaginable web and Sue Way for painstakingly analysing some of the data. The following kindly provided helpful comments on earlier drafts: Jennifer Ruesnick, Bob Paine, Stuart Pimm, Phil Warren, Mark Huxham, Sharon Lawler and Louis Bersier.

REFERENCES

Anderson, R. C. (1965). *Cerebrospinae nematodiasis* in North American cervids. *Transactions of the North American Wildlife Conference and Natural Resources*, **30**, 156–167.

Balciunas, D. & Lawler, S. P. (in press). Effects of basal resources, predation, and alternative prey in microcosm food chains. *Ecology*.

Bengtson, J. (1994). On comparative analyses in community ecology: confounding variables and independent observations in food web studies. *Ecology*, **75**, 1282–1288.

Briand, F. (1983). Environmental control of food web structure. *Ecology*, **64**, 253–263.

Briand, F. & Cohen, J. E. (1984). Community food webs have scale-invariant structure. *Nature*, **307**, 264–266.

Cohen, J. E. (1977). Ratio of prey to predators in community food webs. *Nature*, **270**, 165–167.

Cohen, J. E. (1978). *Food Webs and Niche Space*. Princeton University Press, Princeton, NJ.

Cohen, J. E. (1993). Concluding remarks. *Mutualism and Community Organization* (Ed. by H. Kawanabe, J. E. Cohen & K. Iwasaki), pp. 412–415. Oxford University Press, Oxford.

Cohen, J. E. & Briand, F. (1984). Trophic links of community food webs. *Proceedings of the National Academy of Sciences, USA*, **81**, 4105–4109.

Cohen, J. E., Briand, F. & Newman, C. M. (1986). A stochastic theory of community webs. II. Individual webs. *Proceedings of the Royal Society of London, B*, **228**, 317–353.

Cohen, J. E., Beaver, R. A., Cousins, S. H. (1993a). Improving food webs. *Ecology*, **74**, 252–258.

Cohen, J. E., Briand, F. & Newman, C. M. (1990). *Community Food Webs: Data and Theory*. Springer, London.

Cohen, J. E., Pimm, S. L., Yodzis, P. & Saldana, J. (1993b). Body sizes of animal predators and animal prey in food webs. *Journal of Animal Ecology*, **62**, 67–78.

Connell, J. H. & Sousa, W. P. (1983). On the evidence needed to judge ecological stability or persistence. *American Naturalist*, **121**, 789–824.

Cramp, S. & Simmons, K. E. L. (1977–83). *Handbook of the Birds of Europe, the Middle East and North America*. Vols I–III. Oxford University Press, Oxford.

De Ruiter, P. C., Neutel, A.-M. & Moore, J. C. *et al.* (1995). Energetics, patterns and interaction strengths and stability in real ecosystems. *Science*, **269**, 1257–1260.

Fox, L. R. & Morrow, P. A. (1981). Specialization: species property or local phenomenon? *Science*, **211**, 887–893.

Hairston, N. G., Smith, F. E. & Slobodkin, L. B. (1960). Community structure, population control and competition. *American Naturalist*, **94**, 421–425.

Hall, S. J. & Raffaelli, D. G. (1991). Food web patterns: lessons from a species-rich web. *Journal of Animal Ecology*, **60**, 823–839.

Hall, S. J. & Raffaelli, D. G. (1993). Food webs: theory and reality. *Advances in Ecological Research*, **24**, 187–237.

Havens, K. E. (1994). Experimental perturbation of a freshwater plankton community: a test of hypotheses regarding the effects of stress. *Oikos*, **69**, 147–153.

Hutchinson, G. E. (1959). Homage to Santa Rosalia, or why are there so many kinds of animals? *American Naturalist*, **93**, 145–159.

Huxham, M., Raffaelli, D. G. & Pike, A. W. (1994). The influence of *Cryptocotyle lingua* infections on the survival and fecundity of *Littorina littorea*: an ecological approach. *Journal of Experimental Marine Biology and Ecology*, **168**, 223–238.

Huxham, M., Raffaelli, D. G. & Pike, A. W. (1995a). The effect of larval trematodes on the growth and burrowing behaviour of *Hydrobia ulvae* (Gastropoda: Prosobranchia) in the Ythan estuary, Northeast Scotland. *Journal of Experimental Marine Biology and Ecology*, **1985**, 1–17.

Huxham, M., Raffaelli, D. G. & Pike, A. W. (1995b). Parasites and food web patterns. *Journal of Animal Ecology*, **64**, 168–176.

Huxham, M., Beeney, S. & Raffaelli, D. G. (1996). Do parasites reduce the chances of triangulation in a real food web? *Oikos*, **76**, 284–300.

Jenkins, B., Kitching, R. L. & Pimm, S. L. (1992). Productivity, disturbance and food web structure at a local spatial scale in experimental container habitats. *Oikos*, **65**, 249–255.

Laukner, G. (1988). Ecological effects of trematode infestations in a littoral marine invertebrate population. *International Journal for Parasitology*, **17**, 391–398.

Law, R. & Blackford, J. C. (1992). Self-assembling food webs: a global viewpoint of co-existence of species Lotka–Volterra communities. *Ecology*, **73**, 567–578.

Lawton, J. H. (1989). Food webs. *Ecological Concepts. The Contribution of Ecology to an Understanding of the Natural World* (Ed. by J. M. Cherrett), pp. 43–78. Blackwell Scientific Publications, Oxford.

Lawton, J. H. (1992). Feeble links in food webs. *Nature*, **355**, 19–20.

Lawton, J. H., Naeem, S., Woodfin, R. M., Brown, V. K., Gange, A., Godfray, H. J. C., Heads, P. A., Lawler, S., Magda, D., Thomas, C. D., Thompson, L. J. & Young, S. (1993). The Ecotron:

a controlled environment facility for the investigation of population and ecosystem processes. *Philosophical Transactions of the Royal Society of London, B.*, **341**, 181–194.

Martinez, N. (**1991**). Artifacts or attributes? Effects of resolution on the Little Rock Lake food web. *Ecological Monographs*, **61**, 367–392.

Martinez, N. (**1993**). Effect of scale on food web structure. *Science*, **269**, 242–243.

May, R. M. (**1973**). *Stability and Complexity in Model Ecosystems*. Princeton University Press, Princeton, NJ.

Menge, B. A. (**1995**). Indirect effects in rocky intertidal interaction webs: patterns and importance. *Ecological Monographs*, **65**, 21–74.

Morin, P. J. & Lawler, S. P. (**1993**). Food web architecture and population dynamics in laboratory microcosms of protists. *American Naturalist*, **141**, 675–686.

Morin, P. J. & Lawler, S. P. (**1996**). Effects of food chain length and omnivory on population dynamics in experimental food webs. *Food Webs: Patterns and Process* (Ed. by G. Polis & K. Winemiller), pp. 218–230. Chapman & Hall, London.

Morin, P. J. & Lawler, S. P. (**in press**). Food web architecture and population dynami_s: theory and empirical evidence. *Annual Review of Ecology and Systematics*.

Paine, R. T. (**1980**). Food webs: linkage, interaction strength and community infrastructure. *Journal of Animal Ecology*, **49**, 667–685.

Paine, R. T. (**1988**). On food webs: road maps of interactions or the grist for theoretical development? *Ecology*, **69**, 1648–1654.

Paine, R. T. (**1992**). Food web analyses through field measurements of per capita interaction strength. *Nature*, **350**, 669–674.

Paine, R. T. (**1994**). *Marine Rocky Shores and Community Ecology: An Experimentalist's Perspective*. Ecology Institute, Oldendorf.

Pimm, S. L. (**1982**). *Food Webs*. Chapman & Hall, London.

Pimm, S. L. (**1984**). The complexity and stability of ecosystems. *Nature*, **307**, 321–326.

Pimm, S. L. & Kitching, R. L. (**1987**). The determinants of food chain lengths. *Oikos*, **50**, 302–307.

Pimm, S. L. & Lawton, J. H. (**1978**). On feeding on more than one trophic level. *Nature*, **275**, 542–545.

Pimm, S. L., Lawton, J. H. & Cohen, J. E. (**1991**). Food web patterns and their consequences. *Nature*, **350**, 669–674.

Polis, G. A. (**1991**). Complex trophic interactions in deserts: an empirical critique of food web theory. *American Naturalist*, **138**, 123–155.

Polis, G. A. (**1994**). Food webs, trophic cascades and community structure. *Australian Journal of Ecology*, **19**, 121–136.

Power, M. (**1992**). Top-down and bottom-up forces in food webs. Do plants have primacy? *Ecology*, **73**, 733–746.

Raffaelli, D. G. & Hall, S. J. (**1996**). Assessing the relative importance of trophic links in food webs. *Food Webs: Patterns and Process* (Ed. by G. Polis & K. Winemiller), pp. 185–191. Chapman & Hall, London.

Schoener, T. W. (**1993**). On the relative importance of direct versus indirect effects in ecological communities. *Mutualism and Community Organization* (Ed. by H. Kawanabe, J. E. Cohen & K. Iwasaki), pp. 365–411. Oxford University Press, Oxford.

Schoener, T. W. & Spiller, D. A. (**1987**). Effect of lizards on spider populations: manipulative reconstruction of a natural experiment. *Science*, **236**, 949–952.

Sprules, W. G. & Bowerman, J. E. (**1988**). Omnivory and food chain lengths in zooplankton food webs. *Ecology*, **69**, 418–426.

Stenseth, N. C. (**1983**). A co-evolutionary theory for communities and food web configurations. *Oikos*, **41**, 487–495.

Warren, P. H. (**1989**). Spatial and temporal variation in the structure of a freshwater food web. *Oikos*, **55**, 299–311.

Warren, P. H. (1994). Making connections in food webs. *Trends in Ecology and Evolution*, 9, 136–141.

Warren, P. H. & Lawton, J. H. (1987). Invertebrate predator–prey body size relationships: an explanation for upper triangularity food webs and patterns in food web structure? *Oecologia*, 74, 231–235.

Warren, P. H. & Spencer, M. (in press). Community and food web responses to the manipulation of energy input and disturbance in small ponds. *Oikos*.

Yodzis, P. (1984). Energy flow and the vertical structure of real ecosystems. *Oecologia*, 65, 86–88.

Yodzis, P. (1989). *An Introduction to Theoretical Ecology*, Harper & Row, New York.

CONCLUDING REMARKS

ROBERT M. MAY

Here and elsewhere (de Ruiter *et al.* 1995), Moore and his colleagues have shown that soil communities exhibit pronounced patterns in the way resources are used by the constituent species, which arguably enhance the stability of the system. In other words, they argue that a discussion of the energetic organization of interactions among species in the soil, including a variety of decomposer species, necessarily forms the basis for discussion of the stability of such ecosystems. This work is of particular interest in making direct connections between patterns of energy flow (the basis of one school of thought in ecology) and community dynamics (the basis of a conventionally different school of thought). Such connections, with their emphasis upon the role of detritus and decomposers, undercut a good deal of work which seeks for understanding of the way ecosystems work by making catalogues of food webs (but without decomposers and detritus), and then seeks for patterns within these catalogues.

FUNCTIONAL GROUPS AND PATTERN-SEEKING IN COMMUNITY STRUCTURE

The remarks in the previous paragraph notwithstanding, I believe that much has been learned—and much remains to be learned—by phenomenological searches for patterns in catalogues of food webs.

In any such activity, one of the first problems is characterizing the entities in the web. Must they be individually identified species? Or can we use 'functional groups', appropriately defined one way or another? There has been much critical discussion of these questions, and some of it has been reviewed here by Oksanen *et al.*, Moore and de Ruiter, and Hall and Raffaelli.

Hall and Raffaelli also reviewed a variety of patterns, and, arguably more important, variations within patterns, that have been suggested as being seen in multitrophic food webs. Their review, and particularly their examination of how manipulative experiments can be used to test ideas that arise from such analyses of food webs, is useful. But it must be observed that this review omits a large class of extremely interesting and sometimes enigmatic patterns which have been well documented for food webs by Cohen (1994), Sugihara *et al.* (1989) and others. Admittedly, some of these patterns are of a character which does not lend itself to easy description: intervality, rigid circuits and other patterns in the topology of the web. I make no attempt here to recapitulate this intriguing, but technically complicated, body of work. These patterns do, however, point to kinds of structure which is almost universal, and which is largely independent of degrees of aggregation into functional groupings. It has long been my belief that more work needs to be done to try to understand the origins of these more abstract structures. It is a pity that the otherwise excellent and embracing contribution referred to above did not include these aspects.

SPATIAL DIMENSIONS

Up to now, most work on multitrophic interactions (animal or otherwise) has dealt with temporal dimensions of how ecosystems work, in the sense of asking questions about persistence and ability to resist disturbance. Spatial aspects have, as in so many other areas of ecology, arrived on the scene relatively late. Thus, I think Holt's paper is particularly interesting in its suggestion that recently fashionable ideas about metapopulation dynamics and structure may be inescapable, when considering the analysis of food webs. Metapopulation dynamics, until now, has dealt largely with the way individual populations persist by virtue of a shifting balance between local extinction and recolonization within a mosaic of intercommunicating patches. Holt's work suggests that such considerations are likely to be important for an understanding of multitrophic interactions, and the persistence of complicated communities. This looks like a growth area to me.

TOWARDS A UNIFIED UNDERSTANDING OF COMPLICATED COMMUNITIES?

In his Tansley Lecture, Paine (1980) observed that an approach to ecosystems based on considerations of energy flow and plumbing diagrams tells us a lot, but can miss interactions which carry little energy yet which are absolutely crucial to the functioning of the system (pollinators, or some low-density-population seed dispersers, are obvious examples). Conversely, approaches which emphasize the connections among species, and thence the topology of the food web, can tell us interesting things, but have the fault that they do not always distinguish the beams which hold the building up from minor connections which are essentially architectural embellishment. Paine called for more work which combines the strength of these two differing approaches. This is easier to ask for than to get, as the slowness of the subsequent evolution of the subject has shown. I think the present volume, however, reveals encouraging progress, particularly in work such as Moore and de Ruiter's with its inclusion of decomposers and detritus, and its conclusion that the compartmentalization of resource utilization is an important ingredient in ecosystem persistence.

Of one thing I am sadly confident. We are not going to understand ultimately why there are the number of species we have on earth today, or how many species we can lose and still have the biosphere operate to sustain life, unless we also understand how communities and ecosystems are put together and how they work. And so I see continuing efforts to find a method of describing multitrophic interactions, in ways which combine the strengths of the 'energy flows' approach with the strengths of the 'connections' approach as being intimately entwined with our attempts to understand the causes and consequences of biological diversity today.

REFERENCES

Cohen, J. E. (1994). Marine and continental food webs — three paradoxes. *Philosophical Transactions Royal Society of London*, Series B343, 57–69.

Paine, R. T. (1980). Food Webs: linkage, interaction strength and community infrastructure. *Journal of Animal Ecology*, 49, 667–685.

de Ruiter, P. C., Neutel, A. M. & Moore, J. C. (1995). Energetics, patterns of interaction strengths, and stability in real ecosystems. *Science*, 269, 1257–1260.

Sugihara, G., Schoenly, K. & Trombla, A. (1989). Scale-invariance in food web properties. *Science*, 245 48–52.

AUTHOR INDEX

SUBJECT INDEX